Lloyd R Smith
1978

Nonlinear Parameter Estimation

Nonlinear Parameter Estimation

YONATHAN BARD
International Business Machines Corporation
Cambridge, Massachusetts

1974

ACADEMIC PRESS New York San Francisco London
A Subsidiary of Harcourt Brace Jovanovich, Publishers

ACADEMIC PRESS, INC.
111 Fifth Avenue, New York, New York 10003

United Kingdom Edition published by
ACADEMIC PRESS, INC. (LONDON) LTD.
24/28 Oval Road, London NW1

Library of Congress Cataloging in Publication Data

Bard, Yonathan.
Nonlinear parameter estimation.

Bibliography: p.
1. Estimation theory. I. Title.
QA276.8.B37 519.5'4 72-13616
ISBN 0-12-078250-2

AMS(MOS) 1970 Subject classifications: 62F10, 62J05, 90C30

PRINTED IN THE UNITED STATES OF AMERICA

Contents

Preface ix

Chapter I **Introduction**

1-1. Curve Fitting *1* 1-2. Model Fitting *2* 1-3. Estimation *3*
1-4. Linearity *5* 1-5. Point and Interval Estimation *6* 1-6. Historical Background *6* 1-7. Notation *7*

Chapter II **Problem Formulation**

A DETERMINISTIC MODELS

2-1. Basic Concepts *11* 2-2. Structural Model *12* 2-3. Parameter Evaluation *13* 2-4. Reduced Model *13* 2-5. Application Areas *14*

B DATA

2-6. Experiments and Data Matrix *17*

C PROBABILISTIC MODELS AND LIKELIHOOD

2-7. Randomness in Data *18* 2-8. The Normal Distribution *18*
2-9. The Uniform Distribution *21* 2-10. Distribution of Errors *22*
2-11. Stochastic Form of the Model *24* 2-12. Likelihood–Standard Reduced Model *26* 2-13. Likelihood–Structural Models *27* 2-14. An Example *29*
2-15. Utility of Distribution Assumptions *32*

D PRIOR INFORMATION AND POSTERIOR DISTRIBUTION

2-16. Prior Information *32* 2-17. Prior Distribution *33* 2-18. Informative and Noninformative Priors *34* 2-19. Bayes' Theorem *36* 2-20. Problems *37*

Chapter III **Estimators and Their Properties**

A STATISTICAL PROPERTIES

3-1. The Sampling Distribution *39* 3-2. Properties of the Sampling Distribution *40*
3-3. Evaluation of Statistical Properties *45*

B MATHEMATICAL PROPERTIES

3-4. Optimization *47* 3-5. Unconstrained Optimization *48* 3-6. Equality Constraints *49* 3-7. Inequality Constraints *51* 3-8. Problems *53*

Chapter IV **Methods of Estimation**

4-1. Residuals *54*

A LEAST SQUARES

4-2 Unweighted Least Squares *55* 4-3. Weighted Least Squares *56* 4-4. Multiple Linear Regression *58*

B MAXIMUM LIKELIHOOD

4-5. Definition *61* 4-6. Likelihood Equations *62* 4-7. Normal Distribution *63* 4-8. Unknown Diagonal Covariance *64* 4-9. Unknown General Covariance *65* 4-10. Independent Variables Subject to Error *67* 4-11. Exact Structural Models *68* 4-12. Data Requirements *69* 4-13. Some Other Distributions *70*

C BAYESIAN ESTIMATION

4-14. Definition *72* 4-15. Mode of the Posterior Distribution *73* 4-16. Minimum Risk Estimates *74*

D OTHER METHODS

4-17. Minimax Deviation *77* 4-18. Pseudomaximum Likelihood *78* 4-19. Linearizing Transformations *78* 4-20. Minimum Chi-Square Method *80* 4-21. Problems *80*

Chapter V **Computation of the Estimates I: Unconstrained Problems**

5-1. Introduction *83* 5-2. Iterative Scheme *84* 5-3. Acceptability *85* 5-4. Convergence *87* 5-5. Steepest Descent *88* 5-6. Newton's Method *88* 5-7. Directional Discrimination *91* 5-8. The Marquardt Method *94* 5-9. The Gauss Method *96* 5-10. The Gauss Method as a Sequence of Linear Regression Problems *99* 5-11. The Implementation of the Gauss Method *101* 5-12. Variable Metric Methods *106* 5-13. Step Size *110* 5-14. Interpolation–Extrapolation *111* 5-15. Termination *114* 5-16. Remarks on Convergence *115* 5-17. Derivative Free Methods *117* 5-18. Finite Differences *117* 5-19. Direct Search Methods *119* 5-20. The Initial Guess *120* 5-21. A Single-Equation Least Squares Problem *123* 5-22. Adding Prior Information *131* 5-23. A Two-Equation Maximum Likelihood Problem *133* 5-24. Problems *139*

Chapter VI **Computation of the Estimates II: Problems with Constraints**

A INEQUALITY CONSTRAINTS

6-1. Penalty Functions *141* 6-2. Projection Methods *146* 6-3. Projection with Bounded Parameters *151* 6-4. Transformation of Variables *153* 6-5. Minimax Problems *154*

B EQUALITY CONSTRAINTS

6-6. Exact Structural Models *154* 6-7. Convergence Monitoring *156*
6-8. Some Special Cases *157* 6-9. Penalty Functions *159* 6-10. Linear Equality Constraints *160* 6-11. Least Squares Problem with Penalty Functions *160*
6-12. Least Squares Problem—Projection Method *162* 6-13. Independent Variables Subject to Error *163* 6-14. An Implicit Equations Model *167*
6-15. Problems *168*

Chapter VII **Interpretation of the Estimates**

7-1. Introduction *170* 7-2. Response Surface Techniques *171* 7-3. Canonical Form *174* 7-4. The Sampling Distribution *175* 7-5. The Covariance Matrix of the Estimates *176* 7-6. Exact Structural Model *179* 7-7. Constraints *180*
7-8. Principal Components *183* 7-9. Confidence Intervals *184*
7-10. Confidence Regions *187* 7-11. Linearization *189* 7-12. The Posterior Distribution *191* 7-13. The Residuals *192* 7-14. The Independent Variables Subject to Error *196* 7-15. Goodness of Fit *198* 7-16. Tests on Residuals *199*
7-17. Runs and Outliers *201* 7-18. Causes of Failure *202* 7-19. Prediction *204*
7-20. Parameter Transformation *205* 7-21. Single-Equation Least Squares Problem *206* 7-22. A Monte Carlo Study *210* 7-23. Independent Variables Subject to Error *212* 7-24. Two-Equation Maximum Likelihood Problem *213*
7-25. Problems *216*

Chapter VIII **Dynamic Models**

8-1. Models Involving Differential Equations *218* 8-2. The Standard Dynamic Model *221* 8-3. Models Reducible to Standard Form *223* 8-4. Computation of the Objective Function and Its Gradient *225* 8-5. Numerical Integration *230*
8-6. Some Difficulties Associated with Dynamic Systems *231* 8-7. A Chemical Kinetics Problem *233* 8-8. Linearly Dependent Equations *238*
8-9. Problems *242*

Chapter IX **Some Special Problems**

9-1. Missing Observations *244* 9-2. Inhomogeneous Covariance *246*
9-3. Sequential Reestimation *248* 9-4. Computational Aspects *249*
9-5. Stochastic Approximation *251* 9-6. A Missing Data Problem *251*
9-7. Further Problem with Missing Data *253* 9-8. A Sequential Reestimation Problem *255* 9-9. Problems *257*

Chapter X **Design of Experiments**

10-1. Introduction *258* 10-2. Information and Uncertainty *261* 10-3. Design Criterion for Parameter Estimation *262* 10-4. Design Criterion for Prediction *265*

10-5. Design Criterion for Model Discrimination *266* 10-6. Termination Criteria *269* 10-7. Some Practical Considerations *271* 10-8. Computational Considerations *273* 10-9. Computer Simulated Experiments *276* 10-10. Design for Decision Making *283* 10-11. Problems *286*

Appendix A **Matrix Analysis**

A-1. Matrix Algebra *287* A-2. Matrix Differentiation *293* A-3. Pivoting and Sweeping *296* A-4. Eigenvalues and Vectors of a Real Symmetric Matrix *302* A-5. Spectral Decompositions *303*

Appendix B **Probability** *310*

Appendix C **The Rao–Cramer Theorem** *313*

Appendix D **Generating a Sample from a Given Multivariate Normal Distribution** *316*

Appendix E **The Gauss–Markov Theorem** *318*

Appendix F **A Convergence Theorem for Gradient Methods** *320*

Appendix G **Some Estimation Programs** *323*

References *325*

Author Index 333
Subject Index 337

Preface

This book is intended primarily for use by the scientist or engineer who is concerned with fitting mathematical models to numerical data, and for use in courses on data analysis which deal with that subject. Such fitting is frequently done by the method of least squares, with no regard paid to previous knowledge concerning the values of the parameters (coefficients), nor to the statistical nature of the measurement errors. In Chapters II–IV we show how the problem can be formulated so as to take all these factors into account. In Chapters V–VI we discuss the computational methods used to solve the problem, once its formulation has been completed. Chapter VII is devoted to the question of what conclusions can be drawn, after the estimates have been computed, concerning the validity of the estimates, or of the model which has been fitted. In Chapter VIII we discuss the important special case of models which are stated in the form of differential equations. Other special problems are treated in Chapter IX. Finally, in Chapter X we suggest methods for planning the experiments in such a way that the data will shed the greatest possible light on the model and its parameters. We cannot stress too strongly the point that if data are to be gathered for the purpose of establishing a mathematical model, then the experiments should be designed with this purpose in mind. Hence the importance of Chapter X.

A practical, rather than theoretical point of view has been taken throughout this book. We describe computational algorithms which have performed well on a variety of problems, even if their convergence has not been proven, and even if they have failed on some other problems. We have as yet no foolproof, efficient methods for solving nonlinear problems; hence we cannot afford to throw away useful tools just because they are not perfect.

The presentation uses matrix algebra and probability theory on a very elementary level. Reviews of the needed concepts and proofs of some important theorems will be found in the appendixes. Some supplementary material has been included in the form of problems at the ends of chapters. Problems requiring actual computation have not been included; the reader is likely to have his own data to compute with, and additional data may be found in many of the cited references. Several numerical problems have, however, been worked out in great detail in separate sections at the ends of Chapters V–IX for the purpose of illustrating the methods discussed in those chapters.

The author is deeply indebted to the IBM Corporation, and in particular to the managements of the New York and Cambridge Scientific Centers, who have supported the writing of this book and provided all the necessary resources. The author is also grateful to Professor L. Lapidus of Princeton University, and to his colleagues J. L. Greenstadt, P. G. Comba, H. Eisenpress, K. Spielberg, and P. Backer, for many helpful discussions, and for reviewing portions of the manuscript.

Chapter

I

Introduction

1-1. Curve Fitting

A scientist who has compiled tables of data wishes to reduce them to a more convenient and comprehensible form. He accomplishes this by representing the data in graphical or functional form. In the first case, he plots his data points, and then draws some curve through them. In the second case, he selects a class of functions, and chooses from this class the one that best fits his data. This is called *curve fitting.*

In the simplest case, the data consist of values $y_1, y_2, \ldots, y_n$ of a dependent variable y measured for various values $x_1, x_2, \ldots, x_n$ of an independent variable x. A frequently chosen class of functions is the set of all polynomials of order not exceeding m

$$y = \theta_0 + \theta_1 x + \theta_2 x^2 + \cdots + \theta_m x^m \tag{1-1-1}$$

The values of the *parameters* $\theta_0, \theta_1, \ldots, \theta_m$ are chosen so as to get the best possible fit to the data. The most commonly used technique for accomplishing this is the least squares method, in which those values of the θ_i are selected which minimize the sum of squares of the *residuals*, i.e.,

$$S = \sum_{\mu=1}^{n} \left(y_\mu - \sum_{\alpha=0}^{m} \theta_\alpha x_\mu^{\alpha} \right)^2 \tag{1-1-2}$$

Curve fitting procedures are characterized by two degrees of arbitrariness. First, the class of functions used is arbitrary, being dictated only to a minor extent by the physical nature of the process from which the data came. Second, the best fit criterion is arbitrary, being independent of statistical considerations. This arbitrariness can be exploited to make the fitting job easy. Choosing equations which, like Eq. (1),‡ are linear functions of the parameters; using orthogonal or Fourier polynomials (in place of ordinary

‡ This reference is to the first equation of the current section, i.e., Eq. (1-1-1).

polynomials) as the functions to fit; employing the least squares criterion—all these contribute to making the computation of the parameters a mathematically easy job. On the other hand, due to their arbitrary nature, the equations that we get are useful only for summarizing the data and for interpolating between tabulated values. They cannot be used to extrapolate, i.e., to predict the outcome of experiments removed from the region of already available data. Also, the equations and the parameters occurring in them shed little insight on the nature of the process being measured, except to answer such questions as to whether variable x has an influence on variable y.

Curve fitting techniques have widespread applications in situations that go far beyond the simple y vs. x table. An example is the identification of dynamic systems by means of rational transfer functions or Volterra series. Most multiple linear regression, analysis of variance, and econometric time-series problems are also of a curve fitting nature, since the equations used are not derived from "laws of nature." In most of these applications, however, assumptions are made concerning the statistical behavior of the errors, thereby elevating them at least partly to the status of estimation problems as discussed in Section 1-3.

1-2. Model Fitting

Often the scientist is, to a certain extent, familiar with the laws which govern the behavior of the physical system under observation. He can then derive equations describing the relationships among the observed quantities. For instance, the fraction y of a radioactive isotope remaining x seconds after the isotope's formation is given by

$$y = e^{-\theta x} \tag{1-2-1}$$

where the parameter θ is a physical constant proportional to the instantaneous rate of decay of the isotope. The magnitude of θ is unknown, but we wish to assign to it a value which makes Eq. (1) fit the data (y_1, x_1), (y_2, x_2), ..., (y_n, x_n) as well as possible, e.g., by the least squares criterion.

An equation such as Eq. (1) which is derived from theoretical considerations is called a *model*, and the procedure just described constitutes *model fitting*. In principle, model fitting is not much different from curve fitting, except that we can no longer guide the selection of a functional form by considerations of computational convenience. For instance, Eq. (1) is not a linear function of the parameter, and because of this the computation of the "best fit" is more difficult than the computation of the θ_i in Eq. (1-1-1).

1-3. Estimation

A new consideration arises in model fitting that does not exist in curve fitting. The parameters occurring in a model, e.g., θ in Eq. (1-2-1), usually represent quantities that have physical significance. If the model is a correct one, then it is meaningful to ask what is the true value of θ in nature. Because of the generally imprecise nature of measurements we can never hope to determine the true values with absolute certainty. Also, due to the random nature of the errors in measurements, the value of θ that best fits one series of measurements differs from the value that fits another series, even though both series are performed on the same isotope. However, we can look for procedures to obtain values of the parameters that not only fit the data well, but also come on the average fairly close to the true values, and do not vary excessively from one set of experiments to the next. The process of determining parameter values with these statistical considerations in mind is termed *model estimation*.

The classical problem of *statistical estimation* differs somewhat from the model estimation problem that we have just defined. The statistician observes a sequence of values ("realizations") that a random variable assumes. For instance, he may obtain a sequence of numbers such as 1, 5, 6, 3, . . . denoting successive throws of a die. The statistician assumes a "model" in the form of a probability distribution which may depend on some unknown parameters. In our case, the statistician who suspects the die may be loaded assigns probabilities $[\theta_1, \theta_2, \theta_3, \theta_4, \theta_5, 1 - \sum_{i=1}^{5} \theta_i]$ to the six possible outcomes of a throw. He then attempts to estimate the θ_i from the observed values of the random variable. Here he will probably use the estimate

$$\theta_i = n_i / \sum_{j=1}^{6} n_j \qquad (1\text{-}3\text{-}1)$$

where n_i is the number of throws on which the number i showed up ($i = 1, 2, \ldots, 6$).

As a further example, the observed value of the random variable may be the height h of adults in a community. If we assume that this variable has normal (Gaussian) distribution with mean h_0 and standard deviation σ, then the probability density function is given by

$$p(h) = [1/(2\pi)^{1/2}\sigma] \exp[-(1/2\sigma^2)(h - h_0)^2] \qquad (1\text{-}3\text{-}2)$$

If we measure the heights $h_1, h_2, \ldots, h_n$ of n randomly chosen individuals from the community, we form the usual estimates:

$$h_0 = (1/n) \sum_{\mu=1}^{n} h_\mu \qquad (1\text{-}3\text{-}3)$$

$$\sigma^2 = [1/(n-1)] \sum_{\mu=1}^{n} (h_\mu - h_0)^2 \qquad (1\text{-}3\text{-}4)$$

The model estimation problem can be embedded in the statistical estimation problem in the following way: It is reasonable to suppose that the outcome y of a measurement taken at time x_μ (we shall phrase our discussion in terms of the radioactive decay model of Section 1-2) is a random variable whose mean value is given by Eq. (1-2-1) as $\exp(-\theta x_\mu)$. If many measurements were to be taken at the same x_μ we would discover that the observed values y_μ fluctuate around their mean value with standard deviation σ. Suppose these fluctuations have a normal probability distribution. The probability density function for y_μ would then have the form similar to Eq. (2)

$$p(y_\mu) = [1/(2\pi)^{1/2}\sigma] \exp\{-(1/2\sigma^2)[y_\mu - \exp(-\theta x_\mu)]^2\} \tag{1-3-5}$$

In fact we only take one measurement at any specific x_μ. What we have are realizations $y_1, y_2, \ldots, y_n$, each of a different random variable whose distribution depends on the parameter x_μ which varies from one variable to the next, and on some other parameters (θ, σ) which are common to all these distributions. The *parameter estimation* problem which is the primary concern of this book is the problem of estimating these common parameters.

At first glance, the parameter estimation problem appears more general than the classical statistical estimation problem, since in the latter all samples are taken from the same distribution. The distinction between the two problems disappears if we choose to regard all the data as being a single multivariate sample from the joint distribution of all the observations made in the course of the series of experiments. It follows that many of the statistical estimation methods can be applied to our parameter estimation problems. The single sample point of view is, however, rather awkward when one examines, say, the asymptotic properties of these estimates (see Chapter III for definitions) since it requires that the entire set of experiments be repeated over and over again.

Parameter estimation techniques may be applied as computational tools to pure curve fitting problems. One must remember, however, that the statistical properties of these estimates (e.g., those described in Chapters III and VII) sometimes lose their meaning in the curve fitting context.

Clearly, parameter estimation is a more difficult operation than curve fitting, calling for more sophisticated analysis and more extensive computation. The effort is worthwhile since a well established model and precisely estimated physical parameters are much more versatile tools, both for illuminating the present situation and for prediction in new situations, than arbitrarily fitted curves can ever be. To bring home this point, one need only observe that a physical parameter estimated from one model can always be used in another model to which it is relevant. For instance, the viscosity of a liquid estimated from viscometer data can be used to predict the required pumping load for a piping system being designed.

There are other mathematical problems which may be solved by means of parameter estimation or curve fitting techniques. These techniques may be regarded as attempts to solve (as best one can) an overdetermined (more equations than unknowns) system of simultaneous equations. Solving a system of n equations in n unknowns may, therefore, be regarded as fitting to n data points a model involving n unknown parameters. Two-point boundary value problems in ordinary differential equations may be treated as models in which the known terminal conditions are the data, and the missing initial conditions are the unknown parameters. Some optimal control problems may be solved by regarding the control actions as unknown parameters, and the desired trajectory of the system as the data to be fitted. Similarly, some engineering design problems may be posed as requiring parameter values which induce the systems to meet prescribed conditions as closely as possible.

1-4. Linearity

To understand what we mean by the term "nonlinear estimation" we must first make the following definitions: An expression is said to be *linear* in a set of variables $\phi_1, \phi_2, \ldots, \phi_n$ if it has the form $a_0 + \sum_{i=1}^n a_i \phi_i$, where the coefficients a_i $(i = 0, 1, \ldots, n)$ are not functions of the ϕ_i. An expression is *quadratic* in the ϕ_i if it has the form $a_0 + \sum_{i=1}^n a_i \phi_i + \sum_{i,j=1}^n b_{ij} \phi_i \phi_j$, again with all coefficients not depending on the ϕ_i. If we differentiate a quadratic expression with respect to one of the ϕ_i, we obtain a linear expression.

Linear estimation problems are ones in which the model equations are linear expressions in the unknown parameters, e.g., Eq. (1-1-1). When the model equations are not linear, as in Eq. (1-2-1), we speak of *nonlinear estimation*. However even some apparently linear problems are essentially nonlinear. This is so because in order to estimate the parameters we usually minimize some function, such as the sum of squares of residuals. To find the minimum, we equate the derivatives of the function to zero and solve for the values of the parameters. Now when the model equations are linear, the sum of squares function is quadratic, and the derivatives are again linear. The estimates are obtained, therefore, by solving a set of simultaneous linear equations, and all is well. But if some other functions which are not quadratic are chosen to be minimized, then the equations to be solved are no longer linear, even when the model equations are linear. Such problems should also be regarded as nonlinear estimation problems. Examples of such problems are given in Sections 4-8–4-9.

1-5. Point and Interval Estimation

There exist many methods (e.g., least squares) which calculate specific numbers representing estimates for the parameter values. Such numbers are called *point estimates.* A point estimate for the parameters θ, σ appearing in Eq. (1-3-5) may take the form

$$\theta^* = 4, \qquad \sigma^* = 0.1 \tag{1-5-1}$$

A point estimate standing alone is not very satisfactory. Random errors are present in all measurements, and no mathematical model accounts for all facets of a physical situation. Therefore we cannot hope to obtain point estimates exactly equal to the true values of the parameters (if such exist). Nor can we expect point estimates calculated from different data samples to be equal, even if the samples were obtained under similar conditions. Therefore we need to augment the point estimate with some information on its variability. For instance, in place of Eq. (1) we wish to have a statement such as

$$\theta^* = 4 \pm 0.2, \qquad \sigma^* = 0.1 \pm 0.02 \tag{1-5-2}$$

The numbers 0.2 and 0.02 are meant to represent the standard deviations of the variability of the estimates for θ, σ.

The information contained in Eq. (2) may be translated into a statement of the type‡ "We are 75% sure that θ is between 3.6 and 4.4, and we are 75% sure that σ is between 0.06 and 0.14." This statement constitutes an *interval estimate* for our parameters.

Interval estimates can be computed directly, without first calculating point estimates and their variability. In fact, many statisticians prefer interval estimates, because they feel one is not justified in picking out one specific preferred value to be used as a point estimate. We feel, however, that the needs of the scientist or engineer are best served by point estimates with measures of their reliability, so we will not discuss any direct interval estimation procedures. The calculation of interval estimates (called *confidence intervals* in this context) from point estimates is discussed in Sections 7-9–7-10.

1-6. Historical Background

Legendre (1805) was the first to suggest in print the use of the least squares criterion for estimating coefficients in linear curve fitting. Gauss (1809) laid the statistical foundation for parameter estimation by showing that least squares estimates maximized the probability density for a normal (Gaussian)

‡ The statement is derived from Eq. (2) using the Bienaymé–Chebyshev inequality with $k = 2$. See Eq. (7-9-11).

distribution of errors. In this, Gauss anticipated the maximum likelihood method. Gauss and his contemporaries seemed to prefer, however, purely heuristic justifications for the least squares method. Further work in the 19th and early 20th centuries, by Gauss himself, Cauchy, Bienaymé, Chebyshev, Gram, Schmidt, and others‡ concentrated on computational aspects of linear least squares curve fitting, including the introduction of orthogonal polynomials.

The development of statistical estimation methods received its impetus from the work of Karl Pearson around the turn of the century and R. A. Fisher in the 1920s and 1930s. The latter revived the maximum likelihood method and studied estimator properties such as consistency, efficiency, and sufficiency [see the collection of Fisher's (1950) papers]. The development of decision theory by Wald and others has, in the post-World War II years, introduced a new basis for selecting estimation criteria. The practical impact of these methods in the area of nonlinear parameter estimation has so far been slight, except for causing increased awareness of the uses of prior distributions.

The first modern applications of statistical estimation theory to model estimation were made in the field of econometrics by Koopmans and others, starting in the 1930s. Their work is summarized in the Cowles Commission Reports (Hood and Koopmans, 1953). The main contributions to the application of statistical techniques in the construction and estimation of mathematical models in the physical sciences have come from professor G. E. P. Box and his coworkers at Princeton University and the University of Wisconsin.

The computation of estimates for nonlinear models usually requires finding the maximum or minimum of a nonlinear function. Computational methods bearing the names of Newton, Gauss, and Cauchy have been known for a long time, but their extensive application to practical problems had to await the arrival of the electronic computer. The first general purpose computer program for solving nonlinear least squares problems was written by Booth and Peterson (1958) in collaboration with Box. The program employed a modified Gauss method. It has since been followed by many other programs, some more general in nature and some dealing with more specific estimation problems. A list of such programs can be found in Appendix G.

1-7. Notation

Matrix and vector notation are used throughout this book.

A boldface capital letter denotes a matrix: $\mathbf{A}$, $\boldsymbol{\Gamma}$.

A boldface lower case letter denotes a column vector: $\mathbf{a}$, $\boldsymbol{\gamma}$.

‡ References to this work, along with a more detailed historical survey are given by Seal (1967).

The (i, j) element, appearing in the ith row and jth column of $\mathbf{A}$ is denoted A_{ij} or $[\mathbf{A}]_{ij}$.

The ith element of $\mathbf{a}$ is denoted a_i or $[\mathbf{a}]_i$.

$\mathbf{A}_\mu$ is the μth in a sequence of matrices $\mathbf{A}_1, \mathbf{A}_2, \mathbf{A}_3, \ldots$. The (i, j) element of $\mathbf{A}_\mu$ is denoted $A_{\mu ij}$ or $[\mathbf{A}_\mu]_{ij}$. Analogously for vectors.

$\mathbf{A}^{\mathrm{T}}$ is the transpose of $\mathbf{A}$, i.e., $[\mathbf{A}^{\mathrm{T}}]_{ij} = [\mathbf{A}]_{ji}$.

$\mathbf{a}^{\mathrm{T}}$ is the row vector with the same elements as $\mathbf{a}$.

$\mathbf{A}^{-1}$ is the inverse of $\mathbf{A}$ if such exists.

$\mathbf{A}^{+}$ is the pseudoinverse of $\mathbf{A}$.

$\det(\mathbf{A})$ is the determinant of $\mathbf{A}$.

$\mathrm{Tr}(\mathbf{A}) = \sum_i A_{ii}$ is the trace of $\mathbf{A}$.

$\mathbf{A}$ is said to be $m \times n$ if it has m rows and n columns. A column vector is $m \times 1$ and a row vector $1 \times n$.

$\mathbf{I}$ is the identity matrix, i.e.,

$$I_{ij} = \delta_{ij} = \begin{cases} 1 & (i = j) \\ 0 & (i \neq j) \end{cases}$$

$\mathbf{I}_m$ is the $m \times m$ identity matrix.

$\mathbf{A} = \mathrm{diag}(\mathbf{a})$ means that $\mathbf{A}$ is a matrix with elements $A_{ij} = a_i \delta_{ij}$.

Suppose α is a function of the vectors $\mathbf{a}$ and $\mathbf{b}$ and the matrix $\mathbf{A}$. Then:

$$\begin{aligned} &\partial\alpha/\partial\mathbf{a} \quad \text{is the column vector} \quad [\partial\alpha/\partial\mathbf{a}]_i = \partial\alpha/\partial a_i \\ &\partial\alpha/\partial\mathbf{A} \quad \text{is the matrix} \quad [\partial\alpha/\partial\mathbf{A}]_{ij} = \partial\alpha/\partial A_{ij} \\ &\partial^2\alpha/\partial\mathbf{a}\,\partial\mathbf{b} \quad \text{is the matrix} \quad [\partial^2\alpha/\partial\mathbf{a}\,\partial\mathbf{b}]_{ij} = \partial^2\alpha/\partial a_i\,\partial b_j \end{aligned}$$

Suppose $\mathbf{a}$ is a vector function of the scalar β and the vector $\mathbf{b}$. Then:

$$\begin{aligned} &\partial\mathbf{a}/\partial\beta \quad \text{is the column vector} \quad [\partial\mathbf{a}/\partial\beta]_i = \partial a_i/\partial\beta \\ &\partial\mathbf{a}/\partial\mathbf{b} \quad \text{is the matrix} \quad [\partial\mathbf{a}/\partial\mathbf{b}]_{ij} = \partial a_i/\partial b_j \end{aligned}$$

Suppose $\mathbf{A}$ is a matrix function of the scalar α. Then:

$$\partial\mathbf{A}/\partial\alpha \quad \text{is the matrix} \quad [\partial\mathbf{A}/\partial\alpha]_{ij} = \partial A_{ij}/\partial\alpha$$

Derivatives of matrices with respect to vectors and matrices, or of vectors with respect to matrices, give rise to arrays with more than two dimensions. Rules for differentiating vector and matrix expressions are given in Section A-2 Appendix A.

We also make use of some notation associated with probability concepts.

$\Pr(A)$ is the probability of event A.

If x is a random variable, then

$p(x)$ is the probability density function of x.

$p(x|A)$ is the probability density of x given that A occurred.

$E(x)$ is the expected value of x.

$E(x|A)$ is the expected value of x given that A occurred.

$V_x = E\{[x - E(x)]^2\}$ is the variance of x.

$\sigma_x = V_x^{1/2}$ is the standard deviation of x.

The notation $p(x|y)$ is meant to indicate that the probability density of x is also a function of the variable y.

The reader totally unfamiliar with matrix and probability theories is urged to study texts on these subjects. The reader who merely wishes to refresh his memory may consult Appendixes A and B which contain skeleton definitions of the terms involved and the operations applying to them.

Other notation:

$A \equiv B$ means that A equals B by definition.

$A \approx B$ means that A equals B approximately, or to within the order of approximation being considered (e.g., up to second-order terms in a Taylor series).

$\log x$ is the natural logarithm of x.

$\exp(x) \equiv e^x$.

$N_k(\mathbf{a}, \mathbf{V})$ is the k-dimensional normal distribution with mean $\mathbf{a}$ and covariance matrix $\mathbf{V}$.

Unless otherwise stated, the notation $x = a \pm b$ denotes that x is a random variable or estimate with mean a and standard deviation b.

The estimated value of some quantity x is denoted x^*, and its true (though unknown) value is denoted $\hat{x}$.

Formulas and equations are numbered by chapter and section. For instance, Eq. (5-3-6) is the sixth equation in Section 5-3. The chapter and section numbers are omitted from references to equations within the same section.

Subscripts:

$a, b, c, \ldots$ refer to model equations or dependent variables. The usual range is 1 to m.

Example: in $y_a = f_a(\mathbf{x}, \boldsymbol{\theta})$ the ath dependent variable y_a is a function f_a of $\mathbf{x}$ and $\boldsymbol{\theta}$.

$\alpha, \beta, \gamma, \ldots$ refer to parameters. The usual range is 1 to l.

Example: $q_\alpha = \partial\Phi/\partial\theta_\alpha$ is the αth component of the gradient of Φ with respect to $\boldsymbol{\theta}$.

μ, η, ϕ refer to experiments. The usual range is 1 to n.

Example: $\mathbf{y}_\mu$ is the vector of dependent variables measured in the μth experiment. Its ath component is $y_{\mu a}$.

i frequently (but not always) refers to iteration number.

Example: $\boldsymbol{\theta}_i$ is the vector $\boldsymbol{\theta}$ appearing in the ith iteration. Its αth component is $\theta_{i\alpha}$.

Chapter

II

Problem Formulation

A. Deterministic Models

2-1. Basic Concepts

The scientist often expresses his theories in the form of mathematical relationships among certain quantities. Similarly, the engineer derives equations that describe the properties of his structures or the workings of his processes. We refer to the relations which supposedly describe a certain physical situation, as a *model*. Typically, a model consists of one or more equations. The quantities appearing in the equations we classify into *variables* and *parameters*. The distinction between these is not always clear cut, and it frequently depends on the context in which the variables appear. Usually a model is designed to explain the relationships that exist among quantities which can be measured independently in an experiment; these are the variables of the model. To formulate these relationships, however, one frequently introduces "constants" which stand for inherent properties of nature (or of the materials and equipment used in a given experiment). These are the parameters.

We illustrate by means of an example: A cylindrical vessel of cross-sectional area A is filled with a liquid of density ρ and viscosity μ. It is allowed to drain through a capillary tube of radius R and length L. Let h and h_0 denote the depth of the liquid in the vessel at times t and t_0, respectively. The equations of laminar flow yield, for this case, the relation

$$\log(h_0/h) = (\pi g R^4/8A\phi L)(t - t_0) \qquad (2\text{-}1\text{-}1)$$

where g is the acceleration due to gravity, and $\phi = \mu/\rho$ is the kinematic viscosity of the liquid. If we interpret Eq. (1) as a relationship between the height of the liquid and the time, then we shall regard h, h_0, t, and t_0 as the variables, and g, A, R, L, and ϕ as the parameters. Among the latter, the first is a constant of nature, the next three reflect the properties of the apparatus, and the last one a property of the material used. If we performed experiments on several different vessels, we might add R, A, and L to the list of variables, leaving g and ϕ as the sole parameters.

On the other hand, suppose our instrument is to be used as a viscometer. We place two marks on the vessel, at heights h_0 and h from the bottom, and measure the time Δt that it takes for the surface of the liquid to pass from the higher to the lower mark. The kinematic viscosity of the liquid can then be calculated from the following rearrangement of Eq. (1)

$$\phi = \alpha \, \Delta t \tag{2-1-2}$$

where $\alpha = \pi g R^4/8AL \log(h_0/h)$. We calibrate the instrument with liquids whose viscosities are known. For the purposes of calibration, then, Eq. (2) contains the variables Δt (directly measurable) and ϕ (which can be found in published tables), and the parameter α (in whose physical significance we are not at the moment interested).

The values of some of the parameters which appear in a model may be known with great precision (e.g., the gravitational constant g in Eq. (1)). The role of such parameters does not differ, at least for our purposes, from that of purely numerical constants, such as π or 8 in Eq. (1). We exclude such parameters from further considerations.

2-2. Structural Model

The models we have so far considered take the general functional form

$$\mathbf{g}(\mathbf{z}, \boldsymbol{\theta}) = \mathbf{0} \tag{2-2-1}$$

where:

$$\mathbf{g} = \{g_1, g_2, \ldots, g_m)\}^{\mathrm{T}}$$

is an m-dimensional vector of functions.

$$\mathbf{z} = \{z_1, z_2, \ldots, z_k\}^{\mathrm{T}}$$

is a k-dimensional vector of variables.

$$\boldsymbol{\theta} = \{\theta_1, \theta_2, \ldots, \theta_l\}^{\mathrm{T}}$$

is an l-dimensional vector of parameters whose values are not precisely known.

Equations (1) are referred to as the *structural equations* of the model. Looking at the model represented by Eq. (2-1-1), we find that there is only one equation, hence $m = 1$; there are four variables $z_1 = h$, $z_2 = h_0$, $z_3 = t$, and $z_4 = t_0$; and there are four unknown parameters $\theta_1 = A$, $\theta_2 = R$, $\theta_3 = L$, $\theta_4 = \phi$. Eq. (1) then takes the form

$$g_1(\mathbf{z}, \boldsymbol{\theta}) \equiv \log(z_2/z_1) - (\pi g/8)(\theta_2{}^4/\theta_1\theta_3\theta_4)(z_3 - z_4) = 0 \qquad (2\text{-}2\text{-}2)$$

A model for which $m = 1$ is called a *single equation* model.

We refer to a model as *linear* if each one of the model equations has the form

$$g_i(\mathbf{z}, \boldsymbol{\theta}) = B_{i0}(\mathbf{z}) + \sum_{j=1}^{l} B_{ij}(\mathbf{z})\theta_j = 0 \qquad (i = 1, 2, \ldots, m) \qquad (2\text{-}2\text{-}3)$$

where B_{ij} $(i = 0, 1, \ldots, m; j = 1, 2, \ldots, k)$ are known functions of the $\mathbf{z}$. Models which are not linear are referred to as *nonlinear*. Equation (2-1-2) is a linear model (with α as the parameter), whereas Eq. (2) is nonlinear.

2-3. Parameter Evaluation

A model whose form corresponds to Eq. (2-2-1) is called a *deterministic model*, since all the quantities appearing in it are assumed to be well determined, at least in principle. The model can be of little practical value, however, unless the values of its parameters are known. There are two principal methods by which we may establish the values of the parameters:

1. Calculate the value of a parameter by applying established laws of nature to already known quantities. For example, if R, A, L, h, and h_0 have been measured, we can compute $\alpha = \pi g R^4/8AL \log(h_0/h)$ as the value of the parameter to be used in Eq. (2-1-2).
2. Measure the values of the model variables that occur in actual physical situations, and then seek parameter values which cause the model equations to be satisfied, at least approximately. We are concerned here with the implementation of this second procedure.

2-4. Reduced Model

The structural Eqs. (2-2-1) are suitable for checking the validity of the model. If values can be found for the parameters such that the equations are at least approximately satisfied, then we do not reject the model. The most important practical use to which the model may be put is that of *prediction*. For this purpose, the variables $\mathbf{z}$ are classified into two groups:

1. The r variables $\mathbf{y} = y_1, y_2, \ldots, y_r$ whose values we wish to predict. These we call the *dependent variables*.

2. The s variables $\mathbf{x} = x_1, x_2, \ldots, x_s$ on the basis of which we wish to do the prediction. We call these the *independent variables*.

The problem of prediction, then, is that of determining in advance the values that the dependent variables will take for given values of the independent variables.

Rewriting the structural equations with **x** and **y** replacing **z**

$$\mathbf{g}(\mathbf{x}, \mathbf{y}, \boldsymbol{\theta}) = \mathbf{0} \tag{2-4-1}$$

We see that reasonable prediction is possible if all of the following conditions hold:

1. The model is reasonably correct.
2. The values of the parameters are known to a good approximation.
3. The structural equations can be solved for the dependent variables, yielding the *reduced* equations

$$\mathbf{y} = \mathbf{f}(\mathbf{x}, \boldsymbol{\theta}) \tag{2-4-2}$$

where $\mathbf{f} = f_1, f_2, \ldots, f_r$ is an r-dimensional vector of functions.

Since the number of structural equations is m, we can usually solve for the values of up to m dependent variables, leaving $s = k - m$ independent variables.

A *linear reduced model* is one in which the functions **f** are linear in the $\boldsymbol{\theta}$. A linear structural model may result in a nonlinear reduced model. For instance, the linear structural model $\log y + \theta x = 0$ reduces to the nonlinear model $y = e^{-\theta x}$.

Strictly speaking, we should refer to the "structural form" or "reduced form" of the same model. In practice, however, we shall attach the designation "model" to whatever set of equations we happen to be dealing with at the moment.

2-5. Application Areas

There is nothing in Eqs. (2-2-1) or (2-4-2) to imply that we need have explicit analytic expressions for the functions **g** and **f**. All that is required is that given the values of their arguments (**z** and $\boldsymbol{\theta}$, or **x** and $\boldsymbol{\theta}$), one can calculate the values of the functions. This may require solution of a system of differential equations, or a complicated system simulation. When the structural equations cannot be solved explicitly, we may still obtain predicted values of the dependent variables by solving the equations numerically.

The following example of a model requiring the solution of differential equations is taken from the field of chemical reaction kinetics. Consider a chemical reaction in which molecules of a certain species (compound) A decompose spontaneously into molecules of B and C. In chemical notation, the reaction would be written as

$$A \rightarrow B + C \tag{2-5-1}$$

The law of mass action states that the rate of decomposition is, at any moment, proportional to the concentration of A at that moment. This leads to the differential equation

$$dy_A/dt = -k_1 y_A \tag{2-5-2}$$

where y_A is the concentration of A at time t, and k_1 is the so-called reaction rate constant. Eq. (2) may be integrated explicitly to yield

$$y_A = x_A \exp(-k_1 t) \tag{2-5-3}$$

where x_A is the concentration of A at zero time. This is a reduced equation, with y_A the dependent variable, x_A and t the independent variables, and k_1 a parameter. While in this case the differential equation could be solved explicitly, it is not uncommon to find models where the integration can only be performed numerically. Such models are treated in detail in Chapter VIII.

We cannot show here how mathematical models are derived in the various branches of science, but we can cite a few examples to demonstrate that the utility of parameter estimation methods is not confined to the field of chemical reaction kinetics.

(a) Nuclear Physics. Scattering data have been used to estimate parameters referring to nuclear structure or nuclear–nuclear forces [see Melkanoff *et al.* (1966); Arndt and MacGregor (1966)].

(b) Geophysical Exploration. Geophysical surveys are often conducted by flying over the region of interest and recording measured values of variables such as magnetic and gravimetric field intensities. These records are then scanned for anomalies which may indicate the underground presence of valuable ore deposits. Assuming the ore deposit to have given shape, size and location, it is possible to derive expressions for the magnetic and gravimetric fields along the flight paths [see Grant and West (1965)] Although these expressions are very complicated, they can be used (Eisenpress and Surkan, 1966) to estimate ore deposit parameters from aerial survey data.

(c) Biophysics. To study the manner in which substances are transported from one part of an organism to another, biologists conceive of the body as consisting of compartments separated by semipermeable membranes. A tracer substance is injected into one compartment, and its concentration in the other compartments is subsequently measured at various points in time. These data may be used to estimate intercompartmental transport rate parameters (Berman *et al.*, 1962; Turner *et al.*, 1963; Beauchamp and Cornell, 1966).

Another interesting application is the determination of the dipole moments of various sections of the heart from measurements of skin potential (Bellman, Collier, Kagiwada, Kalaba, and Selvester, 1964).

(d) Probability. Given many samples of a random variable having a given probability distribution, we wish to determine parameters (e.g., mean, standard deviation, etc.) appearing in the distribution. This is the classical estimation problem in statistics. A "curve fitting" approach to the problem is to construct a histogram from the data, and fit to it the expression for the probability density function.

(e) Econometrics. Econometricians attempt to construct mathematical models for the national economy or certain segments of it. These models describe the dynamic relationships among variables such as income, sales, production and employment. Parameters appearing in the model may be estimated from past data, and used to predict future trends (Johnston, 1963).

(f) Orbit Calculations. The orbit of a satellite can be expressed as a function of parameters which describe the heavenly bodies that attract the satellite. These parameters can be estimated from the observed orbits (Kelley and Denham, 1966).

All these are examples in which the parameters to be estimated possessed (more or less) a physical significance, and the model equations attempted to represent true cause and effect relationships. The following examples are of a different kind. We attempt to determine design parameters which will confer desirable properties on a device to be constructed.

(g) A smoothing filter is to be installed in an electrical circuit. We calculate the ideal transfer function for the filter by solving the Wiener–Hopf equation (Wiener, 1949). The filter must be constructed from passive elements (resistors, capacitors, and inductors) so that its transfer function can only be a rational function, i.e., the ratio of two polynomials. Our task is to determine the coefficients in the two polynomials so that their ratio approximates the Wiener–Hopf solution as closely as possible.

(h) Designers of artificial limbs attempt to reproduce the observed kinematics of natural limbs. They must estimate the design parameters so as to best approximate the observed motions (Freudenstein and Woo, 1968).

B. Data

2-6. Experiments and Data Matrix

Parameter estimation is based on data, and the data consist of observed or measured values of the model variables. One may obtain the data by observing situations occurring in nature, or one may set up experiments in which conditions are controlled so as to favor the process of observation. In Chapter X we shall go into the question of what experiments should be performed for estimating a given model. For the present, however, it is immaterial where and how the data were obtained, except inasmuch as the measurement process affects the errors in the observations.

In most cases the data gathering process possesses a certain structure. Performing an *experiment* consists of recording the observed values of a set of variables under a given set of *experimental conditions*. Sometimes this means that the dependent variables are observed for given values of the independent variables. Sometimes, however, the experimental conditions themselves are not among the variables of the model. We may, for instance, wish to relate height and weight of individuals in a population. In this case, the individual chosen can be considered the "set of experimental conditions," whereas the height and weight are the model variables.

Frequently, in the course of an investigation, several experiments are performed, each under a different set of experimental conditions. A variable subscripted by a letter μ, η, or ϕ denotes the value of that variable as measured in the corresponding experiment

$$\mathbf{z}_\mu = [z_{\mu 1}, z_{\mu 2}, \ldots, z_{\mu k}]^{\mathrm{T}}$$

are the values of the model variables observed in the μth experiment. A function subscripted with one of these letters denotes that function computed for the values observed in the corresponding experiment

$$\mathbf{g}_\mu(\boldsymbol{\theta}) \equiv \mathbf{g}(\mathbf{z}_\mu, \boldsymbol{\theta}) \tag{2-6-1}$$

We shall use corresponding capital letters to designate the *data matrix*, i.e., the matrix whose μth row consists of the data vector for the μth experiment. Thus, $\mathbf{Z}$ and $\mathbf{G}$ are the matrices whose μth rows are $\mathbf{z}_\mu^{\mathrm{T}}$ and $\mathbf{g}_\mu^{\mathrm{T}}$ respectively, e.g.,

$$\mathbf{Z} \equiv \begin{bmatrix} z_{11} & z_{12} & \cdots & z_{1k} \\ \vdots & & & \\ z_{n1} & z_{n2} & \cdots & z_{nk} \end{bmatrix} \tag{2-6-2}$$

where n is the number of experiments. The definitions of $\mathbf{x}_\mu$, $\mathbf{y}_\mu$, $\mathbf{X}$, and $\mathbf{Y}$ are obvious.

In practice it happens frequently that not all variables are measured in every experiment, or even that the set of dependent variables measured differs completely from one set of experiments to the next. In most cases this will raise no undue difficulties; we simply use the appropriate set of model equations for each experiment. Some of the problems that do arise in this connection are discussed in Section 9-1.

C. Probabilistic Models and Likelihood

2-7. Randomness in Data

Deterministic models describe reality only in an idealized sense. If the values of all the variables were known exactly, and if no forces other than those explicitly considered were at work, then and only then could we expect to find parameter values that cause the model equations to be satisfied exactly.

In practice, we know that measurement techniques possess limited accuracy, that repeated measurements of one and the same quantity yield different values, that the conditions for which the model was derived are never quite attainable, and that disturbances which could not be predicted or taken into account in the model always occur. Yet these unpredictable disturbances are as much parts of physical reality as are the underlying exact quantities which appear in the model. The model is not complete, then, unless it also describes in an appropriate manner these random elements of the situation.

The appropriate description of random phenomena is through probability statements. The following sections will demonstrate the manner in which the deterministic model can be imbedded in the probabilistic description of the data, but first we digress somewhat to describe some probability distributions that are applicable to experimental errors.

2-8. The Normal Distribution

The importance of the normal distribution (defined below) derives from several reasons.

(a) It has been found to approximate closely the behavior of many measurements in nature.

(b) It is the limit which many other distributions approach when the sample size is increased beyond bound. In particular, we have so-called *central limit theorems* (Feller, 1966) which state that, under fairly general conditions, the distribution of the sum of n independent random variables approaches the normal distribution as n is made sufficiently large. Central limit theorems are often used to explain the widespread occurrence of this distribution in nature: if the observed value of the random variable is the resultant of many additive, independent, effects, the resulting distribution is likely to be normal.

In some cases the normal distribution applies not to the variable itself, but to some function of it. For instance, if a given effect is built up over a period of time as the sum of many random effects, each of which has a standard deviation proportional to the magnitude of the overall effect at the time, then the distribution of the logarithm of the overall effect is likely to be normal. This phenomenon is observed in situations relating to the growth of individuals (Cramér, 1946, p. 220).

(c) By specifying the distribution of a random variable, we convey a certain amount of information concerning the values assumed by the variable. A suitable measure of the information contained in the distribution whose probability distribution function (pdf) is $p(\mathbf{x})$ is given by

$$I(p) \equiv E(\log p) = \int p(\mathbf{x}) \log p(\mathbf{x})\, d\mathbf{x} \tag{2-8-1}$$

(Shannon, 1948; see also Section 10-2). Consider the following situation: A scientist knows that the measuring errors of some apparatus have mean μ and standard deviation σ. For certain reasons (which should become clear in the sequel) the scientist is compelled to assign a pdf $p(x)$ to the measurement errors. This pdf will later be used to make inferences concerning the true values of the measured variable. In the absence of any further information, what function $p(x)$ should be chosen?

The function $p(x)$ must satisfy the following conditions:

1. It is a pdf, i.e., $p(x) \geqslant 0$, and

$$\int_{-\infty}^{\infty} p(x)\, dx = 1 \tag{2-8-2}$$

2. Its mean is as specified, i.e.,

$$\int_{-\infty}^{\infty} xp(x)\, dx = \mu \tag{2-8-3}$$

3. Its variance is as specified, i.e.,

$$\int_{-\infty}^{\infty} (x-\mu)^2 p(x)\, dx = \sigma^2 \tag{2-8-4}$$

It is reasonable to select from those functions $p(x)$ satisfying these conditions *the one whose information content is least.* By doing so, we are adding the smallest possible additional information over and above what we legitimately know (i.e., the values of μ and σ).

Finding $p(x)$ such that $I(p)$ is minimized and Eqs. (2)–(4) are satisfied is an exercise in the calculus of variations. Following standard procedures, we form the Lagrangian functional

$$\Lambda(p) \equiv \int_{-\infty}^{\infty} [p \log p + \lambda_1 p + \lambda_2 xp + \lambda_3 (x-\mu)^2 p]\, dx \qquad (2\text{-}8\text{-}5)$$

where the λ_i are Lagrange multipliers (see Section 3-6). The Euler equation that p must satisfy to make $\Lambda(p)$ stationary is obtained by differentiating with respect to p the expression under the integral sign

$$\log p + 1 + \lambda_1 + \lambda_2 x + \lambda_3 (x - \mu)^2 = 0 \qquad (2\text{-}8\text{-}6)$$

Hence

$$p(x) = \exp[-1 - \lambda_1 - \lambda_2 x - \lambda_3 (x - \mu)^2] \qquad (2\text{-}8\text{-}7)$$

The values of the λ_i can be determined by substituting Eq. (7) in Eqs. (2)–(4). Using the relation $\int_{-\infty}^{\infty} \exp(-\lambda u^2)\, du = (\pi/\lambda)^{1/2}$, we find ultimately that

$$\lambda_1 = \tfrac{1}{2} \log 2\pi + \log \sigma - 1, \qquad \lambda_2 = 0, \qquad \lambda_3 = (1/2\sigma^2) \qquad (2\text{-}8\text{-}8)$$

Hence

$$p(x) = [1/(2\pi)^{1/2}\sigma] \exp[-(1/2\sigma^2)(x - \mu)^2] \qquad (2\text{-}8\text{-}9)$$

This is the *univariate normal distribution* with mean μ and variance σ^2. We designate this distribution $N_1(\mu, \sigma^2)$.

When $\mathbf{x}$ is an n-dimensional vector random variable with mean $\boldsymbol{\mu}$ and nonsingular covariance matrix $\mathbf{V}$, we find by similar arguments that the least informative pdf has the form

$$p(\mathbf{x}) = (2\pi)^{-n/2} \det^{-1/2} \mathbf{V} \exp[-\tfrac{1}{2}(\mathbf{x} - \boldsymbol{\mu})^{\mathrm{T}} \mathbf{V}^{-1} (\mathbf{x} - \boldsymbol{\mu})] \qquad (2\text{-}8\text{-}10)$$

which is the *multivariate normal distribution* with mean $\boldsymbol{\mu}$ and covariance matrix $\mathbf{V}$. We designate this distribution $N_n(\boldsymbol{\mu}, \mathbf{V})$.

To summarize: when we specify only the mean and variance of a random variable, we have not determined the entire distribution. If an entire distribution is demanded, however, then by specifying the normal distribution we assume the least possible amount of extraneous information.

(d) The normal distribution is particularly tractable mathematically. Many results can be worked out explicitly only for this distribution. Therefore, it is frequently convenient to assume a normal distribution where no specific justification for it exists. This is unlikely to cause much harm, except when the estimation method selected is very sensitive to the shape of the tails of the distributions.

A normal distribution of an n-dimensional random vector $\mathbf{x}$ is completely characterized by the mean $\boldsymbol{\mu}$ and the covariance matrix $\mathbf{V}$, as shown by Eq. (10). We assumed that $\mathbf{V}$ was nonsingular; otherwise $\mathbf{V}^{-1}$ could not have been formed. When $\mathbf{V}$ is singular, i.e., $\det \mathbf{V} = 0$, we speak of a *singular normal distribution.* If $m < n$ is the rank of $\mathbf{V}$, then there exist m linear combinations of the $\mathbf{x}$ which possess a nonsingular normal distribution.‡

In a normal distribution, uncorrelated variables are independent. The mean, mode, and median coincide. The following are additional useful properties of the normal distribution. We assume throughout that $\mathbf{x}$ is $N_n(\mathbf{0}, \mathbf{V})$ (this is just another way of saying that $\mathbf{x}$ is an n-dimensional normally distributed random vector with mean $\mathbf{0}$ and covariance $\mathbf{V}$). Then

1. $\mathbf{Ax}$ is $N_m(\mathbf{0}, \mathbf{AVA}^{\mathrm{T}})$, where $\mathbf{A}$ is any $m \times n$ matrix.
2. Let $\mathbf{C}$ be an $n \times n$ matrix, such that $\mathbf{C}^{\mathrm{T}}\mathbf{C} = \mathbf{V}^{-1}$. Then $\mathbf{y} \equiv \mathbf{Cx}$ is $N_n(\mathbf{0}, \mathbf{I})$, i.e., the elements of $\mathbf{y}$ are n independent normal variables with zero means and unit variances. Such variables are called *standard normal deviates.*
3. If $\mathbf{y}$ is $N_n(\mathbf{0}, \mathbf{I})$, then $\mathbf{y}^{\mathrm{T}}\mathbf{y}$ is a random variable whose distribution is called chi-square with n degrees of freedom, designated χ_n^2. In other words, χ_n^2 is the distribution of the sum of squares of n independent standard normal deviates.
4. We have $\mathbf{y}^{\mathrm{T}}\mathbf{y} = \mathbf{x}^{\mathrm{T}}\mathbf{C}^{\mathrm{T}}\mathbf{Cx} = \mathbf{x}^{\mathrm{T}}\mathbf{V}^{-1}\mathbf{x}$. Hence, $\mathbf{x}^{\mathrm{T}}\mathbf{V}^{-1}\mathbf{x}$ is χ_n^2.
5. If $\mathscr{A}$ and $\mathscr{B}$ are independent random variables with distributions χ_p^2 and χ_q^2, respectively, then $q\mathscr{A}/p\mathscr{B}$ is a random variable whose distribution is designated $F_{p,q}$.

The χ^2 and F distributions play an important role in establishing confidence intervals for estimated parameters, and in testing the goodness of fit of the model to the data (see Chapter VII).

2-9. The Uniform Distribution

The uniform distribution (also called *rectangular*) is one in which the range of possible values of each variable is confined to a finite interval, and all values in the interval are equally likely. Thus, if $\mathbf{a}$ and $\mathbf{b}$ are n-dimensional

‡ These are the principal components corresponding to nonzero eigenvalues of $\mathbf{V}$. See Section 7-8.

vectors with $b_i > a_i$ for $i = 1, 2, \ldots, n$, then the uniform distribution within the n dimensional rectangle $a_i \leqslant x_i \leqslant b_i$ is given by:

$$p(\mathbf{x}) = [(b_1 - a_1)(b_2 - a_2) \cdots (b_n - a_n)]^{-1} \qquad \text{for } a_i \leqslant x_i \leqslant b_i,\ i = 1, 2, \ldots, n \tag{2-9-1}$$

$$p(\mathbf{x}) = 0 \qquad \text{otherwise}$$

For this distribution we have:

$$\bar{\mathbf{x}} = \tfrac{1}{2}(\mathbf{a} + \mathbf{b})$$

$$V_{ii} = (1/12)(b_i - a_i)^2 \tag{2-9-2}$$

$$V_{ij} = 0 \qquad (i \neq j) \tag{2-9-3}$$

The components of the vector $\mathbf{x}$ are independent.

→ The uniform distribution frequently describes the errors of measurement due to the limited number of significant digits that can be read on a scale, because all intermediate values between scale marks are equally probable.

The assumption of a uniform distribution with known bounds for all the errors in a model implies that all these errors are restricted in magnitude. This means that the model must be rejected if no parameter values can be found that keep all the errors within the permitted bounds. The use of this distribution can be justified only if one is willing to accept such drastic conclusions. One must be quite certain that the measurements really differ from the true values of the variables by no more than the specified error bounds, and that no other random factors have been overlooked in the model. In contrast, the normal distribution assigns nonzero (though small) probabilities to any error, no matter how large. It is more forgiving towards inadequacies in the model, and does not break down upon the appearance of an occasional unexpectedly large error. This objection to the uniform distribution does not apply if the upper bound on the error magnitude is not known in advance.

2-10. Distribution of Errors

Attempting to relate the deterministic model to the data gathered from n experiments, we are led to the set of equations

$$\mathbf{g}_\mu(\boldsymbol{\theta}) = \mathbf{0} \qquad (\mu = 1, 2, \ldots, n) \tag{2-10-1}$$

The total number of equations in Eq. (1) usually far exceeds the number of unknown parameters $\boldsymbol{\theta}$. Only under exceptional circumstances do there exist values of $\boldsymbol{\theta}$ which cause all the equations in Eq. (1) to be satisfied. Indeed we cannot expect all these equations to be satisfied, since

1. The measured values of the variables do not always represent their true values.
2. The model is not exactly accurate, various effects having been neglected in its formulation.

To account for errors of type 1, we scan the list of all the measured quantities $\mathbf{Z}$, and break it up into two sets: quantities $\mathbf{U}$ which are believed to be free of significant error, and quantities $\mathbf{W}$ whose measured values may differ significantly, in a random manner, from their true underlying values, which we designate $\hat{\mathbf{W}}$. The difference between the measured and true values of a variable we call the *error*

$$\mathbf{E} \equiv \mathbf{W} - \hat{\mathbf{W}} \tag{2-10-2}$$

that is

$$\boldsymbol{\varepsilon}_\mu \equiv \mathbf{w}_\mu - \hat{\mathbf{w}}_\mu \qquad (\mu = 1, 2, \ldots, n)$$

We now assume that each $w_{\mu a}$ is a realization of a random variable $\omega_{\mu a}$, or, equivalently, that $\mathbf{W}$ is a realization of a matrix random variable $\boldsymbol{\Omega}$. This means that $\mathbf{W}$ is one *sample* out of all possible results of our series of n experiments. Furthermore, we assume that the random variables $\omega_{\mu a}$ possess a joint pdf which depends on the true values $\hat{w}_{\mu a}$, as well as on some parameters $\boldsymbol{\psi}$, whose values may or may not be known. Thus, the pdf has the form $p(\boldsymbol{\Omega} | \hat{\mathbf{W}}, \boldsymbol{\psi})$. It is usually the case that the pdf depends explicitly on the $\boldsymbol{\Omega}$ and $\hat{\mathbf{W}}$ only through their difference, i.e., it has the form $p(\mathbf{E} | \boldsymbol{\psi})$. It is also frequently the case that the errors in different experiments are statistically independent. That means that we have a pdf $p_\mu(\boldsymbol{\varepsilon}_\mu | \boldsymbol{\psi}_\mu)$ associated with the errors $\boldsymbol{\varepsilon}_\mu$ in the μth experiment, and the joint pdf for all experiments is given by

$$p(\mathbf{E} | \boldsymbol{\psi}_1, \boldsymbol{\psi}_2, \ldots, \boldsymbol{\psi}_n) = \prod_{\mu=1}^{n} p_\mu(\boldsymbol{\varepsilon}_\mu | \boldsymbol{\psi}_\mu) \tag{2-10-3}$$

To illustrate, assume that the errors in the μth experiment are distributed as $N_r(\mathbf{0}, \mathbf{V}_\mu)$. Then

$$p_\mu(\boldsymbol{\varepsilon}_\mu | \mathbf{V}_\mu) = (2\pi)^{-r/2} \det{}^{-1/2} \mathbf{V}_\mu \exp(-\tfrac{1}{2}\boldsymbol{\varepsilon}_\mu{}^{\mathrm{T}}\mathbf{V}_\mu^{-1}\boldsymbol{\varepsilon}_\mu) \tag{2-10-4}$$

Hence, the joint pdf is given by

$$p(\mathbf{E} | \mathbf{V}_1, \mathbf{V}_2, \ldots, \mathbf{V}_n) = (2\pi)^{-nr/2} \prod_{\mu=1}^{n} \det{}^{-1/2} \mathbf{V}_\mu \exp\left(-\tfrac{1}{2} \sum_{\mu=1}^{n} \boldsymbol{\varepsilon}_\mu{}^{\mathrm{T}}\mathbf{V}_\mu^{-1}\boldsymbol{\varepsilon}_\mu\right) \tag{2-10-5}$$

The vector of distribution parameters $\boldsymbol{\psi}$ here consists of the elements of the matrices $\mathbf{V}_\mu$.

It must be remembered that when we speak here of random variables we are referring to the results of the measurements, not to the choice of experimental conditions. In many cases, the experimental conditions are selected randomly, e.g., by drawing individuals at random from a population. This does not concern us here; once the individual has been chosen, he ceases to be random. What we are interested in are the random differences that may arise between repeated measurements on the same individual.

The values of $\mathbf{w}_\mu$ and $\mathbf{w}_\eta$ for $\mu \neq \eta$ are usually realizations of different vector random variables $\boldsymbol{\omega}_\mu$ and $\boldsymbol{\omega}_\eta$. Only in the case when experiments μ and η are replications of each other are $\mathbf{w}_\mu$ and $\mathbf{w}_\eta$ realizations of one and the same random variable.

2-11. Stochastic Form of the Model

How can the deterministic model be modified so as to account for the variability in the data and model? There are several ways in which this can be done. The specific form chosen should depend on what we know about the system being described. Do we have strong confidence in the model but not in the data? Do we trust the data but not the model? Perhaps both are subject to significant errors? The type of model that is appropriate in a given situation depends on the answers to these questions. We list below some of the forms that the model may take. A typical example which illustrates the conditions under which these forms are appropriate follows in Section 2-14.

(a) Suppose the data are subject to measurement errors, but the model equations are thought to apply exactly to the true (though unknown) values of the variables

$$\mathbf{g}(\mathbf{u}_\mu, \hat{\mathbf{w}}_\mu, \boldsymbol{\theta}) = \mathbf{0} \qquad (\mu = 1, 2, \ldots, n) \tag{2-11-1}$$

We refer to Eq. (1) as an *exact structural model*. The measurement errors $\mathbf{E} = \mathbf{W} - \hat{\mathbf{W}}$ are assumed to have a joint pdf $p(\mathbf{E} \mid \boldsymbol{\psi})$.

(b) Suppose the model equations apply only approximately even to the true values of the variables. The error in the model at the μth experiment is assumed to be a random variable $\boldsymbol{\gamma}_\mu$

$$\mathbf{g}(\mathbf{u}_\mu, \hat{\mathbf{w}}_\mu, \boldsymbol{\theta}) = \boldsymbol{\gamma}_\mu \qquad (\mu = 1, 2, \ldots, n) \tag{2-11-2}$$

We refer to Eq. (2) as an *inexact structural model*. In conformity with our usual notation, we let $\boldsymbol{\Gamma}$ be the matrix whose μth row is $\boldsymbol{\gamma}_\mu^{\mathrm{T}}$. The $\boldsymbol{\gamma}_\mu$ are supposed to account for forces that were neglected in the formulation of the model. One usually assumes that the γ_μ have zero means, and that they are statistically independent of the measurement errors. Then the overall pdf applicable to this model has the form $p(\mathbf{E}|\boldsymbol{\psi})p(\boldsymbol{\Gamma}|\boldsymbol{\psi}')$, where $\boldsymbol{\psi}'$ is an additional set of distribution parameters.

(c) A special case of (b) occurs when all variables are measured precisely so that **w** is vacuous. Then

$$\mathbf{g}(\mathbf{u}_\mu, \boldsymbol{\theta}) = \boldsymbol{\gamma}_\mu \qquad (\mu = 1, 2, \ldots, n) \tag{2-11-3}$$

The relevant pdf is simply $p(\boldsymbol{\Gamma}|\boldsymbol{\psi}')$. Let us introduce an artificial variable $\mathbf{y}_\mu$ to which we assign the "observed" value zero, and let us define $\boldsymbol{\varepsilon}_\mu \equiv -\boldsymbol{\gamma}_\mu$. Then Eq. (3) is equivalent to $\mathbf{y}_\mu = \mathbf{g}(\mathbf{u}_\mu, \boldsymbol{\theta}) + \boldsymbol{\varepsilon}_\mu$, which has the same form as the reduced model Eq. (9) discussed below.

(d) In some applications, particularly in the field of econometrics, it has been found appropriate not to introduce the "true value" $\hat{\mathbf{w}}_\mu$ explicitly, but rather to treat the model equations as applying approximately to the measured values

$$\mathbf{g}(\mathbf{u}_\mu, \mathbf{w}_\mu, \boldsymbol{\theta}) = \boldsymbol{\gamma}_\mu \qquad (\mu = 1, 2, \ldots, n) \tag{2-11-4}$$

where $\boldsymbol{\gamma}_\mu$ is an error term which is not treated as a random variable in its own right. Rather, one assumes that $\boldsymbol{\omega}_\mu$ is a random variable distributed in such a way that $\boldsymbol{\gamma}_\mu = \mathbf{g}(\mathbf{u}_\mu, \boldsymbol{\omega}_\mu, \boldsymbol{\theta})$ (regarded as a function of $\boldsymbol{\omega}_\mu$) has a given pdf $\tilde{p}(\boldsymbol{\gamma}_\mu)$. The pdf for the original variable $\boldsymbol{\omega}_\mu$ is then obtained according to the rules for transforming variables in probability distributions

$$p(\boldsymbol{\omega}_\mu) = \tilde{p}(\boldsymbol{\gamma}_\mu)\,|\det(\partial \mathbf{g}_\mu/\partial \boldsymbol{\omega}_\mu)| \tag{2-11-5}$$

The quantity $\det(\partial \mathbf{g}_\mu/\partial \mathbf{w}_\mu)$ is the *Jacobian* of the transformation from $\mathbf{w}_\mu$ to $\mathbf{g}_\mu$, and the dimension of **w** must be the same as that of **g**. The econometricians refer to the **w** as the *endogenous* variables.

Example The following two-equation production model is due to Bodkin and Klein [1967, Eqs. (12) and (18)]

$$\mathbf{g}_\mu \equiv \begin{cases} g_{\mu 1} \equiv w_{\mu 1} - \theta_3\, \theta_2^{u_{\mu 1}} w_{\mu 2}^{(1-\theta_1)} = \gamma_{\mu 1} \\ g_{\mu 2} \equiv w_{\mu 1} - u_{\mu 2}/\theta_1 = \gamma_{\mu 2} \end{cases} \tag{2-11-6}$$

where w_1 is the ratio of real production output to labor input, w_2 the ratio of capital input to labor input, u_1 the time, and u_2 the ratio of wage rate to price of output. Here

$$\det(\partial \mathbf{g}_\mu/\partial \mathbf{w}_\mu) = \begin{vmatrix} 1 & -(1-\theta_1)\theta_3\, \theta_2^{u_{\mu 1}} w_{\mu 2}^{-\theta_1} \\ 1 & 0 \end{vmatrix} = (1-\theta_1)\theta_3\, \theta_2^{u_{\mu 1}} w_{\mu 2}^{-\theta_1} \tag{2-11-7}$$

If $[\gamma_{\mu 1}, \gamma_{\mu 2}]$ are assumed distributed as $N_2(\mathbf{0}, \mathbf{V})$, then $[w_{\mu 1}, w_{\mu 2}]$ have the pdf

$$p(\mathbf{w}_\mu) = |(1-\theta_1)\theta_3 \theta_2^{u_{\mu 1}} w_{\mu 2}^{-\theta_1}|(2\pi)^{-1}(\det \mathbf{V})^{-1/2} \exp(-\tfrac{1}{2}\mathbf{g}_\mu{}^T \mathbf{V}^{-1} \mathbf{g}_\mu) \quad (2\text{-}11\text{-}8)$$

(e) Suppose the dimension of **w** is equal to the dimension of **g**, and suppose further that the structural equations $\mathbf{g}(\mathbf{u}, \mathbf{w}, \boldsymbol{\theta}) = \mathbf{0}$ can be solved for **w**. Then we obtain the reduced model $\mathbf{w} = \mathbf{f}(\mathbf{u}, \boldsymbol{\theta})$. In the μth experiment there may be errors of two kinds; errors $\boldsymbol{\varepsilon}_{\mu 1}$ in the measurement of **w**, and errors $\boldsymbol{\varepsilon}_{\mu 2}$ in the model equations. In conformity with usual practice when dealing with reduced models, we write **y** in place of **w** and **x** in place of **u**. The model now takes the form

$$\mathbf{y}_\mu = \mathbf{f}(\mathbf{x}_\mu, \boldsymbol{\theta}) + \boldsymbol{\varepsilon}_\mu = \mathbf{f}_\mu(\boldsymbol{\theta}) + \boldsymbol{\varepsilon}_\mu \quad (2\text{-}11\text{-}9)$$

where $\boldsymbol{\varepsilon}_\mu \equiv \boldsymbol{\varepsilon}_{\mu 1} + \boldsymbol{\varepsilon}_{\mu 2}$. If we define $\hat{\mathbf{y}}_\mu \equiv \mathbf{y}_\mu - \boldsymbol{\varepsilon}_\mu$, then we may write the model as

$$\hat{\mathbf{y}}_\mu = \mathbf{f}(\mathbf{x}_\mu, \boldsymbol{\theta}) = \mathbf{f}_\mu(\boldsymbol{\theta}) \quad (2\text{ }11\text{-}10)$$

The quantity $\hat{\mathbf{y}}_\mu$ cannot legitimately be thought of as a "true" value of $\mathbf{y}_\mu$ unless $\boldsymbol{\varepsilon}_{\mu 2}$ is negligible. The relevant pdf for the model Eq. (9) has the form $p(\boldsymbol{\varepsilon}_{\mu 1}, \boldsymbol{\varepsilon}_{\mu 2})$, but in practice the dual nature of the errors is usually ignored, and the pdf is written simply as $p(\boldsymbol{\varepsilon}_\mu)$. The joint pdf for all the errors has the form $p(\mathbf{E}|\boldsymbol{\psi})$.

We refer to a reduced model in which the independent variables **x** are measured precisely as a *standard reduced model*. The appropriate representation of such a model is given by Eq. (9) or, equivalently, by Eq. (10). Of all nonlinear models, this is the one for which the calculation of the estimates is easiest. For this reason, it is tempting to neglect errors in $\mathbf{x}_\mu$ in any reduced model, regardless of whether this is justified in physical fact. The resulting errors in the estimates are difficult to predict, and a Monte Carlo study would be appropriate (see Section 3-3). Be that as it may, the vast majority of all nonlinear estimation calculations have in practice been undertaken on the implicit assumption that the model was in standard reduced form.

2-12. Likelihood–Standard Reduced Model

The standard reduced model Eq. (2-11-9) can be put in the more concise form

$$\mathbf{Y} = \mathbf{F}(\mathbf{X}, \boldsymbol{\theta}) + \mathbf{E} \quad (2\text{-}12\text{-}1)$$

Suppose the model is specified, along with the joint pdf $p(\mathbf{E}|\boldsymbol{\psi})$ and with the data **Y**, **X**. For any given values of the parameters $\boldsymbol{\theta}$ we can compute the *residuals*

$$\mathbf{E}(\boldsymbol{\theta}) \equiv \mathbf{Y} - \mathbf{F}(\mathbf{X}, \boldsymbol{\theta}) \quad (2\text{-}12\text{-}2)$$

i.e., the differences $e_{\mu a}(\boldsymbol{\theta})$ between the observed values $y_{\mu a}$ and the "computed" values $f_a(\mathbf{x}_\mu, \boldsymbol{\theta})$ of the dependent variables. If $\boldsymbol{\theta}$ is close to the true value $\hat{\boldsymbol{\theta}}$, then $\mathbf{E}(\boldsymbol{\theta})$ should be close to the true errors $\mathbf{E}$. In the joint pdf, let us replace the errors by the expressions for the residuals. The resulting expression, which is a function of $\boldsymbol{\theta}$ and $\boldsymbol{\psi}$ alone, is called the *likelihood function* of the sample

$$L(\boldsymbol{\theta}, \boldsymbol{\psi}) \equiv p(\mathbf{E}(\boldsymbol{\theta}) \,|\, \boldsymbol{\psi}) = p(\mathbf{Y} - \mathbf{F}(\mathbf{X}, \boldsymbol{\theta}) \,|\, \boldsymbol{\psi}) \tag{2-12-3}$$

Note that since $\mathbf{X}$ and $\mathbf{Y}$ are known quantities, they do not appear as variables among the arguments of the likelihood function.

As an example, suppose the pdf is given by Eq. (2-10-5), i.e., the errors in the μth experiment are distributed as $N_m(\mathbf{0}, \mathbf{V}_\mu)$, and errors in different experiments are independent. The likelihood is obtained by substituting $\mathbf{y}_\mu - \mathbf{f}(\mathbf{x}_\mu, \boldsymbol{\theta})$ for $\boldsymbol{\varepsilon}_\mu$ in Eq. (2-10-5)

$$L(\boldsymbol{\theta}, \mathbf{V}_1, \mathbf{V}_2, \ldots, \mathbf{V}_n) = (2\pi)^{-nm/2} \prod_{\mu=1}^{n} \det{}^{-1/2} \mathbf{V}_\mu$$

$$\times \exp\left\{ -\tfrac{1}{2} \sum_{\mu=1}^{n} [\mathbf{y}_\mu - \mathbf{f}(\mathbf{x}_\mu, \boldsymbol{\theta})]^{\mathrm{T}} \mathbf{V}_\mu^{-1} [\mathbf{y}_\mu - \mathbf{f}(\mathbf{x}_\mu, \boldsymbol{\theta})] \right\} \tag{2-12-4}$$

The likelihood function can be defined in more generality as follows: take the joint pdf of the deviations or errors, and substitute for all random variables their sample values in the form of expressions involving measured variables and unknown parameters; the resulting expression is the likelihood function. In the next section we carry out this procedure for several additional models.

2-13. Likelihood–Structural Models

(a) Exact Structural Model. Referring to Eq. (2-11-1) we find that the $\hat{\mathbf{w}}_\mu$ appear as additional unknown parameters in the model. We define the residuals here as the differences between the measured values $\mathbf{w}_\mu$ and any particular assumed values for $\hat{\mathbf{w}}_\mu$, i.e.,

$$\mathbf{E}(\hat{\mathbf{W}}) \equiv \mathbf{W} - \hat{\mathbf{W}} \tag{2-13-1}$$

Hence the likelihood function, as derived from the pdf $p(\mathbf{E} \,|\, \boldsymbol{\psi})$, has the form

$$L(\hat{\mathbf{W}}, \boldsymbol{\psi}) \equiv p(\mathbf{W} - \hat{\mathbf{W}} \,|\, \boldsymbol{\psi}) \tag{2-13-2}$$

The parameters $\hat{\mathbf{W}}$ are not free to assume any values whatsoever; they are constrained to satisfy the structural Eqs. (2-11-1).

As an example, when the joint pdf is given by Eq. (2-10-5), all we have to do is substitute $\mathbf{w}_\mu - \hat{\mathbf{w}}_\mu$ for $\boldsymbol{\varepsilon}_\mu$ to form the likelihood

$$L(\mathbf{V}_1, \mathbf{V}_2, \ldots, \mathbf{V}_n, \hat{\mathbf{w}}_1, \hat{\mathbf{w}}_2, \ldots, \hat{\mathbf{w}}_n) = (2\pi)^{-nr/2} \prod_{\mu=1}^{n} \det^{-1/2}\mathbf{V}_\mu \times \exp\left[-\tfrac{1}{2}\sum_{\mu=1}^{n} (\mathbf{w}_\mu - \hat{\mathbf{w}}_\mu)^{\mathrm{T}}\mathbf{V}_\mu^{-1}(\mathbf{w}_\mu - \hat{\mathbf{w}}_\mu)\right] \tag{2-13-3}$$

Note again that $\mathbf{w}_1, \mathbf{w}_2, \ldots, \mathbf{w}_n$ are known vectors (being the measured data). Hence they do not appear among the arguments of L.

Since an exact structural model requires a large number of additional unknown parameters $\hat{\mathbf{W}}$, it is desirable to transform it to reduced form when possible.

(b) Inexact Structural Model. Let the model be described by Eq. (2-11-2). Once more, the residuals are defined by Eq. (1), and they take the place of $\mathbf{E}$ in the pdf $p(\mathbf{E}\,|\,\boldsymbol{\psi})p(\boldsymbol{\Gamma}\,|\,\boldsymbol{\psi}')$. An expression for $\boldsymbol{\Gamma}$ is obtained simply by evaluating Eq. (2-11-2) for specific values of $\hat{\mathbf{W}}$ and $\boldsymbol{\theta}$, i.e., by substituting $\mathbf{G}(\mathbf{U}, \hat{\mathbf{W}}, \boldsymbol{\theta})$ for $\boldsymbol{\Gamma}$. Thus

$$L(\hat{\mathbf{W}}, \boldsymbol{\theta}, \boldsymbol{\psi}, \boldsymbol{\psi}') \equiv p(\mathbf{W} - \hat{\mathbf{W}}\,|\,\boldsymbol{\psi})p[\mathbf{G}(\mathbf{U}, \hat{\mathbf{W}}, \boldsymbol{\theta})\,|\,\boldsymbol{\psi}'] \tag{2-13-4}$$

In this case, no restrictions apply to the $\boldsymbol{\theta}$ and $\hat{\mathbf{W}}$. As an example, suppose the errors in $\mathbf{W}$ are again distributed as Eq. (2-10-5) and that the $\boldsymbol{\gamma}_\mu$ are similarly distributed as $N_m(\mathbf{0}, \mathbf{Q}_\mu)$. We obtain the likelihood function by substituting $\mathbf{w}_\mu - \hat{\mathbf{w}}_\mu$ for $\boldsymbol{\varepsilon}_\mu$, and $\mathbf{g}_\mu$ for $\boldsymbol{\gamma}_\mu$

$$L(\boldsymbol{\theta}, \mathbf{V}_1, \mathbf{V}_2, \ldots, \mathbf{V}_n, \mathbf{Q}_1, \mathbf{Q}_2, \ldots, \mathbf{Q}_n, \hat{\mathbf{w}}_1, \hat{\mathbf{w}}_2, \ldots, \hat{\mathbf{w}}_n) = (2\pi)^{-(n/2)(r+m)} \prod_{\mu=1}^{n} (\det^{-1/2}\mathbf{Q}_\mu \det^{-1/2}\mathbf{V}_\mu) \times \exp\left\{-\tfrac{1}{2}\sum_{\mu=1}^{n} [(\mathbf{w}_\mu - \hat{\mathbf{w}}_\mu)^{\mathrm{T}}\mathbf{V}_\mu^{-1}(\mathbf{w}_\mu - \hat{\mathbf{w}}_\mu) + \mathbf{g}_\mu^{\mathrm{T}}(\mathbf{u}_\mu, \hat{\mathbf{w}}_\mu, \boldsymbol{\theta})\mathbf{Q}_\mu^{-1}\mathbf{g}_\mu(\mathbf{u}_\mu, \hat{\mathbf{w}}_\mu, \boldsymbol{\theta})]\right\} \tag{2-13-5}$$

Again, $\mathbf{u}_\mu$ and $\mathbf{w}_\mu$, being known vectors, do not appear as variables among the arguments of L.

(c) For the econometric models discussed under (d) in Section 2-11, the likelihood function is found by multiplying the terms Eq. (2-11-5) for all values of μ, i.e.,

$$L(\boldsymbol{\theta}) = \prod_{\mu=1}^{n} p(\mathbf{w}_\mu) = \prod_{\mu=1}^{n} \tilde{p}(\mathbf{g}_\mu)\,|\det(\partial\mathbf{g}_\mu/\partial\mathbf{w}_\mu)| \tag{2-13-6}$$

For the model of Eqs. (2-11-6) this turns out to be

$$L(\boldsymbol{\theta}, \mathbf{V}) = [(1 - \theta_1)\theta_3(2\pi)^{-1}(\det \mathbf{V})^{-1/2}]^n$$

$$\times\, \theta_2^{\sum_{\mu=1}^{n} u_{\mu 1}} \left[\prod_{\mu=1}^{n} w_{\mu 2}\right]^{-\theta_1} \exp\left[-\tfrac{1}{2}\sum_{\mu=1}^{n} \mathbf{g}_\mu{}^{\mathrm{T}}\mathbf{V}^{-1}\mathbf{g}_\mu\right] \qquad \text{(2-13-7)}$$

The reducibility of an exact model $\mathbf{g}(\mathbf{u}_\mu, \hat{\mathbf{w}}_\mu, \boldsymbol{\theta}) = \mathbf{0}$ to the form $\hat{\mathbf{w}} = \mathbf{f}(\mathbf{u}_\mu, \boldsymbol{\theta})$ depends primarily on the relation between the number m of equations and the number r of random variables per experiment. We distinguish three cases:

1. $r = m$. Except in certain singular cases (vanishing Jacobian) the equations may, in principle, be solved for the $\hat{\mathbf{w}}_\mu$ (although the solution may not be unique). Even when the solution cannot be exhibited in explicit form, it can be computed numerically. Thus, at least in principle, the $\hat{\mathbf{w}}_\mu$ may be eliminated from the likelihood function, which remains a function of $\boldsymbol{\theta}$ and $\boldsymbol{\psi}$ alone.
2. $r > m$. Not all the $\hat{\mathbf{w}}_\mu$ can be eliminated. We can, however, choose m of the $\hat{\mathbf{w}}_\mu$, solve for those, and substitute in the likelihood function. This leaves us with only $r - m$ unknown $\hat{\mathbf{w}}_\mu$ per experiment, and these are unrestricted in value.
3. $r < m$. In this case, we may solve r of the equations for the $\hat{\mathbf{w}}_\mu$, and use these to eliminate the $\hat{\mathbf{w}}_\mu$ from the likelihood and from the remaining $m - r$ model equations. For each experiment there remain $m - r$ equations, containing only $\mathbf{u}_\mu$ and $\boldsymbol{\theta}$. Therefore, if n is the number of experiments, the total number of equations is $(m - r)n$, and the model must contain at least $(m - r)n$ unknown parameters $\boldsymbol{\theta}$ for there to be a solution. In most cases we find that the number of random variables per experiment must at least equal the number of equations.

2-14. An Example

The following example should clarify the conditions under which the various types of model are appropriate. A sphere of radius r and mass m is dropping freely through an incompressible Newtonian fluid of viscosity μ. The force of gravity acting on the sphere is $g(m - m_0)$, where m_0 is the mass of the fluid displaced by the sphere. According to Stokes's law, the drag opposing the motion of the sphere (when the motion is slow) is $6\pi r\mu v$, where v is the velocity of the sphere. Newton's first law of motion takes the form

$$m\dot{v} = g(m - m_0) - 6\pi r\mu v \qquad \text{(2-14-1)}$$

where $\dot{v} \equiv dv/dt$ is the acceleration. We may rewrite this as

$$\dot{v} + \alpha v = \beta \tag{2-14-2}$$

where

$$\alpha \equiv 6\pi r\mu/m, \qquad \beta \equiv g(m - m_0)/m \tag{2-14-3}$$

Assuming that the sphere was initially at rest, we can integrate Eq. (2) to find

$$v(t) = (\beta/\alpha)(1 - e^{-\alpha t}) \tag{2-14-4}$$

The distance s traveled by the sphere since the inception of its motion is

$$s = \int_0^t v(\tau)\, d\tau = (\beta/\alpha)t - (\beta/\alpha^2)(1 - e^{-\alpha t})$$

which can be translated into the "model"

$$g(s, t, \alpha, \gamma) \equiv s - \gamma[t - (1/\alpha)(1 - e^{-\alpha t})] = 0 \tag{2-14-5}$$

where

$$\gamma \equiv \beta/\alpha = g(m - m_0)/6\pi r\mu \tag{2-14-6}$$

Suppose we have measurements $s_1, s_2, \ldots, s_n$ recorded when the clock indicated times $t_1, t_2, \ldots, t_n$. We are interested in estimating some of the physical constants appearing in the model. At the outset it is clear that the model equation contains only two independent parameters. Hence only two of the physical constants g, r, m, m_0, μ appearing in the model can be estimated independently. Since all the information contained in Eq. (5) relative to these constants is derivable from the values of α and γ, we shall assume that these are the parameters to be estimated. We examine the following cases, in all of which it is assumed that errors in different measurements are statistically independent. The parameters of the error distributions are represented as ψ.

(a) The model equation Eq. (5) is exact, i.e., any systematic deviations from it are negligible compared to measurement errors.

1. The measurements of t_μ are precise, but those of s_μ are subject to errors with pdf $p(\varepsilon|\psi)$. Eq. (5) becomes in reduced form

$$\hat{s}_\mu = \gamma[t_\mu - (1/\alpha)(1 - \exp(-\alpha t_\mu))] \qquad (\mu = 1, 2, \ldots, n) \tag{2-14-7}$$

and the likelihood is

$$L(\alpha, \gamma, \psi) = \prod_{\mu=1}^{n} p\{s_\mu - \gamma[t_\mu - (1/\alpha)(1 - \exp(-\alpha t_\mu))]|\psi\} \tag{2-14-8}$$

2. Both t_μ and s_μ are subject to measurement errors ε_t and ε_s, respectively, with pdf $p(\varepsilon_t, \varepsilon_s)|\psi)$. We have the exact structural model

$$\hat{s}_\mu - \gamma[\hat{t}_\mu - (1/\alpha)(1 - \exp(-\alpha\hat{t}_\mu))] = 0 \qquad (\mu = 1, 2, \ldots, n) \tag{2-14-9}$$

where $\hat{s}_\mu$ and $\hat{t}_\mu$ are the "true" values of s and t at the μth measurement. The likelihood is

$$L[\hat{s}_\mu, \hat{t}_\mu(\mu = 1, 2, \ldots, n), \psi] = \prod_{\mu=1}^{n} p\{t_\mu - \hat{t}_\mu, s_\mu - \hat{s}_\mu | \psi\} \qquad \text{(2-14-10)}$$

with $\hat{t}_\mu$ and $\hat{s}_\mu$ constrained by Eq. (9). Alternately, we can substitute $\hat{s}_\mu$ from Eq. (9) into Eq. (10) to obtain

$$L(\hat{t}_\mu(\mu = 1, 2, \ldots, n), \alpha, \gamma, \psi)$$
$$= \prod_{\mu=1}^{n} p\{t_\mu - \hat{t}_\mu, \gamma[\hat{t}_\mu - (1/\alpha)(1 - \exp(-\alpha\hat{t}_\mu))] | \psi\} \qquad \text{(2-14-11)}$$

with no constraints applying to the t_μ.

3. Only t_μ is subject to significant errors. If Eq. (5) could be solved for t_μ we would have a standard reduced model. Since this is impossible, we again adopt an exact structural model, except that $\hat{s}_\mu$ is replaced by s_μ in Eq. (9), and all references to $\hat{s}_\mu$ and s_μ are deleted from Eq. (10).

(b) The model Eq. (5) is inexact. For instance, if the sphere is sufficiently small, then the drag force is randomly perturbed due to the impact of individual molecules. A Brownian motion is thereby superimposed on the falling motion of the sphere. Eq. (2) must be amended to read

$$\dot{v} + \alpha v = \beta + \eta \qquad \text{(2-14-12)}$$

where η is a random variable whose distribution may be derived from the laws of statistical mechanics. When the equations are integrated, there will arise a perturbation on s, i.e., Eq. (5) will take the form

$$s - \gamma[t - (1/\alpha)(1 - e^{-\alpha t})] = \phi \qquad \text{(2-14-13)}$$

where ϕ is also a random variable. Let $p(\phi_1, \phi_2, \ldots, \phi_n | \alpha, \gamma, \psi)$ be the joint pdf of the ϕ_μ. If the measurements of s_μ and t_μ are precise relative to the standard deviation of ϕ, then we have the likelihood

$$L(\alpha, \gamma, \psi) = p\{s_1 - \gamma[t_1 - (1/\alpha)(1 - \exp(-\alpha t_1))],$$
$$s_2 - \gamma[t_2 - (1/\alpha)(1 - \exp(-\alpha t_2))], \ldots, | \alpha, \gamma, \psi\} \qquad \text{(2-14-14)}$$

We have used a joint pdf instead of the product of individual pdf's because the assumption of independence between observations is tenable here only if the experiment is restarted from rest for each measurement. Otherwise, if the disturbances up to time t_2 have conspired to make s_2 larger than expected from Eq. (5), then s is likely to remain too large in succeeding periods. See Problem 4 of Section 8-9.

2-15. Utility of Distribution Assumptions

The problem of estimating the parameters $\boldsymbol{\theta}$ has now been augmented, inasmuch as we must also estimate the parameters $\boldsymbol{\psi}$, and possibly $\hat{\mathbf{W}}$. We let $\boldsymbol{\phi}$ denote the entire set of unknown parameters, i.e.,

$$\boldsymbol{\phi} \equiv \{\boldsymbol{\theta}, \boldsymbol{\psi}, \hat{\mathbf{W}}\} \tag{2-15-1}$$

Those $\hat{\mathbf{W}}$ which could be eliminated (see Section 2-13) are excluded. A scientist may be reluctant to erect the entire probabilistic superstructure, only to find himself with a larger problem than he started with. It is true that some parameter estimation procedures may be applied without making any probabilistic assumptions. The resulting estimates, however, are rather meaningless. They may suffice for curve fitting, but nothing will be known about the relationship between the estimated and true values of the parameters.

Frequently people make implicit assumptions concerning the probability distribution without realizing the fact. This happens, for instance, when weights are assigned in the least squares procedure (see Section 4-3). By recognizing the role of such weights as parameters $\boldsymbol{\psi}$ of a distribution, we are able to *estimate* the weights rather than *assign* them. Thus we are able to shift a burden from ourselves to the computer (see Sections 4-8–4.9).

It should be noted that in some cases (particularly with linear models) it suffices to specify the covariance matrix of the distribution without committing oneself to any specific form of the density functions.

D. Prior Information and Posterior Distribution

2-16. Prior Information

The scientist usually has some ideas concerning the values of his parameters even before any data have been gathered. He is frequently able to exclude entirely some values. For instance, the rate constant in a chemical reaction or the viscosity of a liquid must be positive. An estimation procedure that came up with negative values for such parameters should be entirely unacceptable. His physical intuition may lead the scientist to reject some other values as being entirely implausible, even though they are strictly speaking not impossible. Even among the admissible values the scientist may regard some as more plausible than others. For instance, suppose a chemist knows with great precision the viscosities η_6 of *n*-hexane and η_8 of *n*-octane, and he is trying to determine η_7, the viscosity of *n*-heptane. Experience with the properties of homologous series of organic compounds will lead him to reject entirely values of η_7 such that $\eta_7 \leqslant \eta_6$ or $\eta_7 \geqslant \eta_8$. Among the remaining values, he will prefer those near $(\eta_6 + \eta_8)/2$ to those near η_6 or η_8.

2-17. Prior Distribution

The scientist may summarize his prior information in what is called the *prior distribution* of the parameters. The prior distribution may be characterized by means of the *prior density* function $p_0(\boldsymbol{\phi})$. The prior density function is required to be nonnegative, and to possess the property that if $\boldsymbol{\phi}_1$ and $\boldsymbol{\phi}_2$ are any two values of $\boldsymbol{\phi}$, then $p_0(\boldsymbol{\phi}_1)/p_0(\boldsymbol{\phi}_2)$ represents the ratio of the plausibility‡ of $\boldsymbol{\phi}_1$ to that of $\boldsymbol{\phi}_2$. Note that we do not require the normalization condition $\int p_0(\boldsymbol{\phi})\, d\boldsymbol{\phi} = 1$. In fact, we do not even require the integral $\int p_0(\boldsymbol{\phi})\, d\boldsymbol{\phi}$ to exist. Thus, we are permitted to assign the *uniform priority density* $p_0(\boldsymbol{\phi}) = 1$ to describe the case when all values of $\boldsymbol{\phi}$ are equally plausible. We always assign $p_0(\boldsymbol{\phi}) = 0$ to all values of $\boldsymbol{\phi}$ which are to be entirely excluded.

Controversy still rages around the question of whether or not the prior distribution may be regarded as a true probability distribution. For the statistician belonging to the frequentist school a probability distribution is meaningful only when applied to a random variable. When the parameters represent physical constants, their values are perfectly definite (although unknown), and they cannot be regarded as random variables. Proponents of subjective probability (e.g., Savage, 1954) and decision theory (e.g., Raiffa and Schlaifer, 1961; Ferguson, 1967), however, do indeed admit "degrees of belief" and subjective choices of plausibility as probability densities. Not only do they allow one to postulate prior densities, they actually insist that one do so. They believe that any sensible subsequent action (e.g., parameter estimation) must be based on some choice of a prior distribution.

We shall not attempt to resolve the controversy here. For a concise discussion of the problem we refer the reader to Cornfield (1967). We feel that it is up to the scientist to decide for himself the extent of his commitment to a prior distribution. He should remember that introducing a prior distribution biases the results of the estimation process so as to favor parameter values for which $p_0(\boldsymbol{\phi})$ is relatively large. This bias diminishes as the number of experiments is increased. In other words, when the amount of data available is sufficiently large, the effect of the prior distribution on the parameter estimates is negligible, except that values of $\boldsymbol{\phi}$ for which $p_0(\boldsymbol{\phi}) = 0$ remain excluded.

‡ This is an intuitive concept, which we do not attempt to define here. Some authors have given definitions based on bets that the scientist is willing to lay on each value of $\boldsymbol{\phi}$.

There are several cases for which the use of a prior distribution is noncontroversial:

(a) Assigning $p_0(\boldsymbol{\phi}) = 0$ to physically impossible values of $\boldsymbol{\phi}$.

(b) If $\boldsymbol{\phi}$ is truly a random variable, its pdf should be used as the prior density. For instance, $\boldsymbol{\phi}$ may represent the physical properties of a batch of chemicals. If these properties are known to vary randomly from one batch to the next according to some pdf $p(\boldsymbol{\phi})$, it is entirely proper to use $p_0(\boldsymbol{\phi}) = p(\boldsymbol{\phi})$ when attempting to estimate the properties of one specific batch.

(c) Suppose a number of relevant experiments has already been conducted, but additional experiments are being planned. As will be seen shortly, the information on the parameters contained in the data may be expressed in the form of a so-called posterior distribution. It is entirely proper to use the posterior distribution from the already completed experiments as the prior distribution for the experiments yet to be conducted.

2-18. Informative and Noninformative Priors

In case (c) above the choice of the prior distribution is obvious. The same is true in case (b), provided $p(\boldsymbol{\phi})$ is known. How to choose a distribution in other cases? We distinguish three situations:

(a) The *noninformative* case occurs when we really have no marked preferences for some values of $\boldsymbol{\phi}$ over others, at least within the relevant region in $\boldsymbol{\phi}$-space.‡ The simplest solution, and a very satisfactory one in practice,§ is to assume no prior distribution at all, and to use a parameter estimation method which does not require one. This avenue, however, is closed to practitioners who are irrevocably committed to decision theory and Bayesian statistics. They are forced to choose a prior distribution, and are likely to assume a uniform prior density. This is logically unsatisfactory, since if ϕ has a uniform prior distribution, any nontrivial function of ϕ (say ϕ^3) has a nonuniform distribution. As we could have written the model in terms of ϕ^3 rather than ϕ, we find ourselves in a situation where the choice of parametrization affects the outcome of the estimation.

‡ By the relevant region we mean that region in which the likelihood function is far from vanishing.

§ Author's opinion.

Alternative procedures are available. Raiffa and Schlaifer (1961) introduce the concept of the conjugate distribution. This is a distribution which has the same mathematical form as the likelihood function derived from some hypothetical sample. A suitable noninformative prior distribution can sometimes be obtained by finding the function to which the conjugate distribution tends as the sample size (now assumed a continuous variable) tends to zero. Another suggestion, due to Jeffreys (1961), is to use $1/\phi$ as the noninformative prior density for nonnegative variables. The justification given is that this distribution is unaffected when ϕ is replaced by ϕ^n.

(b) The *informative* case occurs when we do prefer some values of $\boldsymbol{\phi}$ to others within the relevant region. Generally the precise form of the prior distribution is immaterial as long as it has approximately the right shape. The method of conjugate distributions can be used here too; the scientist postulates what seems to him a likely data set, and the likelihood function corresponding to it is used as a prior density. The advantage this method offers is that the prior density, the likelihood function, and the posterior density all have the same mathematical form, which sometimes simplifies formal manipulations. In the case of nonlinear models, however, all these functions are complicated. Numerical evaluation replaces formal manipulation, and this method no longer offers any advantage. One is probably better off assuming a normal or other simple prior density with suitably chosen means and variances. Another approach is to attempt the graphical construction of a suitable pdf or cumulative distribution function. Winkler (1967) has carried out experiments in which students were made to construct prior densities to represent their beliefs concerning the values of certain parameters, using several of the above mentioned methods. The feasibility of constructing such functions was demonstrated, but the question of whether they would be of practical value to the parameter estimator remains unanswered.

In practice, our prior information on a parameter θ often takes the form of a value $\theta_0 \pm \alpha$ reported in the literature. The number α may be a standard deviation, in which case we assign the prior density $N_1(\theta_0, \alpha^2)$; or, $\pm\alpha$ may represent absolute bounds on the deviation from θ_0, so that the uniform distribution over the interval $\theta_0 - \alpha \leqslant \theta \leqslant \theta_0 + \alpha$ is appropriate. In both cases, the chosen distribution is the least informative one among all distributions satisfying the given conditions.

(c) It may happen that $\boldsymbol{\phi}$ is truly a random variable with pdf $p(\boldsymbol{\phi})$ (as in case (b), Section 2-17), yet the function $p(\boldsymbol{\phi})$ is not known. The *empirical Bayes'* method of Robbins (1955, and 1964) (see also Neyman, 1962) provides an approach to the estimation of $p(\boldsymbol{\phi})$ on the basis of available data.

2-19. Bayes' Theorem

We have summarized the information contained in the data by means of the likelihood function L, and the prior information by means of the prior density $p_0(\boldsymbol{\phi})$. We combine the two in the so-called *posterior density* which is proportional to their product

$$p^*(\boldsymbol{\phi}) = cL(\boldsymbol{\phi})p_0(\boldsymbol{\phi}) \tag{2-19-1}$$

with

$$c = \left[\int L(\boldsymbol{\phi})p_0(\boldsymbol{\phi})\, d\boldsymbol{\phi}\right]^{-1} \tag{2-19-2}$$

provided the integral exists.‡

If $p_0(\boldsymbol{\phi})$ is regarded as the probability density ascribed to $\boldsymbol{\phi}$ before the experiments were performed, then $p^*(\boldsymbol{\phi})$ is the density we must ascribe to $\boldsymbol{\phi}$ after the data were obtained. This follows from Bayes' (Bayes, 1763) theorem, which may be stated as follows:

Bayes Theorem. Let A and B be two events whose probabilities of occurrence are $P(A)$ and $P(B) \neq 0$ respectively. Let $P(A\,|\,B)$ denote the conditional probability that A occurs, given that B has occurred, and let $P(B\,|\,A)$ be defined analogously. Then

$$P(A\,|\,B) = P(B\,|\,A)P(A)/P(B) \tag{2-19-3}$$

Proof. The proof follows immediately from the definition of conditional probability

$$P(A \;\&\; B) = P(A\,|\,B)P(B) \tag{2-19-4}$$

where $P(A \;\&\; B)$ is the probability of A and B both occurring. But also

$$P(A \;\&\; B) = P(B\,|\,A)P(A) \tag{2-19-5}$$

Dividing Eq. (5) by Eq. (4) and solving for $P(A\,|\,B)$ yields Eq. (3) directly.

In our case we define A to be the event "the true value of $\boldsymbol{\phi}$ is within a hypercube of volume§ $d\boldsymbol{\phi}$ centered at $\boldsymbol{\phi}_0$" and B to be the event "the true value of $\boldsymbol{\Omega}$ is within a hypercube of volume $d\boldsymbol{\Omega}$ centered at $\mathbf{W}$."

‡ In some applications the value of c is immaterial, and we can proceed even when the integral does not exist. See Section 4-15.

§ We use the notation $d\boldsymbol{\phi}$ as a shorthand for $d\phi_1\, d\phi_2 \ldots d\phi_l$ and $d\boldsymbol{\Omega}$ for $d\omega_{11}\, d\omega_{12} \ldots d\omega_{1r}\, d\omega_{21} \ldots d\omega_{nr}$.

By the definitions of the pdf, the prior distribution, and the likelihood function, we have

$$P(A) = p_0(\boldsymbol{\phi}_0)\, d\boldsymbol{\phi} \tag{2-19-6}$$

$$P(B\,|\,A) = L(\boldsymbol{\phi}_0)\, d\boldsymbol{\Omega} \tag{2-19-7}$$

The value of $P(B)$ is obtained by summing $P(A \;\&\; B) = P(B\,|\,A)P(A)$ over all possible A's, i.e.,

$$P(B) = \left[\int L(\boldsymbol{\phi}_0)p_0(\boldsymbol{\phi}_0)\, d\boldsymbol{\phi}\right] d\boldsymbol{\Omega} \tag{2-19-8}$$

Substitution in Eq. (3) yields

$$P(A\,|\,B) = L(\boldsymbol{\phi}_0)p_0(\boldsymbol{\phi}_0)\, d\boldsymbol{\phi} \Big/ \int L(\boldsymbol{\phi}_0)p_0(\boldsymbol{\phi}_0)\, d\boldsymbol{\phi} \tag{2-19-9}$$

But $P(A\,|\,B)$ is the probability of A occurring given that the experiments yielded the data **W**. By definition, then

$$P(A\,|\,B) = p^*(\boldsymbol{\phi}_0)\, d\boldsymbol{\phi} \tag{2-19-10}$$

from which Eq. (1) follows immediately.

Note that $p^*(\boldsymbol{\phi})$ is a meaningful pdf only if $p_0(\boldsymbol{\phi})$ was one. The frequentist who in a given case does not accept a prior distribution will not accept a posterior distribution either.

When the prior distribution is uniform, the posterior density is proportional to the likelihood function. If the results of two series of experiments are statistically independent, the joint likelihood function is the product of the two individual likelihoods. Formally it is then possible to regard the likelihood from the first series as the prior for the second series, and then the posterior for the second series equals (except for a constant factor) the joint likelihood function. This is the basis for the assertion made under case (c) in Section 2-17.

The posterior distribution (equal to the likelihood in the absence of prior information) combined with any constraints that may be applicable, embodies all four elements that enter into the parameter estimation problem, namely the model, the data, the probability distribution of the errors, and the prior information on the parameters. Formulating a parameter estimation problem is in many cases equivalent to writing down the posterior distribution.

2-20. Problems

1. Write down explicit expressions for all the likelihood functions appearing in Section 2-14, assuming that all error distributions are normal with zero means and given variances.

2. The gamma distribution with parameters α and ν has the pdf

$$\Gamma_{\alpha,\nu}(\theta) \equiv \begin{cases} \Gamma^{-1}(\nu)\alpha^{\nu}\theta^{\nu-1}e^{-\alpha\theta} & (\theta \geqslant 0) \\ 0 & (\theta < 0) \end{cases}$$

where

$$\Gamma(\nu) \equiv \int_0^{\infty} x^{\nu-1}e^{-x}\,dx$$

is the gamma function. Show that $E(\theta) = \nu/\alpha$ and $V(\theta) = \nu/\alpha^2$.

3. Suppose an object is measured n times to determine its length θ. The measurements are denoted w_μ $(\mu = 1, 2, \ldots, n)$. The model equation is $\hat{w}_\mu = \theta$. Assume the errors are distributed as $N_1(0, \sigma^2)$. Suppose σ is known and θ is assigned the prior distribution $N_1(\theta_0, \sigma_0{}^2)$. Write down the likelihood and the posterior pdf $p^*(\theta)$. Show that $p^*(\theta)$ is normal, and find its mean and variance.

4. As in the previous problem, but assume θ is known and σ is to be estimated. Let $\tau \equiv 1/\sigma^2$ and assume that τ has the prior distribution Γ_{α_0, ν_0}. Show that the posterior distribution of τ is also gamma, and find its parameters.

5. Investigate the shape of the gamma distribution for various ranges of its parameter values. Under what circumstances is the gamma distribution a suitable prior for a parameter?

6. Show that the examples of Problems 3 and 4 contain conjugate distributions. Show how the parameters of the prior distribution are related to the sizes of hypothetical samples. Investigate the behavior of the prior distribution as the hypothetical sample size is reduced to zero.

Chapter

III

Estimators and Their Properties

A. Statistical Properties

3-1. The Sampling Distribution

A *point estimation method* is a procedure which enables one to compute an estimate $\boldsymbol{\phi}^*$ for the parameter vector $\boldsymbol{\phi}$, given the data matrix $\mathbf{W}$. The estimation method defines (at least implicitly) a vector valued function $\mathbf{h}$

$$\boldsymbol{\phi}^* = \mathbf{h}(\mathbf{W}) \tag{3-1-1}$$

We shall use the expression "the estimator $\mathbf{h}$" to mean "the estimation procedure defined by the function $\mathbf{h}$."

If the experiments which yielded our data were to be repeated, we would obtain different values of the $\mathbf{W}$, i.e., different realizations of the random variables $\boldsymbol{\Omega}$. Application of the estimator $\mathbf{h}$ to the new data would yield different values of $\boldsymbol{\phi}^*$. We see then that the estimates $\boldsymbol{\phi}^*$ are themselves random variables, possessing a certain probability distribution, which depends both on the nature of $\mathbf{h}$ and on the distribution of the $\boldsymbol{\Omega}$. We refer to this distribution as the *sampling distribution* of the estimator $\mathbf{h}$, and denote its pdf (if one exists) $p_{\mathbf{h}}(\boldsymbol{\phi}^*)$.

Note the fundamental difference between the sampling distribution and the posterior distribution. The sampling distribution refers to the estimate $\boldsymbol{\phi}^*$, which is truly a random variable. The sampling distribution is defined only once the estimation procedure is defined; different estimation procedures when applied to the same data generally give rise to different sampling distributions. On the other hand, the posterior distribution is independent of the estimation procedure. It applies to the true values $\boldsymbol{\phi}$, and its interpretation in cases where these are not random is, therefore, controversial.

A glance at Eq. (1) reveals that the sampling distribution of $\boldsymbol{\phi}^*$ depends on the actual distribution of $\boldsymbol{\Omega}$. This distribution however, depends on the true values $\hat{\boldsymbol{\phi}}$, which are generally unknown (or we would not be trying to estimate them). Therefore, even when we can derive a formula for the sampling distribution, we can evaluate only the approximation obtained by substituting the estimated parameter values for the true ones. Still, one can frequently deduce some important properties of the distribution, as shown by the following example: Suppose we measure an object n times to determine its length θ. The measurements will be denoted by w_μ ($\mu = 1, 2, \ldots, n$). The model equations take the form

$$\hat{w}_\mu = \theta \tag{3-1-2}$$

Assume the measurements to be independent, and normally distributed with variance σ^2 and mean $\hat{\theta}$. Consider the estimator

$$\theta^* = (1/n) \sum_{\mu=1}^{n} w_\mu \tag{3-1-3}$$

i.e., θ^* is the mean of the observations. It is well known that θ^* is normally distributed with mean $\hat{\theta}$ and variance σ^2/n. Thus we have found that the mean of the sampling distribution is equal to the true value $\hat{\theta}$, and its variance decreases as $1/n$ when the sample size is increased.

The above example introduced the concepts of the mean and variance of the sampling distribution, also referred to as the mean and variance of the estimate. In the general case, the mean $\bar{\boldsymbol{\phi}}$ and covariance matrix $\mathbf{V}_{\boldsymbol{\phi}}$ of the estimate are given by:

$$\bar{\boldsymbol{\phi}} \equiv E(\boldsymbol{\phi}^*) = \int \boldsymbol{\phi}^* p_{\mathbf{h}}(\boldsymbol{\phi}^*)\, d\boldsymbol{\phi}^* = \int \mathbf{h}(\mathbf{W}) p(\mathbf{W} \mid \hat{\boldsymbol{\phi}})\, d\mathbf{W} \tag{3-1-4}$$

$$\mathbf{V}_{\boldsymbol{\phi}} \equiv E(\boldsymbol{\phi}^* - \bar{\boldsymbol{\phi}})(\boldsymbol{\phi}^* - \bar{\boldsymbol{\phi}})^{\mathrm{T}} = \int (\mathbf{h}(\mathbf{W}) - \bar{\boldsymbol{\phi}})(\mathbf{h}(\mathbf{W}) - \bar{\boldsymbol{\phi}})^{\mathrm{T}} p(\mathbf{W} \mid \hat{\boldsymbol{\phi}})\, d\mathbf{W} \tag{3-1-5}$$

3-2. Properties of the Sampling Distribution

It is clearly desirable to have an estimator whose sampling distribution is concentrated in the neighborhood of the true values of the parameters. More formally, we define the following properties of estimators:

(a) The *bias* of an estimator is the difference between the expected value of the estimator and the true value of the parameter, i.e., $\mathbf{b} \equiv \bar{\boldsymbol{\phi}} - \hat{\boldsymbol{\phi}}$. An estimator is *unbiased* if its bias vanishes, i.e., if $\bar{\boldsymbol{\phi}} = \hat{\boldsymbol{\phi}}$. In the example of Section 3-1 we saw that the estimate Eq. (3-1-3) was unbiased. Clearly, we desire estimators with small (in absolute value) bias, but total unbiasedness is mostly unobtainable. Nor is unbiasedness particularly important, since the bias is not the only error in any given estimator. Furthermore, if an estimate is unbiased for some parameter ϕ, it is generally biased for nontrivial functions of ϕ. That is, even if $E(\phi^*) = \hat{\phi}$, it need not be true that, say, $E(\phi^{*2}) = \hat{\phi}^2$. Thus, the presence or absence of bias in an estimator is affected by a change in parametrization.

(b) While the bias is a measure of the "systematic" error in an estimator, the variance measures its random error. The Rao–Cramer theorem (see Appendix C) establishes a theoretical lower bound on the attainable covariance matrix $\mathbf{V}_{\boldsymbol{\phi}}$ of an estimator.

We see from Eq. (3-1-4) that $\overline{\boldsymbol{\phi}}$ is a function of $\hat{\boldsymbol{\phi}}$. If this function is differentiable, then we can form the matrix

$$\mathbf{P} \equiv \partial\overline{\boldsymbol{\phi}}/\partial\hat{\boldsymbol{\phi}} = \int \mathbf{h}(\mathbf{W})(\partial p/\partial\hat{\boldsymbol{\phi}})^{\mathrm{T}}\, d\mathbf{W} \tag{3-2-1}$$

We also define a matrix $\mathbf{R}$ by

$$\begin{aligned}\mathbf{R} &= E((\partial \log p/\partial\hat{\boldsymbol{\phi}})(\partial \log p/\partial\hat{\boldsymbol{\phi}})^{\mathrm{T}}) \\ &= \int (\partial \log p/\partial\hat{\boldsymbol{\phi}})(\partial \log p/\partial\hat{\boldsymbol{\phi}})^{\mathrm{T}} p\, d\mathbf{W} \end{aligned}\tag{3-2-2}$$

The theorem asserts that the matrix $\mathbf{V}_{\boldsymbol{\phi}} - \mathbf{P}\mathbf{R}^{-1}\mathbf{P}^{\mathrm{T}}$ is positive semidefinite, and that it is null if and only if there exists a matrix $\mathbf{A}$ (whose elements may be functions of $\hat{\boldsymbol{\phi}}$) such that

$$\boldsymbol{\phi}^* - \hat{\boldsymbol{\phi}} = \mathbf{A}(\hat{\boldsymbol{\phi}})\partial \log p/\partial\hat{\boldsymbol{\phi}} \tag{3-2-3}$$

The proof is given in Appendix C. The matrix $\mathbf{P}\mathbf{R}^{-1}\mathbf{P}^{\mathrm{T}}$ is called the *minimum variance bound* (MVB).

Since the diagonal elements of a positive semidefinite matrix must be nonnegative, we have for the variance of each ϕ_α^*

$$V_{\boldsymbol{\phi}\alpha\alpha} \geqslant \sum_{\beta,\gamma} P_{\alpha\beta}[\mathbf{R}^{-1}]_{\beta\gamma} P_{\alpha\gamma} \tag{3-2-4}$$

with equality holding only when Eq. (3) is satisfied.

An estimate is called *efficient*‡ if its variance is the lowest theoretically attainable, i.e., when

$$\mathbf{V}_{\boldsymbol{\phi}} = \mathbf{P}\mathbf{R}^{-1}\mathbf{P}^{\mathrm{T}} \tag{3-2-5}$$

When the estimate is unbiased, we have $\overline{\boldsymbol{\phi}} = \hat{\boldsymbol{\phi}}$, and hence $\mathbf{P} = \mathbf{I}$. The Rao–Cramer theorem reduces to the statement that $\mathbf{V}_{\boldsymbol{\phi}} - \mathbf{R}^{-1}$ is positive semidefinite, and an efficient unbiased estimate has the covariance

$$\mathbf{V}_{\boldsymbol{\phi}} = \mathbf{R}^{-1} \tag{3-2-6}$$

‡ Some authors use the term "MVB estimate" instead, and reserve the term "efficient" for what we call "asymptotically efficient" here.

In the example of Section 3-1 we had an unbiased estimate for the single parameter θ. Its variance was found to be $V_\theta = \sigma^2/n$.
The likelihood function is

$$p(\mathbf{W}, \hat{\theta}) = (2\pi)^{-n/2}\sigma^{-n} \exp\left[-(1/2\sigma^2)\sum_{\mu=1}^{n}(w_\mu - \hat{\theta})^2\right] \tag{3-2-7}$$

whence

$$\partial \log p/\partial\hat{\theta} = (1/\sigma^2)\sum_{\mu=1}^{n}(w_\mu - \hat{\theta}) \tag{3-2-8}$$

and

$$R = E(\partial \log p/\partial\hat{\theta})^2 = (1/\sigma^4)E\left\{\left[\sum_{\mu=1}^{n}(w_\mu - \hat{\theta})\right]^2\right\}$$

$$= (1/\sigma^4)E\left[\sum_{\mu=1}^{n}\sum_{\eta=1}^{n}(w_\mu - \hat{\theta})(w_\eta - \hat{\theta})\right] \tag{3-2-9}$$

Now the variables w_μ ($\mu = 1, 2, \ldots, n$) were assumed independent with means $\hat{\theta}$ and variances σ^2. Hence

$$E(w_\mu - \hat{\theta})(w_\eta - \hat{\theta}) = \sigma^2\delta_{\mu\eta} \tag{3-2-10}$$

and

$$R = (1/\sigma^4)\sum_{\mu=1}^{n} E(w_\mu - \hat{\theta})^2 = n\sigma^2/\sigma^4 = n/\sigma^2 \tag{3-2-11}$$

In this case, then, $V_\theta = R^{-1}$, and the estimate is efficient. We could have deduced this fact also from Eq. (3-1-3), which yields

$$\theta^* - \hat{\theta} = (1/n)\sum_{\mu=1}^{n} w_\mu - \hat{\theta} = (1/n)\sum_{\mu=1}^{n}(w_\mu - \hat{\theta}) = (\sigma^2/n)\,\partial \log p/\partial\hat{\theta} \tag{3-2-12}$$

so that Eq. (3) is satisfied with $\mathbf{A} = \sigma^2/n$.

As with unbiasedness, efficiency can be attained only in a small class of relatively simple models. It should be pointed out that in some cases, although no efficient estimate exists, among those estimates that do exist or among estimates of a certain class there may be one whose variance is least. For instance, if all errors in a linear model are identically and independently distributed, then least squares estimates have, among all linear unbiased estimates, the smallest variance. Yet, they are not necessarily efficient.

(c) While unbiased and efficient estimates cannot generally be found for samples of finite size, the situation changes drastically when the number of experiments increases beyond bound. Under this condition, the actual value of many estimates converges with probability one to the true parameter values. We refer to such estimates as *consistent* or *asymptotically unbiased.* The variances of most of the estimates that we shall deal with approach zero as $1/n$ when n increases. Hence, if the estimator is consistent for ϕ it is consistent for any well-behaved function of ϕ. Thus consistency is a more significant concept then unbiasedness.

(d) Both $\mathbf{V}_\phi$ and $\mathbf{R}^{-1}$ tend to zero as $1/n$ for most relevant estimators. Hence we call a consistent estimator *asymptotically efficient* if with probability one

$$\lim_{n \to \infty} n(\mathbf{V}_\phi - \mathbf{R}^{-1}) = \mathbf{0} \tag{3-2-13}$$

(e) Our discussion of estimation criteria would not be complete without the mention of sufficient statistics. A *statistic* $\boldsymbol{\rho}$ of a sample is any function computed from the values of the sample for the purpose of extracting relevant information

$$\boldsymbol{\rho} = \mathbf{r}(\mathbf{W}) \tag{3-2-14}$$

In particular, any estimate $\boldsymbol{\phi}^*$ is a statistic defined by Eq. (3-1-1). A statistic $\boldsymbol{\rho}$ is deemed *sufficient* for the parameters $\boldsymbol{\phi}$ if the value of $\boldsymbol{\rho}$ conveys as much information concerning the value of $\boldsymbol{\phi}$ as did the original sample $\mathbf{W}$. In other words, we may compute the value of $\boldsymbol{\rho}$ from the sample, and then discard the data $\mathbf{W}$ without losing any information relevant to the estimation of $\boldsymbol{\phi}$.

This statement should be interpreted as follows: A sample can contain information concerning the value of $\hat{\boldsymbol{\phi}}$ only if the distribution of the sample is a function of $\hat{\boldsymbol{\phi}}$. To say that $\boldsymbol{\rho}$ is a sufficient statistic for $\hat{\boldsymbol{\phi}}$ implies, then, that once the value of $\boldsymbol{\rho}$ is determined, the distribution of the sample can be represented in terms of $\boldsymbol{\rho}$ alone, with no further dependence on $\hat{\boldsymbol{\phi}}$. Therefore, there must exist a function $q(\mathbf{W}, \boldsymbol{\rho})$ such that

$$p(\mathbf{W} \mid \hat{\boldsymbol{\phi}}, \boldsymbol{\rho}) = q(\mathbf{W}, \boldsymbol{\rho}) \tag{3-2-15}$$

But we have, from the definition of conditional probability

$$p(\mathbf{W} \mid \hat{\boldsymbol{\phi}}) = p(\mathbf{W} \mid \hat{\boldsymbol{\phi}}, \boldsymbol{\rho}) p(\boldsymbol{\rho} \mid \hat{\boldsymbol{\phi}}) \tag{3-2-16}$$

Letting $p(\boldsymbol{\rho} \mid \hat{\boldsymbol{\phi}}) = s(\boldsymbol{\rho}, \hat{\boldsymbol{\phi}})$ (this is the sampling distribution of the statistic $\boldsymbol{\rho}$) we are led to the factorization theorem for sufficient statistics $\boldsymbol{\rho}$

$$p(\mathbf{W} \mid \hat{\boldsymbol{\phi}}) = q(\mathbf{W}, \boldsymbol{\rho}) s(\boldsymbol{\rho}, \hat{\boldsymbol{\phi}}) \tag{3-2-17}$$

Taking logarithms and differentiating both sides with respect to $\hat{\boldsymbol{\phi}}$ we obtain

$$\partial \log p(\mathbf{W} \mid \hat{\boldsymbol{\phi}})/\partial \hat{\boldsymbol{\phi}} = \partial \log s(\boldsymbol{\rho}, \hat{\boldsymbol{\phi}})/\partial \hat{\boldsymbol{\phi}} \equiv t(\boldsymbol{\rho}, \hat{\boldsymbol{\phi}}) \tag{3-2-18}$$

In the example of Section 3-1, we have

$$\begin{aligned} p(\mathbf{W}/\hat{\theta}) &= (2\pi)^{-n/2}\sigma^{-n} \exp\left\{-(1/2\sigma^2)\left(\sum_{\mu=1}^{n} w_\mu{}^2 - 2\hat{\theta} \sum_{\mu=1}^{n} w_\mu + n\hat{\theta}^2\right)\right\} \\ &= \left\{(2\pi)^{-n/2}\sigma^{-n} \exp\left[-(1/2\sigma^2) \sum_{\mu=1}^{n} w_n{}^2\right]\right\}\left\{\exp\left[(1/2\sigma^2)\left(2\hat{\theta} \sum_{\mu=1}^{n} w_\mu - n\hat{\theta}^2\right)\right]\right\} \end{aligned} \tag{3-2-19}$$

which is in the form Eq. (17) with $\rho = \sum_{\mu=1}^{n} w_\mu$. Thus, the sum of the observations is a sufficient statistic in this case.

An estimate which is a function of a sufficient statistic is a *sufficient estimate.* In our example, $\theta^* = (1/n)\rho$, hence θ^* is a sufficient estimate for θ.

Supposing $\boldsymbol{\phi}^*$ is an efficient estimate of $\boldsymbol{\phi}$, then from Eq. (3) we find that

$$\partial \log p/\partial\hat{\boldsymbol{\phi}} = \mathbf{A}^{-1}(\hat{\boldsymbol{\phi}})(\boldsymbol{\phi}^* - \hat{\boldsymbol{\phi}}) \tag{3-2-20}$$

Comparison of Eq. (20) with Eq. (18) shows that $\boldsymbol{\rho} = \boldsymbol{\phi}^*$ is sufficient. Thus we have proved that if an efficient estimate exists, it must also be sufficient. Conversely, if a sufficient estimate exists, some function of it is an efficient estimate of some function of the parameters.

(f) The value of an estimate usually depends on the form of the probability distribution that we assume. We rarely possess exact knowledge of the distribution, and must usually content ourselves with a rough approximation. We desire, therefore, that our estimate be *robust*, i.e., that it be only slightly affected by seemingly unimportant changes in the form of the assumed distribution.

(g) The choice of parameters appearing in a model is often arbitrary. We may replace the original parameters $\boldsymbol{\phi}$ with a different set of parameters $\tilde{\boldsymbol{\phi}}$ which are single valued functions of the $\boldsymbol{\phi}$, e.g., $\tilde{\boldsymbol{\phi}} = \mathbf{s}(\boldsymbol{\phi})$ where $\mathbf{s}$ is a vector of functions. It is desirable that our estimators be *invariant under reparametrization*. That is, if $\boldsymbol{\phi}^*$ and $\tilde{\boldsymbol{\phi}}^*$ are the estimates obtained when the model is represented respectively in terms of $\boldsymbol{\phi}$ and $\tilde{\boldsymbol{\phi}}$, then we expect to find that $\tilde{\boldsymbol{\phi}}^* = \mathbf{s}(\boldsymbol{\phi}^*)$.

(h) An estimation procedure which cannot be implemented on available computing machinery is of little use. An estimate which is readily computable is, from a practical point of view, more valuable than a statistically more efficient estimate which is computable only with an excessive amount of labor. In other words, an adherent of decision theory should make his cost function depend not only on the error of the estimate, but also on the cost of computing the estimate.

(i) A *linear estimate* is one which is a linear function of the data, i.e., $\boldsymbol{\phi}^*$ is a linear estimate if there exists a matrix $\mathbf{A}$ (which may be a function of the $\mathbf{u}_\mu$) such that

$$\boldsymbol{\phi}^* = \mathbf{A}\mathbf{w} \tag{3-2-21}$$

Useful linear estimates valid over a wide range of data values can be found only when the model itself is linear in the parameters.

Among the properties that we have defined, the most important in practice are small bias, small variance, robustness, and computability.

A true measure of the accuracy and precision of an estimate is given by the mean square error relative to the true, rather than mean, value. It is easily verified that

$$E(\boldsymbol{\phi}^* - \hat{\boldsymbol{\phi}})(\boldsymbol{\phi}^* - \hat{\boldsymbol{\phi}})^{\mathrm{T}} = E(\boldsymbol{\phi}^* - \overline{\boldsymbol{\phi}})(\boldsymbol{\phi}^* - \overline{\boldsymbol{\phi}})^{\mathrm{T}} + (\overline{\boldsymbol{\phi}} - \hat{\boldsymbol{\phi}})(\overline{\boldsymbol{\phi}} - \hat{\boldsymbol{\phi}})^{\mathrm{T}} = \mathbf{V}_{\boldsymbol{\phi}} + \mathbf{b}\mathbf{b}^{\mathrm{T}} \tag{3-2-22}$$

where **b** is the bias. The root-mean-square error in the estimate ϕ_i^* of the ith parameter is given, therefore, by $(\sigma_i^2 + b_i^2)^{1/2}$, where $\sigma_i \equiv V_{\boldsymbol{\phi}ii}^{1/2}$ is the standard deviation of the estimate ϕ_i^*. This points out the fact that we gain little in laboring to make a highly biased estimate (large b_i) very efficient (small σ_i), or to eliminate bias in an inefficient estimate.

Ideally, we would like to derive formulas for estimators having the desired properties. This, unfortunately, is not possible except in simple cases such as with linear models (see Appendix E). The best that can be done in practice is to propose reasonable estimators, and to test their properties as described in the next section.

3-3. Evaluation of Statistical Properties

Given an estimation procedure **h(W)**, how do we determine its statistical properties? How do we determine whether its bias and variance are within acceptable limits? How can we compare its performance to that of other estimators? How can we assess robustness? Several approaches to the answering of these questions suggest themselves:

(a) Theoretical Analysis. Once an estimation procedure is properly defined, it may be possible to derive the precise sampling distribution, or at least some of its relevant properties such as the mean and variance. In practice, such an analysis can be carried out only for linear models (see Appendix E), or for the asymptotic distribution when the sample size is increased beyond bound. This unfortunately leaves open the most common situation, i.e., a nonlinear model with moderate sample size.

(b) Replication. If we repeat the whole series of experiments many times, and apply our estimator to each data set in turn, our estimates will form a large sample drawn from the sampling distribution. This sample can be used to estimate the mean, variance, and other properties of the sampling distribution. This procedure possesses some serious drawbacks. First, it is expensive, inasmuch as a very large number of experiments must be performed each time the estimator is applied to a new model. Second, although we may find the mean of the sampling distribution, we cannot determine the bias unless the true values of the parameters are known. Therefore, we must first carry out a series of experiments on a system whose parameters are known. Such a system is not always available.

(c) Computer Simulation (Monte Carlo Method). The objections to the replication method disappear if the experiments are not carried out on a real physical system, but are simulated on a computer instead. To simulate a series of experiments on the computer we proceed as follows:

1. Define the system by prescribing the model equations, the probability distribution of the errors, and, where applicable, a prior distribution. Assign "true" values $\hat{\boldsymbol{\phi}}$ to all the parameters $\boldsymbol{\phi}$.

2. Assign "true" values $\hat{\mathbf{z}}_\mu$ to the variables in the μth experiment ($\mu = 1, 2, \ldots, n$). Choose these values so that the model equations $\mathbf{g}(\hat{\mathbf{z}}_\mu, \hat{\boldsymbol{\theta}}) = \mathbf{0}$ are satisfied. One way of doing this is to choose a set of independent variables, assign arbitrary values to these, and then solve the equations for the values of the remaining variables. The task is most easily performed if the equations are in reduced form.

3. Use the computer to produce a set of errors $\mathbf{e}_\mu$ drawn from the prescribed probability distribution. For most computers there are available routines which generate a stream of numbers having the appearance of being random numbers uniformly distributed in the interval zero to one. They are referred to as *pseudorandom numbers*. From these, suitable transformations may be used to obtain samples from any other desired distribution. In Appendix D we show how a sample from a multivariate normal distribution may be obtained. For a more general treatment the reader is referred to the literature on Monte Carlo methods, e.g., Hammersley and Hanscomb (1964).

Having generated the values of the errors $\mathbf{e}_\mu$, we add them to the $\hat{\mathbf{w}}_\mu$ (previously generated among the $\hat{\mathbf{z}}_\mu$) to obtain the actual "data" $\mathbf{w}_\mu$.

4. The estimation procedure is now applied to the data generated by the computer as though they were obtained in real experiments. This yields an estimate $\boldsymbol{\phi}^*$ for the parameters.

5. Replicate the series of experiments as many times as we please by repeating steps 3 and 4, each time with a new sample of errors.

6. The relevant properties of the sampling distribution are obtained by averaging over all replications. Let $\boldsymbol{\phi}_i^*$ be the estimate of $\boldsymbol{\phi}$ obtained in the ith replication of the series of experiments, and let N be the total number of replications. Then, we estimate the mean of the sampling distribution as

$$\bar{\boldsymbol{\phi}}^* = (1/N) \sum_{i=1}^{N} \boldsymbol{\phi}_i^* \tag{3-3-1}$$

and its covariance matrix as

$$\mathbf{Q}^* = [1/(N-1)] \sum_{i=1}^{N} (\boldsymbol{\phi}_i^* - \bar{\boldsymbol{\phi}}^*)(\boldsymbol{\phi}_i^* - \bar{\boldsymbol{\phi}}^*)^{\mathrm{T}} \tag{3-3-2}$$

These formulas apply, of course, also when the experiments are real, not simulated.

The bias **b** of the estimator is then estimated as

$$\mathbf{b} = \bar{\boldsymbol{\phi}}^* - \hat{\boldsymbol{\phi}} \tag{3-3-3}$$

The flexibility of the simulation method is endless. We can estimate the properties of the sampling distribution for any model, and for any values of the parameters within the model. We may examine the effects of errors in the formulation of the model, by deliberately using a slightly different model in the estimator than was used in the data generator. When the errors come from a distribution that is not the same as the one assumed in the estimation procedure, we obtain a measure of the robustness of the estimator. We can also compare the true sampling distributions to theoretically derived approximations (see Chapter VII). All this can be done on the modern computer at a small fraction of the cost, in time and money, of a comparable set of physical experiments.

The results of an actual Monte Carlo study appear in Section 7-22.

B. Mathematical Properties

3-4. Optimization

In most parameter estimation methods we proceed in two stages:

(a) Define a function $\Phi(\boldsymbol{\phi})$ which is a suitable measure of the departure of the data from the model, i.e., of the "lack of fit". We refer to this function as the *objective function*. For example, in the least squares method, the objective function is the sum of squares of residuals.

(b) Seek those values $\boldsymbol{\phi}^*$ of the parameters $\boldsymbol{\phi}$ at which the objective function attains its minimum or maximum, as appropriate. We accept $\boldsymbol{\phi}^*$ as our estimate for $\boldsymbol{\phi}$. The process of computing $\boldsymbol{\phi}^*$ is called *optimization*.

Chapter IV is devoted to the realization of (a), i.e., to the definition of the objective functions. The remainder of this chapter is devoted to the analytic properties exhibited by the solutions to the optimization problem. Descriptions of the optimization process itself will be found in Chapters V and VI.

When the unknown parameters are free to assume any values whatsoever, we speak of *unconstrained optimization*. Sometimes, only parameter values satisfying certain constraints are permitted. We may have a vector of *equality constraints*

$$\mathbf{g}(\boldsymbol{\phi}) = \mathbf{0} \tag{3-4-1}$$

and/or *inequality constraints*

$$\mathbf{h}(\boldsymbol{\phi}) \geqslant \mathbf{0} \tag{3-4-2}$$

where $\mathbf{h} \geqslant \mathbf{0}$ means that each component $h_i \geqslant 0$. The set of all values of $\boldsymbol{\phi}$ satisfying all the constraints is called the *feasible region*. Feasible points satisfying Eq. (2) with strict inequality constitute the *interior* of the feasible region. A point satisfying $h_j(\boldsymbol{\phi}) = 0$ for some j is said to be *on the jth constraint*, and also on the *boundary* of the feasible region.

We examine conditions that characterize the minima‡ in the various cases that may arise. A point $\boldsymbol{\phi}^*$ is said to be a *local minimum* of $\Phi(\boldsymbol{\phi})$ if in some neighborhood (e.g., sphere) around $\boldsymbol{\phi}^*$ there is no feasible point $\boldsymbol{\phi}^{**}$ such that $\Phi(\boldsymbol{\phi}^{**}) < \Phi(\boldsymbol{\phi}^*)$. A point is a *global minimum* if there is no feasible point $\boldsymbol{\phi}^{**}$ such that $\Phi(\boldsymbol{\phi}^{**}) < \Phi(\boldsymbol{\phi}^*)$. Clearly, any global minimum is also a local minimum. Although we wish to find the global minimum, the conditions at our disposal usually characterize local minima, and there is generally no easy way to tell whether a given local minimum is the global minimum.

The problem of optimization, with or without constraints, is often referred to as the problem of *mathematical programming*. If both objective function and constraints are linear functions of the unknown parameters, we speak of *linear programming*. When constraints are present, but either they or the objective function are nonlinear, we speak of *nonlinear programming*. In the sequel, we shall assume that the objective and all constraint functions are twice differentiable functions of the parameters.

3-5. Unconstrained Optimization

To characterize an unconstrained minimum we use the rules of elementary calculus. We state the results briefly. The following are *necessary* conditions for $\boldsymbol{\phi}^*$ to be a minimum of Φ:

N1. $\boldsymbol{\phi}^*$ is a *stationary* point of Φ, that is, the gradient of Φ vanishes at $\boldsymbol{\phi}^*$

$$\partial\Phi/\partial\boldsymbol{\phi})_{\boldsymbol{\phi}=\boldsymbol{\phi}^*} = \mathbf{0} \tag{3-5-1}$$

N2. Let $\mathbf{H}(\boldsymbol{\phi})$ be the *Hessian matrix* of Φ, i.e., $H_{\alpha\beta} \equiv \partial^2\Phi/\partial\phi_\alpha\,\partial\phi_\beta$. Then $\mathbf{H}(\boldsymbol{\phi}^*)$ must be *positive semidefinite*, i.e., for any nonzero vector $\mathbf{y}$ we must have

$$\mathbf{y}^{\mathrm{T}}\mathbf{H}(\boldsymbol{\phi}^*)\mathbf{y} \geqslant 0 \tag{3-5-2}$$

‡ If a maximum of Φ is required, we seek the minimum of $-\Phi$ instead.

The following conditions are *sufficient* for $\boldsymbol{\phi}^*$ to be a local minimum of Φ:

S1. $\boldsymbol{\phi}^*$ is a stationary point of Φ.
S2. $\mathbf{H}(\boldsymbol{\phi}^*)$ is *positive definite*, i.e., Eq. (2) holds with strict inequality.

If $\mathbf{H}(\boldsymbol{\phi})$ is positive definite for all $\boldsymbol{\phi}$, and $\boldsymbol{\phi}^*$ is stationary, then $\boldsymbol{\phi}^*$ is the unique global minimum of Φ.

When $\boldsymbol{\phi}^*$ satisfies N1 and N2 but not S2, it is impossible to determine whether $\boldsymbol{\phi}^*$ is a local minimum without considering higher order derivatives.

Relations Eq. (1), regarded as equations in the unknown $\boldsymbol{\phi}^*$, are called the *normal equations*. Condition N1 states that the minimum must be a solution of the normal equations. Any solution to the normal equations *may* be a local minimum only if it also satisfies N2, and it *must* be a local minimum if it satisfies S2.

3-6. Equality Constraints

Suppose $\Phi(\boldsymbol{\phi})$ is to be minimized subject to the equality constraints Eq. (3-4-1). If the solution is denoted $\boldsymbol{\phi}^*$, then in the neighborhood of $\boldsymbol{\phi}^*$ the function $\Phi(\boldsymbol{\phi})$ must be stationary for all variations in $\boldsymbol{\phi}$ that stay within the constraints. Let $\delta\boldsymbol{\phi}$ be such a variation, i.e.,

$$\mathbf{g}(\boldsymbol{\phi}^* + \delta\boldsymbol{\phi}) = \mathbf{0} \tag{3-6-1}$$

To a first-order approximation

$$\mathbf{g}(\boldsymbol{\phi}^* + \delta\boldsymbol{\phi}) \approx \mathbf{g}(\boldsymbol{\phi}^*) + (\partial\mathbf{g}/\partial\boldsymbol{\phi})\,\delta\boldsymbol{\phi} \tag{3-6-2}$$

From Eq. (1) and Eq. (2) we conclude that

$$(\partial\mathbf{g}/\partial\boldsymbol{\phi})\,\delta\boldsymbol{\phi} = \mathbf{0} \tag{3-6-3}$$

for all permissible variations $\delta\boldsymbol{\phi}$. Expanding $\Phi(\boldsymbol{\phi})$ around $\boldsymbol{\phi}^*$ we find, to a first-order approximation

$$\Phi(\boldsymbol{\phi}^* + \delta\boldsymbol{\phi}) \approx \Phi(\boldsymbol{\phi}^*) + (\partial\Phi/\partial\boldsymbol{\phi})^{\mathrm{T}}\,\delta\boldsymbol{\phi} \tag{3-6-4}$$

Since $\Phi(\boldsymbol{\phi})$ is to be stationary, we must have

$$(\partial\Phi/\partial\boldsymbol{\phi})^{\mathrm{T}}\,\delta\boldsymbol{\phi} = 0 \tag{3-6-5}$$

for all $\delta\boldsymbol{\phi}$ satisfying Eq. (3). This condition may be paraphrased as follows: The vector $\partial\Phi/\partial\boldsymbol{\phi}$ must be orthogonal to all vectors $\delta\boldsymbol{\phi}$ which are orthogonal to the rows of the matrix $\partial\mathbf{g}/\partial\boldsymbol{\phi}$. It follows that $\partial\Phi/\partial\boldsymbol{\phi}$ must belong to the subspace spanned by the rows of $\partial \mathbf{g}/\partial \boldsymbol{\phi}$, which in turn implies the existence of numbers $\lambda_1^*, \lambda_2^*, \ldots, \lambda_\mu^*$ such that

$$\partial \Phi/\partial \boldsymbol{\phi} = \sum_{i=1}^{p} \lambda_i^* \, \partial g_i/\partial \boldsymbol{\phi} \tag{3-6-6}$$

where p is the number of constraints. The λ_i^* are called *Lagrange multipliers*.

Let us now construct the function

$$\Lambda(\boldsymbol{\phi}, \boldsymbol{\lambda}) \equiv \Phi(\boldsymbol{\phi}) - \sum_{i=1}^{p} \lambda_i g_i(\boldsymbol{\phi}) \tag{3-6-7}$$

which has a stationary point at $\boldsymbol{\phi} = \boldsymbol{\phi}^*$, $\boldsymbol{\lambda} = \boldsymbol{\lambda}^*$ if and only if

$$\partial \Lambda/\partial \boldsymbol{\phi})_{\boldsymbol{\phi}=\boldsymbol{\phi}^*} = \partial \Phi/\partial \boldsymbol{\phi})_{\boldsymbol{\phi}=\boldsymbol{\phi}^*} - \sum_{i=1}^{p} \lambda_i^* \, g_i(\boldsymbol{\phi}^*) = \mathbf{0} \tag{3-6-8}$$

and

$$\partial \Lambda/\partial \lambda_i)_{\boldsymbol{\phi}=\boldsymbol{\phi}^*} = -g_i(\boldsymbol{\phi}^*) = 0 \qquad (i = 1, 2, \ldots, p) \tag{3-6-9}$$

But Eq. (8) and Eq. (9) correspond exactly to Eq. (6) and Eq. (3-4-1), respectively. We thus conclude that $\Phi(\boldsymbol{\phi})$ has a constrained stationary point at $\boldsymbol{\phi} = \boldsymbol{\phi}^*$ if and only if $\Lambda(\boldsymbol{\phi}, \boldsymbol{\lambda})$ has an unconstrained stationary point at $\boldsymbol{\phi} = \boldsymbol{\phi}^*$ and $\boldsymbol{\lambda} = \boldsymbol{\lambda}^*$.

In some problems we have an $n \times m$ matrix of constraints $\mathbf{G}(\boldsymbol{\phi}) = \mathbf{0}$. In this case we need an $m \times n$ matrix of Lagrange multipliers $\mathbf{L}$, and the Lagrangian takes the form

$$\Lambda(\boldsymbol{\phi}, \mathbf{L}) \equiv \Phi(\boldsymbol{\phi}) + \mathrm{Tr}(\mathbf{LG}) \tag{3-6-10}$$

To determine the nature of the stationary point $(\boldsymbol{\phi}^*, \boldsymbol{\lambda}^*)$, expand both Φ and the g_i in Taylor series around $\boldsymbol{\phi}^*$, retaining terms up to second order:

$$\Phi(\boldsymbol{\phi}^* + \delta\boldsymbol{\phi}) \approx \Phi(\boldsymbol{\phi}^*) + (\partial \Phi/^{\mathrm{T}} \partial \boldsymbol{\phi})^{\mathrm{T}} \, \delta\boldsymbol{\phi} + \tfrac{1}{2} \, \delta\boldsymbol{\phi}^{\mathrm{T}} \, (\partial^2 \Phi/\partial \boldsymbol{\phi} \, \partial \boldsymbol{\phi}) \, \delta\boldsymbol{\phi} \tag{3-6-11}$$

$$g_i(\boldsymbol{\phi}^* + \delta\boldsymbol{\phi}) \approx g_i(\boldsymbol{\phi}^*) + (\partial g_i/\partial \boldsymbol{\phi})^{\mathrm{T}} \, \delta\boldsymbol{\phi} + \tfrac{1}{2} \, \delta\boldsymbol{\phi}^{\mathrm{T}} \, (\partial^2 g_i/\partial \boldsymbol{\phi} \, \partial \boldsymbol{\phi}) \, \delta\boldsymbol{\phi}$$
$$(i = 1, 2, \ldots, p) \tag{3-6-12}$$

If $\boldsymbol{\phi}^*$ is a local constrained minimum of Φ, it follows that $\Phi(\boldsymbol{\phi}^* + \delta\boldsymbol{\phi}) \geqslant \Phi(\boldsymbol{\phi}^*)$ for all sufficiently small $\delta\boldsymbol{\phi}$ for which

$$g_i(\boldsymbol{\phi}^* + \delta\boldsymbol{\phi}) = 0 \qquad (i = 1, 2, \ldots, p)$$

Such $\delta\boldsymbol{\phi}$ must therefore satisfy (approximately) the equations

$$(\partial g_i/\partial \boldsymbol{\phi})^{\mathrm{T}} \, \delta\boldsymbol{\phi} + \tfrac{1}{2} \, \delta\boldsymbol{\phi}^{\mathrm{T}} \, (\partial^2 g_i/\partial \boldsymbol{\phi} \, \partial \boldsymbol{\phi}) \, \delta\boldsymbol{\phi} = 0 \qquad (i = 1, 2, \ldots, p) \tag{3-6-13}$$

and the inequality

$$(\partial \Phi/\partial \boldsymbol{\phi})^{\mathrm{T}} \, \delta\boldsymbol{\phi} + \tfrac{1}{2} \, \delta\boldsymbol{\phi}^{\mathrm{T}} \, (\partial^2 \Phi/\partial \boldsymbol{\phi} \, \partial \boldsymbol{\phi}) \, \delta\boldsymbol{\phi} \geqslant 0 \tag{3-6-14}$$

Multiplying each Eq. (13) by λ_i^* and subtracting from Eq. (14) we obtain, in view of Eq. (6)

$$\tfrac{1}{2}\,\delta\boldsymbol{\phi}^{\mathrm{T}}\left[(\partial^2\Phi/\partial\boldsymbol{\phi}\,\partial\boldsymbol{\phi}) - \sum_{i=1}^{p}\lambda_i^*\,(\partial^2 g_i/\partial\boldsymbol{\phi}\,\partial\boldsymbol{\phi})\right]\delta\boldsymbol{\phi} \geqslant 0 \qquad \text{(3-6-15)}$$

This must hold for any $\delta\boldsymbol{\phi}$ satisfying Eq. (14), i.e., by cords joining the solution point $\boldsymbol{\phi}^*$ to neighboring points approximately on the constraint surfaces. Because of continuity, Eq. (15) must, therefore, also be satisfied by the tangents to the constraint surfaces at the solution, these being limits of such cords. These tangents satisfy

$$(\partial\mathbf{g}/\partial\boldsymbol{\phi})\,\delta\boldsymbol{\phi} = \mathbf{0} \qquad \text{(3-6-16)}$$

That is, they are the null vectors of the matrix $\partial\mathbf{g}/\partial\boldsymbol{\phi}$. Now let $\mathbf{B}$ be a matrix whose columns span the null space of $\partial\mathbf{g}/\partial\boldsymbol{\phi}$. That is, every null vector $\delta\boldsymbol{\phi}$ of $\partial\mathbf{g}/\partial\boldsymbol{\phi}$ can be expressed in the form $\delta\boldsymbol{\phi} = \mathbf{Bx}$, where $\mathbf{x}$ is an arbitrary vector whose dimension is that of the null space. If the dimension of $\boldsymbol{\phi}$ is r, and if $\partial\mathbf{g}/\partial\boldsymbol{\phi}$ has p linearly independent rows, then the dimension of $\mathbf{x}$ is $r - p$, and $\mathbf{B}$ is an $r \times (r - p)$ matrix. Letting

$$\mathbf{A} \equiv \frac{\partial^2\Phi}{\partial\boldsymbol{\phi}\,\partial\boldsymbol{\phi}} - \sum_{i=1}^{p}\lambda_i^*\,\frac{\partial^2 g_i}{\partial\boldsymbol{\phi}\,\partial\boldsymbol{\phi}} \qquad \text{(3-6-17)}$$

we know, from Eq. (15), that

$$\mathbf{x}^{\mathrm{T}}\mathbf{B}^{\mathrm{T}}\mathbf{A}\mathbf{B}\mathbf{x} \geqslant 0 \qquad \text{(3-6-18)}$$

for all $r - p$ dimensional vectors $\mathbf{x}$, i.e., $\mathbf{B}^{\mathrm{T}}\mathbf{A}\mathbf{B}$ must be positive semidefinite.

Conversely, if $\mathbf{B}^{\mathrm{T}}\mathbf{A}\mathbf{B}$ is positive definite, it follows by continuity that Eq. (15) holds for all sufficiently small $\delta\boldsymbol{\phi}$ satisfying Eq. (13). Hence, $\boldsymbol{\phi}^*$ is a constrained minimum of Φ.

3-7. Inequality Constraints

Suppose we wish to minimize $\Phi(\boldsymbol{\phi})$ subject to the inequality constraint Eq. (3-4-2). If the minimum occurs at a point $\boldsymbol{\phi}^*$ which is in the interior of the feasible region, that is $h_i(\boldsymbol{\phi}^*) > 0$ for all i, then the constraints are irrelevant as far as the local nature of the minimum is concerned. Therefore, $\boldsymbol{\phi}^*$ must satisfy the conditions for an unconstrained minimum. When $\boldsymbol{\phi}^*$ lies on the boundary of the feasible region, there will be some values of i for which

$$h_i(\boldsymbol{\phi}^*) = 0 \qquad \text{(3-7-1)}$$

We refer to these h_i as the *active constraints*. For the purpose of characterizing the point $\boldsymbol{\phi}^*$, we may disregard all the inactive constraints, and in the sequel we shall consider the vector $\mathbf{h}$ to consist of all the active constraints alone. We denote by t the number of active constraints, and by T the total number of constraints.

At an inequality constrained minimum it is required that the gradient of the objective function should point decisively into the feasible region. We observe that since the constraint functions are positive inside and negative outside the feasible region, their gradients also point into the feasible region. The necessary condition for minimality is, then, that the gradient of the objective function should be a linear combination with *positive coefficients* of the gradients of the constraint functions.

More precisely, John (1948) has proven that $\boldsymbol{\phi}^*$ is a minimum only if there exist nonnegative numbers $\lambda_0, \lambda_1, \lambda_2, \ldots, \lambda_t$ not all zero such that

$$\lambda_0 \, \partial\Phi(\boldsymbol{\phi}^*)/\partial\boldsymbol{\phi} = \sum_{i=1}^{t} \lambda_i[\partial h_i(\boldsymbol{\phi}^*)/\partial\boldsymbol{\phi}] \tag{3-7-2}$$

The more famous Kuhn–Tucker condition (Kuhn and Tucker, 1951) asserts that we may choose $\lambda_0 = 1$ provided the constraints [Eq. (1)] meet a certain qualification,‡ which in practice is almost always satisfied.

Clearly, Eq. (2) is unaffected if we add on the right-hand side terms corresponding to the inactive constraints, provided their multipliers assume the value zero. The Kuhn–Tucker condition can then be stated as follows:

$$\partial\Phi(\boldsymbol{\phi}^*)/\partial\boldsymbol{\phi} = \sum_{i=1}^{T} \lambda_i \, \partial h_i/\partial\boldsymbol{\phi} \tag{3-7-3}$$

$$h_i(\boldsymbol{\phi}^*) \geqslant 0 \qquad (i = 1, 2, \ldots, T) \tag{3-7-4}$$

$$\lambda_i \geqslant 0 \qquad (i = 1, 2, \ldots, T) \tag{3-7-5}$$

$$\lambda_i h_i(\boldsymbol{\phi}^*) = 0 \qquad (i = 1, 2, \ldots, T) \tag{3-7-6}$$

The last equation states that either a constraint is active ($h_i = 0$) or its multiplier vanishes. This is known as the principle of *complementary slackness.*

Sufficient conditions for local optimality have been derived by McCormick (1967) and Fiacco (1968). They require the quantity $\mathbf{y}^T\mathbf{A}\mathbf{y}$ to be positive for all vectors $\mathbf{y}$ which point into the feasible region or are tangent to it at $\boldsymbol{\phi}^*$. Here $\mathbf{A}$ is the Hessian of $\Phi - \sum_i \lambda_i h_i$ (λ_0 having been set to one).

The role of the λ_i here is analogous to that of the Lagrange multipliers in the case of equality constraints. The λ_i are called *dual* variables or *shadow prices.* When both equality and inequality constraints apply, the necessary

‡ The qualification requires that for every vector $\mathbf{u}$ such that $\mathbf{u}^T \, \partial h_i/\partial\boldsymbol{\phi})_{\boldsymbol{\phi}=\boldsymbol{\phi}^*} \geqslant 0$ ($i = 1, 2, \ldots, t$), there should exist a vector of functions $\boldsymbol{\phi}(\tau)$ such that $\boldsymbol{\phi}(0) = \boldsymbol{\phi}^*$, $\boldsymbol{\phi}(\tau)$ is in the feasible region for $0 \leqslant t \leqslant 1$, and $\partial\boldsymbol{\phi}/\partial\tau_{\boldsymbol{\phi}=\boldsymbol{\phi}^*} = \mathbf{u}$. A case where the qualification is not met occurs when $\boldsymbol{\phi}^*$ is at a cusp formed by constraints tangent to each other.

condition for a minimum states that there exist scalars $\mu_1, \mu_2, \ldots$, and non-negative scalars $\lambda_1, \lambda_2, \ldots$ such that

$$\partial\Phi/\partial\boldsymbol{\phi})_{\boldsymbol{\phi}=\boldsymbol{\phi}^*} = \sum_i \mu_i \, \partial g_i/\partial\boldsymbol{\phi})_{\boldsymbol{\phi}=\boldsymbol{\phi}^*} + \sum_i \lambda_i \, \partial h_i/\partial\boldsymbol{\phi})_{\boldsymbol{\phi}=\boldsymbol{\phi}^*} \tag{3-7-7}$$

The subject of optimality conditions is discussed in great detail by Mangasarian (1969).

3-8. Problems

1. Consider the model whose likelihood is given by Eq. (3-2-7). Show that if θ and σ are unknown parameters, then $\sum_{\mu=1}^{n} w_\mu$ and $\sum_{\mu=1}^{n} {w_\mu}^2$ form a pair of statistics which is jointly sufficient for θ and σ.

2. Show that the Lagrange multipliers represent the unit cost of the constraints. That is, if a constraint $g_i(\boldsymbol{\theta}) = 0$ is replaced by $g_i(\boldsymbol{\theta}) = \varepsilon$, then the minimum attainable value of Φ is increased (to a first-order approximation) by an amount $\lambda_i \varepsilon$.

3. As above, but for inequality constraints.

4. Let $\mathbf{A}$ be a square matrix. Using Lagrange multipliers, show that of all vectors $\mathbf{x}$ of unit length (i.e., $\mathbf{x}^T\mathbf{x} = 1$), the ones for which $\mathbf{x}^T\mathbf{A}\mathbf{x}$ is either a minimum or a maximum are eigenvectors of $\mathbf{A}$.

Chapter

IV

Methods of Estimation

4-1. Residuals

In the previous chapter we have discussed, in general terms, what desirable properties an estimator should possess, and how any specific estimator may be tested in order to determine whether it possesses these properties. We now proceed to describe specific estimators, or estimation methods, which are in widespread use.

We have defined the *error* $\varepsilon_{\mu a}$ as the difference between the measured and true values of a variable. In the case of a reduced model $\mathbf{y} = \mathbf{f}(\mathbf{x}, \boldsymbol{\theta})$, if we knew the true value $\hat{\boldsymbol{\theta}}$ of $\boldsymbol{\theta}$ we could compute the error

$$\varepsilon_{\mu a} = y_{\mu a} - f_a(\mathbf{x}_\mu, \hat{\boldsymbol{\theta}}) \tag{4-1-1}$$

We can also compute these differences for any other value of $\boldsymbol{\theta}$. This defines functions

$$e_{\mu a}(\boldsymbol{\theta}) \equiv y_{\mu a} - f_a(\mathbf{x}_\mu, \boldsymbol{\theta}) \tag{4-1-2}$$

to which we refer as the *residuals*. The errors are equal to the residuals evaluated with the true value $\boldsymbol{\theta} = \hat{\boldsymbol{\theta}}$.

The residuals in an inexact structural model $\mathbf{g}(\mathbf{z}, \boldsymbol{\theta}) = \boldsymbol{\gamma}$ are obtained simply by evaluation of the model equations

$$e_{\mu a}(\boldsymbol{\theta}) = g_a(\mathbf{z}_\mu, \boldsymbol{\theta}) \tag{4-1-3}$$

When the structural model is exact, the residuals are the differences between the observed and estimated values of the variables

$$e_{\mu a}(\hat{\mathbf{W}}) = w_{\mu a} - \hat{w}^*_{\mu a} \tag{4-1-4}$$

A. Least Squares

4-2. Unweighted Least Squares

The method of least squares is the oldest and most widely used estimation procedure. At least some of its popularity is due to the fact that it can be applied in an ad hoc manner directly to the deteriminstic model, without any cognizance being taken of the probability distribution of the observations. Needless to say, estimates obtained in such a way may be very unsatisfactory indeed, although one can envision situations in which nothing better can be done. We do not wish to imply that least squares estimates are always merely ad hoc. Quite the contrary is true, and where the observations have certain probability distributions, these estimates even possess optimal statistical properties, which will be described in the sequel.

In cases of pure curve fitting, where the coefficients have no physical significance, the least squares method is usually adequate.

We employ the following notation: A small capital letter denotes the vector formed by adjoining to each other the rows of the matrix denoted by the same letter. Thus

$$\mathrm{E}^{\mathrm{T}} \equiv [E_{11}, E_{12}, \ldots, E_{1m}, E_{21}, \ldots, E_{nm}]$$

The least squares procedure in its simplest form consists of finding the values of $\boldsymbol{\theta}$ which minimize the function

$$\Phi(\boldsymbol{\theta}) \equiv \mathrm{E}^{\mathrm{T}}(\boldsymbol{\theta})\mathrm{E}(\boldsymbol{\theta}) = \mathrm{Tr}[\mathbf{E}^{\mathrm{T}}(\boldsymbol{\theta})\mathbf{E}(\boldsymbol{\theta})] \tag{4-2-1}$$

which is, in component form

$$\Phi(\boldsymbol{\theta}) = \sum_{a=1}^{m} \sum_{\mu=1}^{n} e_{\mu a}^2(\boldsymbol{\theta}) \tag{4-2-2}$$

i.e., we minimize the sum of squares of the residuals. When $m = 1$ we speak of *single equation least squares*. In practice, most estimation problems fall into this category. A typical problem is worked out in detail in Section 5-21.

We derive the normal equations easily

$$\partial\Phi/\partial\boldsymbol{\theta} = 2 \sum_{a=1}^{m} \sum_{\mu=1}^{n} e_{\mu a}\, \partial e_{\mu a}/\partial\boldsymbol{\theta} = \mathbf{0} \tag{4-2-3}$$

In the most common case of a single reduced equation, $e_\mu = y_\mu - f(\mathbf{x}_\mu, \boldsymbol{\theta})$, and we have:

$$\Phi(\boldsymbol{\theta}) = \sum_{\mu=1}^{n} [y_\mu - f(\mathbf{x}_\mu, \boldsymbol{\theta})]^2 \tag{4-2-4}$$

$$\partial\Phi/\partial\theta_\alpha = -2\sum_{\mu=1}^{n} e_\mu\, \partial f(\mathbf{x}_\mu, \boldsymbol{\theta})/\partial\theta_\alpha \qquad (\alpha = 1, 2, \ldots, l) \tag{4-2-5}$$

4-3. Weighted Least Squares

An objective function consisting of a simple sum of squares is often unsatisfactory for the following reasons:

(a) The various quantities $y_{\mu a}$ (or $g_{\mu a}$) may represent entities having different physical dimensions, or measured on different scales. For instance, $y_{\mu 1}$ may be the concentration of a chemical, expressed in mole fractions and falling in the range zero to one; at the same times, $y_{\mu 2}$ may be a temperature measured in degrees centigrade, and falling in the range 500–1000. It clearly makes no sense to sum together squares of numbers of such disparate orders of magnitude; the residuals of the temperatures are likely to dominate those of the mole fractions and any information contained in the latter will be lost.

(b) Some observations may be known to be less reliable than others, and we want to make sure that our parameter estimates will be less influenced by those than by the more accurate ones. (Note that we cannot, after all, escape the statistical structure of the data.)

The solution to both of these problems is one and the same; assign a nonnegative weight factor $b_{\mu a}$ to each $e_{\mu a}(\boldsymbol{\theta})$, and minimize

$$\Phi(\boldsymbol{\theta}) \equiv \sum_{a=1}^{m} \sum_{\mu=1}^{n} b_{\mu a}\, e_{\mu a}^2(\boldsymbol{\theta}) \tag{4-3-1}$$

We choose small $b_{\mu a}$ for $y_{\mu a}$ which are measured on a large scale, or which are highly unreliable, and conversely for large $b_{\mu a}$. A more general formulation is one which assigns weights to cross product terms as well, i.e.,

$$\Phi(\boldsymbol{\theta}) = \sum_{a=1}^{m} \sum_{b=1}^{m} \sum_{\mu=1}^{n} \sum_{\eta=1}^{n} B_{(\mu a)(\eta b)}\, e_{\mu a}(\boldsymbol{\theta}) e_{\eta b}(\boldsymbol{\theta}) \tag{4-3-2}$$

The weights $B_{(\mu a)(\eta b)}$ must be elements of a positive definite or semidefinite matrix, for otherwise $\Phi(\boldsymbol{\theta})$ can be made to approach $-\infty$. Clearly, Eq. (1) is a special case of Eq. (2), with $B_{(\mu a)(\eta b)} = b_{\mu a}\delta_{\mu\eta}\delta_{ab}$, and Eq. (4-2-2) is a special case of Eq. (1), with all $b_{\mu a} = 1$.

Additional important special cases of (2) are the following:

(a) Weighting by Variable. Where $B_{(\mu a)(\eta b)} = 0$ ($\mu \neq \eta$), and the weights are independent of μ

$$\Phi(\boldsymbol{\theta}) = \sum_{a=1}^{m} \sum_{b=1}^{m} B_{ab} \sum_{\mu=1}^{n} e_{\mu a}(\boldsymbol{\theta}) e_{\mu b}(\boldsymbol{\theta}) = \sum_{\mu=1}^{n} \mathbf{e}_{\mu}{}^{\mathrm{T}} \mathbf{B} \mathbf{e}_{\mu} \tag{4-3-3}$$

When $\mathbf{B}$ is a diagonal matrix, this simplifies to

$$\Phi(\boldsymbol{\theta}) = \sum_{a=1}^{m} b_a \sum_{\mu=1}^{n} e_{\mu a}^2(\boldsymbol{\theta}) \tag{4-3-4}$$

(b) Weighting by Experiment. Applicable mostly to the single equation case

$$\Phi(\boldsymbol{\theta}) = \sum_{\mu=1}^{n} b_{\mu} e_{\mu}{}^2(\boldsymbol{\theta}) \tag{4-3-5}$$

(c) Weighting by Experiment and Variable.

$$\Phi(\boldsymbol{\theta}) = \sum_{\mu=1}^{n} \sum_{a=1}^{m} \sum_{b=1}^{m} b_{\mu} B_{ab} e_{\mu a}(\boldsymbol{\theta}) e_{\mu b}(\boldsymbol{\theta}) = \sum_{\mu=1}^{n} b_{\mu} \mathbf{e}_{\mu}{}^{\mathrm{T}} \mathbf{B} \mathbf{e}_{\mu} \tag{4-3-6}$$

Does statistical theory tell us what values should be assigned to the weights, or when we are entitled to use the simpler formulas? The answer is at least partly in the affirmative. We shall see later that if the model equations are linear in the parameters (Section 4-4), or if the number of observations is large and the errors are normally distributed (Section 4-7), then the choice of weights leading to least-variance estimates is given by the elements of the inverse of the covariance matrix of the errors. That is

$$B_{(\mu a)(\eta b)} = (\mathbf{V}^{-1})_{(\mu a)(\eta b)} \tag{4-3-7}$$

where

$$V_{(\mu a)(\eta b)} = E(\varepsilon_{\mu a} \varepsilon_{\eta b}) \tag{4-3-8}$$

Although we cannot prove optimal properties in the general case (non-normal distributions with nonlinear models), it is still reasonable, and approximately optimal, to use weights which are the elements of the inverse of the covariance matrix. When the covariance matrix is not known, one may choose either to guess or to use a method such as maximum likelihood which sometimes enables one to estimate the weights along with the other parameters. Or, one can obtain a direct estimate of the covariance matrix by replicating at least some of the experiments.

4-4. Multiple Linear Regression

When the model is linear, the choice of proper weights in Eq. (4-3-2) ensures optimal statistical properties for the corresponding estimators. The linear model takes the form

$$\mathbf{f}_\mu = \mathbf{f}(\mathbf{x}_\mu, \boldsymbol{\theta}) = \mathbf{B}_\mu(\mathbf{x}_\mu)\boldsymbol{\theta} \tag{4-4-1}$$

where $\mathbf{B}_\mu(\mathbf{x}_\mu)$ is a matrix of given functions (polynomials and trigonometric functions are often used in curve fitting).‡ Adjoining the equations for all values of μ, we obtain, in matrix form

$$\mathbf{F} = \mathbf{B}\boldsymbol{\theta} \tag{4-4-2}$$

where $\mathbf{F}^{\mathrm{T}} \equiv [\mathbf{f}_1^{\mathrm{T}}, \mathbf{f}_2^{\mathrm{T}}, \ldots, \mathbf{f}_n^{\mathrm{T}}]$ and $\mathbf{B}^{\mathrm{T}} \equiv [\mathbf{B}_1^{\mathrm{T}}(\mathbf{x}_1), \mathbf{B}_2^{\mathrm{T}}(\mathbf{x}_2), \ldots, \mathbf{B}_n^{\mathrm{T}}(\mathbf{x}_n)]$. For given data, $\mathbf{B}$ is a constant matrix. Suppose the $\mathbf{x}_\mu$ are measured precisely, and each observation $\mathbf{y}_\mu$ is a sample from a random variable whose mean value is $\mathbf{f}_\mu$, and let the joint covariance matrix of all the elements of $\mathbf{Y}$ be $\mathbf{V}$, i.e.,

$$E(y_{\mu a} - f_{\mu a})(y_{\eta b} - f_{\eta b}) = V_{(\mu a)(\eta b)} \tag{4-4-3}$$

If we determine $\boldsymbol{\theta} = \boldsymbol{\theta}^*$ so as to minimize the function

$$\Phi(\boldsymbol{\theta}) \equiv (\mathbf{Y} - \mathbf{B}\boldsymbol{\theta})^{\mathrm{T}}\, \mathbf{V}^{-1}(\mathbf{Y} - \mathbf{B}\boldsymbol{\theta}) \tag{4-4-4}$$

then $\boldsymbol{\theta}^*$ must satisfy:

$$(\partial\Phi/\partial\boldsymbol{\theta}) = -2\mathbf{B}^{\mathrm{T}}\mathbf{V}^{-1}(\mathbf{Y} - \mathbf{B}\boldsymbol{\theta}^*) = \mathbf{0} \tag{4-4-5}$$

This is equivalent to the *normal equations*

$$\mathbf{B}^{\mathrm{T}}\mathbf{V}^{-1}\mathbf{B}\boldsymbol{\theta}^* = \mathbf{B}^{\mathrm{T}}\mathbf{V}^{-1}\mathbf{Y} \tag{4-4-6}$$

Solving for $\boldsymbol{\theta}^*$, we find, provided $\mathbf{B}^{\mathrm{T}}\mathbf{V}^{-1}\mathbf{B}$ is nonsingular,§

$$\boldsymbol{\theta}^* = (\mathbf{B}^{\mathrm{T}}\mathbf{V}^{-1}\mathbf{B})^{-1}\mathbf{B}^{\mathrm{T}}\mathbf{V}^{-1}\mathbf{Y} \tag{4-4-7}$$

‡ Suppose we are fitting the single equation model $f(x, \boldsymbol{\theta}) = \theta_1 + \theta_2 x + \theta_3 x^2$. Then $\mathbf{B}_\mu$ is the row vector $[1, x, x^2]$ and

$$\mathbf{B} = \begin{bmatrix} 1 & x_1 & x_1^2 \\ 1 & x_2 & x_2^2 \\ \vdots & & \\ 1 & x_n & x_n^2 \end{bmatrix}$$

§ If $\mathbf{B}^T\mathbf{V}^{-1}\mathbf{B}$ is singular, the normal equations possess infinitely many solutions. Of these, the one for which $\boldsymbol{\theta}^{*T}\boldsymbol{\theta}^*$ is minimum is given by

$$\boldsymbol{\theta}^* = (\mathbf{B}^T\mathbf{V}^{-1}\mathbf{B})^+\mathbf{B}^T\mathbf{V}^{-1}\mathbf{Y}$$

where $\mathbf{A}^+$ is the pseudoinverse of $\mathbf{A}$ (see Section A-1).

This is the well-known *multiple linear regression* formula. Clearly $\boldsymbol{\theta}^*$ is a linear estimate, having the form $\boldsymbol{\theta}^* = \mathbf{AY}$. By our assumption, $E(\mathbf{Y}) = \mathbf{B}\hat{\boldsymbol{\theta}}$. Hence, $\mathbf{B}$ and $\mathbf{V}$ being constant, we find from Eq. (7)

$$E(\boldsymbol{\theta}^*) = \hat{\boldsymbol{\theta}} \tag{4-4-8}$$

That is, $\boldsymbol{\theta}^*$ is an unbiased estimate of $\boldsymbol{\theta}$. Now, Eq. (3) is equivalent to

$$E(\mathbf{Y} - \mathbf{B}\hat{\boldsymbol{\theta}})(\mathbf{Y} - \mathbf{B}\hat{\boldsymbol{\theta}})^{\mathrm{T}} = \mathbf{V}$$

Also, it is easily seen that

$$\boldsymbol{\theta}^* - \hat{\boldsymbol{\theta}} = (\mathbf{B}^{\mathrm{T}}\mathbf{V}^{-1}\mathbf{B})^{-1}\mathbf{B}^{\mathrm{T}}\mathbf{V}^{-1}(\mathbf{Y} - \mathbf{B}\hat{\boldsymbol{\theta}})$$

Hence the covariance of the sampling distribution of the estimate $\boldsymbol{\theta}^*$ turns out to be

$$\mathbf{V}_{\boldsymbol{\theta}} \equiv E(\boldsymbol{\theta}^* - \hat{\boldsymbol{\theta}})(\boldsymbol{\theta}^* - \hat{\boldsymbol{\theta}})^{\mathrm{T}} = (\mathbf{B}^{\mathrm{T}}\mathbf{V}^{-1}\mathbf{B})^{-1} \tag{4-4-9}$$

The Gauss–Markov Theorem (proved in Appendix E) asserts that among all linear unbiased estimates, Eq. (7) yields the one whose variance is smallest. If, furthermore, the distribution of the $\boldsymbol{\varepsilon}_\mu$ is normal, the estimate is efficient. In the case where the errors of all the observations are independent and of equal variance σ^2, we have $\mathbf{V} = \sigma^2\mathbf{I}$, and

$$\boldsymbol{\theta}^* = (\mathbf{B}^{\mathrm{T}}\mathbf{B})^{-1}\mathbf{B}^{\mathrm{T}}\mathbf{Y} \tag{4-4-10}$$

which is the usual unweighted linear least squares estimate. The covariance matrix of this estimate is given by $\mathbf{V}_{\boldsymbol{\theta}} = \sigma^2(\mathbf{B}^{\mathrm{T}}\mathbf{B})^{-1}$.

Computational methods for solving linear regression problems are discussed in Section 5-11. A question that often arises in connection with linear regression problems is which variables should be included, and which should be excluded from the model. Stated in another way, the question is which parameters should be left out (assumed to be zero) because they do not contribute significantly to the model. The method of stepwise regression (Section A-3) provides an answer to this question.

Before leaving the subject of linear models, let us examine briefly the question of how the optimal properties of the regression estimate are affected when the assumed model is incorrect.

First, suppose some important terms were omitted from the model. As a result, it is no longer true that $E(\mathbf{Y}) = \mathbf{B}\hat{\boldsymbol{\theta}}$; rather, we have

$$E(\mathbf{Y}) = \mathbf{B}\hat{\boldsymbol{\theta}} + \mathbf{s} \tag{4-4-11}$$

where $\mathbf{s}$ is a fixed vector consisting of the omitted terms. If $\boldsymbol{\theta}^*$ is computed from Eq. (7), we find

$$E(\boldsymbol{\theta}^*) = \hat{\boldsymbol{\theta}} + (\mathbf{B}^{\mathrm{T}}\mathbf{V}^{-1}\mathbf{B})^{-1}\mathbf{B}^{\mathrm{T}}\mathbf{V}^{-1}\mathbf{s} \tag{4-4-12}$$

so that $\boldsymbol{\theta}^*$ is no longer an unbiased estimate. The bias is precisely equal to

$$(\mathbf{B}^T\mathbf{V}^{-1}\mathbf{B})^{-1}\mathbf{B}^T\mathbf{V}^{-1}\mathbf{s}.$$

Secondly, consider the case where an erroneous value has been taken for $\mathbf{V}$. Suppose the true covariance matrix is $\mathbf{U} \neq \mathbf{V}$. Then the covariance matrix of the estimate Eq. (7) is given by

$$\mathbf{V} = (\mathbf{B}^T\mathbf{V}^{-1}\mathbf{B})^{-1}\mathbf{B}^T\mathbf{V}^{-1}\mathbf{U}\mathbf{V}^{-1}\mathbf{B}(\mathbf{B}^T\mathbf{V}^{-1}\mathbf{B})^{-1} \tag{4-4-13}$$

We wish to determine how inefficient this estimate is relative to the best possible estimate in which $\mathbf{V} = \mathbf{U}$. The covariance of the latter estimate is, according to Eq. (9), $(\mathbf{B}^T\mathbf{U}^{-1}\mathbf{B})^{-1}$. We define the relative inefficiency e of Eq. (7) as the ratio of its generalized covariance to the minimum attainable generalized covariance, i.e.,

$$e = \det(\mathbf{B}^T\mathbf{V}^{-1}\mathbf{B})^{-1}\mathbf{B}^T\mathbf{V}^{-1}\mathbf{U}\mathbf{V}^{-1}\mathbf{B}(\mathbf{B}^T\mathbf{V}^{-1}\mathbf{B})^{-1}/\det(\mathbf{B}^T\mathbf{U}^{-1}\mathbf{B})^{-1} \tag{4-4-14}$$

Clearly, $e = 1$ if $\mathbf{V} = \mathbf{U}$. In other cases, it can be shown that e can assume any value in the range given below; its actual value depends on $\mathbf{B}$

$$1 \leqslant e \leqslant (1 + \alpha)^2/4\alpha \tag{4-4-15}$$

where α is the condition number, i.e., the ratio of largest to smallest eigenvalues, of the matrix $\mathbf{V}^{-1/2}\mathbf{U}\mathbf{V}^{-1/2}$. To illustrate, suppose an unweighted least squares estimate is used where in fact the error variances range between 10 and 100. We have, then, $\mathbf{V}^{-1/2} = \mathbf{I}$, and $\mathbf{U} = \text{diag}(\mathbf{u})$, where $\mathbf{u}$ is a vector of numbers in the range 10 to 100. It follows that $\mathbf{V}^{-1/2}\mathbf{U}\mathbf{V}^{-1/2} = \mathbf{U} = \text{diag}(\mathbf{u})$, and $\alpha = 100/10 = 10$. The inefficiency of $\boldsymbol{\theta}^*$ may be as high as $(1 + 10)^2/40 \approx 3$.

While an estimate of the form Eq. (7) or Eq. (10) is the best *unbiased* estimate, it is possible to construct biased estimates whose total expected squared error is less. For instance, in the method of ridge regression (Hoerl, 1962; Hoerl and Kennard, 1970), one substitutes for Eq. (10) the estimate

$$\boldsymbol{\theta}^*(\lambda) = (\mathbf{B}^T\mathbf{B} + \lambda\mathbf{I})^{-1}\mathbf{B}^T\mathbf{Y} \tag{4-4-16}$$

where λ is a positive parameter. It can be shown that the expected square error is

$$\mathbf{U}(\lambda) \equiv E(\boldsymbol{\theta}^*(\lambda) - \hat{\boldsymbol{\theta}})(\boldsymbol{\theta}^*(\lambda) - \hat{\boldsymbol{\theta}})^T = (\mathbf{B}^T\mathbf{B} + \lambda\mathbf{I})^{-1}(\sigma^2\mathbf{B}^T\mathbf{B} + \lambda^2\hat{\boldsymbol{\theta}}\hat{\boldsymbol{\theta}}^T)(\mathbf{B}^T\mathbf{B} + \lambda\mathbf{I})^{-1} \tag{4-4-17}$$

and that the quantity $\det \mathbf{U}(\lambda)$ is minimum when λ satisfies the equation

$$\lambda\hat{\boldsymbol{\theta}}^T(\sigma^2\mathbf{B}^T\mathbf{B} + \lambda^2\hat{\boldsymbol{\theta}}\hat{\boldsymbol{\theta}}^T)^{-1}\hat{\boldsymbol{\theta}} = \text{Tr}(\mathbf{B}^T\mathbf{B} + \lambda\mathbf{I})^{-1} \tag{4-4-18}$$

Since $\hat{\boldsymbol{\theta}}$ is unknown, the optimal λ cannot be determined a priori. Hoerl and Kennard recommend construction of a so-called ridge trace, which is a plot of the components of $\boldsymbol{\theta}^*(\lambda)$ versus λ with λ increasing from zero. One chooses a value of λ where $\boldsymbol{\theta}^*(\lambda)$ ceases to vary rapidly. Note that at $\lambda = 0$ we have the usual least squares estimate. Note also that $\lambda = 0$ does not satisfy Eq. (18). Hence, the least squares estimate is never the linear least squared error estimate.

B. Maximum Likelihood

4-5. Definition

In Sections 2-12–2-13 we have defined the likelihood function $L(\boldsymbol{\phi})$ of the sample as being the joint pdf of the observations, viewed as a function of the unknown parameters $\boldsymbol{\phi}$. These unknown parameters were of three kinds:

1. $\boldsymbol{\theta}$ represents the unknown parameters of the deterministic models.
2. $\hat{\mathbf{W}}$ represents the true values of the observed variables.
3. $\boldsymbol{\psi}$ represents other distribution parameters.

In Section 2-13 we saw that the model equations could be regarded as equality constraints which limit the possible values that the $\boldsymbol{\theta}$ and $\hat{\mathbf{W}}$ could attain. In addition, prior information may impose certain inequality constraints (e.g., nonnegativity) on the parameters.

The *maximum likelihood estimate* (MLE) of $\boldsymbol{\phi}$ is that value of $\boldsymbol{\phi}$ satisfying all the equality and inequality constraints, for which the likelihood function attains its maximum value (if such a value exists).

Under relatively mild conditions on the form of the likelihood function, the MLE is consistent and asymptotically efficient. This is a strong argument for using the MLE when the sample is large. The MLE does not usually possess any optimal properties for small samples. It is generally neither unbiased nor efficient, although it is sufficient when a sufficient statistic exists. Sampling experiments [see, e.g., Chow (1964), Cragg (1967), Carney and Goldwyn (1967)] have shown, however, that the maximum likelihood method produces acceptable estimates in many situations. Whereas better methods may be available in specific cases, a powerful argument for the use of the maximum likelihood method is the generality and relative ease of application.

Since the logarithm is a monotonic increasing function of its argument, the value of $\boldsymbol{\phi}$ that maximizes $L(\boldsymbol{\phi})$ also maximizes $\log L(\boldsymbol{\phi})$. Since $\log L$ is frequently a simpler function than L, it is in terms of maximizing $\log L$ that we shall usually formulate the problem.

The following heuristic argument may make the maximum likelihood method seem plausible: the probability of observing a sample lying in a region $\delta \mathbf{W}$ around the actually observed sample $\mathbf{W}$ is given by $p(\mathbf{W} \mid \boldsymbol{\phi})\, \delta \mathbf{W} = L(\boldsymbol{\phi})\, \delta \mathbf{W}$. The value $\boldsymbol{\phi} = \boldsymbol{\phi}^*$ for which this probability is greatest is the MLE. We say that $\boldsymbol{\phi}^*$ is the *most likely* value of $\boldsymbol{\phi}$. Of all possible values of the parameters, $\boldsymbol{\phi}^*$ is the one having the largest probability of giving rise to a sample within $\delta \mathbf{W}$ of the actually observed one.

4-6. Likelihood Equations

In this and subsequent sections we examine the application of MLE to various cases. We shall proceed as far as we can formally. In most applications, the final computation of the estimates requires numerical methods to be described in the next two chapters.

We first discuss the case where no constraints of any kind apply. This occurs when our model is of the reduced type, discussed in Section 2-12. Then the likelihood is a function of $\boldsymbol{\theta}$ and $\boldsymbol{\psi}$ alone, as shown by Eq. (2-12-3).

We know from Section 3-5 that a free (unconstrained) maximum of the function $\log L(\boldsymbol{\phi})$ must satisfy the set of *likelihood equations*

$$\partial \log L(\boldsymbol{\phi}) / \partial \boldsymbol{\phi} = \mathbf{0} \tag{4-6-1}$$

Since $L(\boldsymbol{\phi}) \equiv p(\mathbf{W} \mid \boldsymbol{\phi})$, we find from Eq. (3-2-18) that if $\boldsymbol{\rho}$ is a sufficient statistic for $\boldsymbol{\phi}$, then Eq. (1) is equivalent to

$$t(\boldsymbol{\rho}, \boldsymbol{\phi}) = 0 \tag{4-6-2}$$

Hence, to calculate the maximum likelihood estimate it is sufficient to know the value of the *sufficient* statistic $\boldsymbol{\rho}$, and we may discard the original data $\mathbf{W}$. Unfortunately, there are few practical cases involving nonlinear models for which a sufficient statistic can be found.

One may approach the problem of finding the estimates in two ways:

(a) By solving the likelihood equations, and then determining whether the solution is indeed a maximum. This is the approach taken when the solution is to be found analytically.

(b) By attempting to find the maximum of the likelihood function directly, paying no regard to the likelihood equations. This is the more fruitful approach when the solution is to be found numerically (see Chapter V). Even in this case, however, the likelihood equations can sometimes be used to eliminate some of the parameters, thus reducing the size of the problem to be solved numerically. This method, known as stagewise maximization, works out particularly well for the elimination of the distribution parameters $\boldsymbol{\psi}$, and specific illustrations are given in Sections 4-8 and 4-9.

4-7. Normal Distribution

We consider the case of a normal distribution. In the most general case, we denote by $T_{(\mu a)(\eta b)}$ the covariance of $\varepsilon_{\mu a}$ with $\varepsilon_{\eta b}$, and by $B_{(\mu a)(\eta b)}$ the elements of the matrix inverse to **T**. The errors $\varepsilon_{\mu a}$ ($\mu = 1, 2, \ldots, n$; $a = 1, 2, \ldots, m$) possess the normal distribution $N_{nm}(\mathbf{0}, \mathbf{T})$. The logarithm of the pdf may be derived from Eq. (2-8-10) as being

$$\log L = -(nm/2)\log 2\pi - \tfrac{1}{2}\log\det\mathbf{T} - \tfrac{1}{2}\sum_{a=1}^{m}\sum_{b=1}^{m}\sum_{\mu=1}^{n}\sum_{\eta=1}^{n} B_{(\mu a)(\eta b)} e_{\mu a}(\boldsymbol{\theta}) e_{\eta b}(\boldsymbol{\theta}) \tag{4-7-1}$$

If all the elements of **T** are known, finding the values of **θ** that maximize Eq (1) is equivalent to minimizing

$$\Phi(\boldsymbol{\theta}) \equiv \sum_{a=1}^{m}\sum_{b=1}^{m}\sum_{\mu=1}^{n}\sum_{\eta=1}^{n} B_{(\mu a)(\eta b)} e_{\mu a}(\boldsymbol{\theta}) e_{\eta b}(\boldsymbol{\theta}) \tag{4-7-2}$$

which is the same as Eq. (4-3-2). Thus, for a normal distribution with known covariance, MLE reduces to weighted least squares, with the weights given by the elements of the inverse of the covariance matrix.

We now turn to the case when the covariance matrix is not known, and must be estimated from the data. Variances and covariances are measures of the magnitudes of the errors. The data themselves can tell us nothing about the magnitude of an error unless we have replications of the same error. If we measure the length of an object once, we can gain no idea of the error in the measurement; if we measure it twice, the difference between the measurements can be used to estimate the error.

In the general case described by Eq. (1) we have no replication: each error $\varepsilon_{\mu a}$ is assumed to have its own variance $T_{(\mu a)(\mu a)}$. In order that estimation of the variances should be feasible, we must assume that several measurements (or quantities derived from them) possess identical variances. The following are typical assumptions:

(a) Errors in different experiments are independent.

(b) Errors in each experiment are distributed with the same covariance matrix **V**.

Both assumptions may be summarized by

$$T_{(\mu a)(\eta b)} = \delta_{\mu\eta} V_{ab}, \qquad V_{ab} = E(\varepsilon_{\mu a}\varepsilon_{\mu b}) \tag{4-7-3}$$

The trace of a matrix **A** is the sum of its diagonal elements, i.e., $\mathrm{Tr}(\mathbf{A}) \equiv \sum_i A_{ii}$. It follows that $\mathrm{Tr}(\mathbf{AB}) = \sum_{i,j} A_{ij} B_{ij}$. It is easily verified then that

$$\sum_{i=1}^{n} \mathbf{e}_\mu^{\mathrm{T}}(\boldsymbol{\theta}) \mathbf{V}^{-1} \mathbf{e}_\mu(\boldsymbol{\theta}) = \mathrm{Tr}[\mathbf{V}^{-1}\mathbf{M}(\boldsymbol{\theta})] \tag{4-7-4}$$

where $\mathbf{M}(\boldsymbol{\theta})$ is the *moment matrix* of the residuals, defined by $M_{ab} \equiv \sum_{\mu=1}^{n} e_{\mu a} e_{\mu b}$, i.e ,

$$\mathbf{M}(\boldsymbol{\theta}) \equiv \sum_{\mu=1}^{n} \mathbf{e}_{\mu}(\boldsymbol{\theta})\mathbf{e}_{\mu}^{\mathrm{T}}(\boldsymbol{\theta}) \tag{4-7-5}$$

Under our assumptions (a) and (b) above, the likelihood function takes the form

$$\log L = -(nm/2) \log 2\pi - (n/2) \log \det \mathbf{V} - \tfrac{1}{2} \operatorname{Tr}[\mathbf{V}^{-1}\mathbf{M}(\boldsymbol{\theta})] \tag{4-7-6}$$

Clearly, maximizing Eq. (6) when $\mathbf{V}$ is a known matrix, is equivalent to minimizing

$$\Phi(\boldsymbol{\theta}) \equiv \tfrac{1}{2} \operatorname{Tr}[\mathbf{V}^{-1}\mathbf{M}(\boldsymbol{\theta})] \tag{4-7-7}$$

Retaining the factor $\frac{1}{2}$ in the above expression (and similar constant coefficients in other objective functions to be derived later) is important in Bayesian estimation, when $\log p_0(\boldsymbol{\theta})$ is added to $-\Phi(\boldsymbol{\theta})$ (see Section 4-15).

A further specialization of Eq. (4) is obtained when one adds the following assumption:

(c) All errors are independent, i.e., $\mathbf{V}$ is a diagonal matrix

$$V_{ab} = \delta_{ab} v_a, \qquad v_a = E(\varepsilon_{\mu a}^2) = \sigma_a{}^2 \tag{4-7-8}$$

in which case

$$\log L = -(nm/2) \log 2\pi - (n/2) \sum_{a=1}^{m} \log v_a - \tfrac{1}{2} \sum_{a=1}^{m} v_a^{-1} M_{aa}(\boldsymbol{\theta}) \tag{4-7-9}$$

In the single equation case

$$\log L = -(n/2) \log 2\pi - n \log \sigma - (1/2\sigma^2) M(\boldsymbol{\theta}) \tag{4-7-10}$$

where

$$M(\boldsymbol{\theta}) = \sum_{\mu=1}^{n} e_{\mu}{}^2(\boldsymbol{\theta}) \tag{4-7-11}$$

Whether or not σ is known, $\log L$ is maximized relative to $\boldsymbol{\theta}$ by minimizing $M(\boldsymbol{\theta})$. Maximum likelihood here is equivalent to unweighted least squares.

4-8. Unknown Diagonal Covariance

We shall treat Eq. (4-7-9) first, and then generalize to Eq. (4-7-6). Assuming then that v_a are unknown, we seek those values of $\boldsymbol{\theta}$ and the v_a that maximize Eq. (4-7-9). We proceed by the method of *stagewise maximization* (Koopmans and Hood, 1953). This consists of finding, for any value of $\boldsymbol{\theta}$, the values of the v_a that maximize $\log L$. These will be some functions of $\boldsymbol{\theta}$, say $\hat{v}_a(\boldsymbol{\theta})$. Substitution of $\tilde{v}_a(\boldsymbol{\theta})$ for v_a in Eq. (4-7-9) reduces $\log L$ to a function $\tilde{\mathscr{L}}$ of

$\boldsymbol{\theta}$ alone, and we seek $\boldsymbol{\theta}^*$ so as to maximize $\tilde{\mathscr{L}}(\boldsymbol{\theta})$. The first step, then, is to differentiate Eq. (4-7-9) with respect to each v_a, and equate the derivatives to zero

$$\partial \log L/\partial v_a = -n/2v_a + (1/2v_a{}^2) \sum_{\mu=1}^{n} e_{\mu a}^2(\boldsymbol{\theta}) = 0 \qquad (a = 1, 2, \ldots, m) \quad \text{(4-8-1)}$$

This equation has the unique finite solution

$$\tilde{v}_a(\theta) = (1/n) \sum_{\mu=1}^{n} e_{\mu a}^2(\boldsymbol{\theta}) \quad \text{(4-8-2)}$$

Substituting Eq. (2) in Eq. (4-7-9) one obtains

$$\begin{aligned}\tilde{\mathscr{L}}(\boldsymbol{\theta}) &= \log L(\boldsymbol{\theta}, \tilde{v}_a(\boldsymbol{\theta})) \\ &= -(nm/2) \log 2\pi - (n/2) \sum_{a=1}^{m} \log\left[(1/n) \sum_{\mu=1}^{n} e_{\mu a}^2(\boldsymbol{\theta})\right] \\ &\quad - \tfrac{1}{2} \sum_{a=1}^{m} \left[\sum_{\mu=1}^{n} e_{\mu a}^2(\boldsymbol{\theta}) \Big/ (1/n) \sum_{\mu=1}^{n} e_{\mu a}^2(\boldsymbol{\theta})\right]\end{aligned}$$

which can be reduced to

$$\tilde{\mathscr{L}}(\boldsymbol{\theta}) = (mn/2)(\log (n/2\pi) - 1) - (n/2) \sum_{a=1}^{m} \log \sum_{\mu=1}^{n} e_{\mu a}^2(\boldsymbol{\theta}) \quad \text{(4-8-3)}$$

Maximizing Eq. (3) is clearly equivalent to minimizing

$$\Phi(\boldsymbol{\theta}) \equiv (n/2) \sum_{a=1}^{m} \log M_{aa}(\boldsymbol{\theta}) \quad \text{(4-8-4)}$$

where $\mathbf{M}(\boldsymbol{\theta})$ is the moment matrix of the residuals defined by Eq. (4-7-5). We refer to $\tilde{\mathscr{L}}(\boldsymbol{\theta})$ as the *concentrated likelihood* function.

To solve our estimation problem, we proceed as follows:

1. Find $\boldsymbol{\theta}^*$ which maximizes $\tilde{\mathscr{L}}(\boldsymbol{\theta})$, or minimizes $\Phi(\boldsymbol{\theta})$.
2. Estimate $v_a{}^* = \tilde{v}_a(\boldsymbol{\theta}^*)$, using Eq. (4-8-2). This estimate for v_a is biased but consistent. The bias may be eliminated approximately (exactly for certain linear models) by replacing $v_a{}^*$ with $nv_a{}^*/(n - ml)$ (see Section 7-13 for further details).

4-9. Unknown General Covariance

The results for the case of a nondiagonal unknown covariance matrix (Eq. 4-7-6) are similar to the ones obtained in the preceding section, but require some additional matrix calculus. Let $f(\mathbf{V})$ be some scalar function of a nonsingular matrix $\mathbf{V}$, and let $\partial f/\partial\mathbf{V}$ denote the matrix of partial derivatives of f with respect to the elements of $\mathbf{V}$, i.e.,

$$(\partial f/\partial\mathbf{V})_{ab} = \partial f/\partial V_{ab} \tag{4-9-1}$$

Then the following formulas hold (see Appendix A-2):

$$\partial \log \det \mathbf{V}/\partial\mathbf{V} = (\mathbf{V}^{\mathrm{T}})^{-1} \tag{4-9-2}$$

$$\partial[\mathrm{Tr}(\mathbf{V}^{-1}\mathbf{M})]/\partial\mathbf{V} = -(\mathbf{V}^{\mathrm{T}})^{-1}\mathbf{M}(\mathbf{V}^{\mathrm{T}})^{-1} \tag{4-9-3}$$

Applying these formulas to Eq. (4-7-6) and remembering that $\mathbf{V}$ is symmetric, i.e., $\mathbf{V}^{\mathrm{T}} = \mathbf{V}$, we obtain

$$\partial \log L/\partial\mathbf{V} = -(n/2)\mathbf{V}^{-1} + \tfrac{1}{2}\mathbf{V}^{-1}\mathbf{M}(\boldsymbol{\theta})\mathbf{V}^{-1} = \mathbf{0} \tag{4-9-4}$$

We may rewrite Eq. (4) as

$$\mathbf{V}^{-1} = (1/n)\mathbf{V}^{-1}\mathbf{M}(\boldsymbol{\theta})\mathbf{V}^{-1} \tag{4-9-5}$$

Premultiplying and postmultiplying Eq. (5) by $\mathbf{V}$, we obtain

$$\tilde{\mathbf{V}}(\boldsymbol{\theta}) = (1/n)\mathbf{M}(\boldsymbol{\theta}) \tag{4-9-6}$$

We now have

$$\mathrm{Tr}[\tilde{\mathbf{V}}^{-1}(\boldsymbol{\theta})\mathbf{M}(\theta)] = n \,\mathrm{Tr}\, \mathbf{I}_m = nm \tag{4-9-7}$$

whence, by substituting Eq. (6) in Eq. (4-7-6) we are led after simplification, to

$$\tilde{\mathscr{L}}(\boldsymbol{\theta}) = (nm/2)(\log(n/2\pi) - 1) - (n/2) \log \det \mathbf{M}(\boldsymbol{\theta}) \tag{4-9-8}$$

Maximizing this is equivalent to minimizing

$$\Phi(\boldsymbol{\theta}) \equiv (n/2) \log \det \mathbf{M}(\boldsymbol{\theta}) \tag{4-9-9}$$

The two steps are:

1. Find $\boldsymbol{\theta}^*$ to maximize $\tilde{\mathscr{L}}(\boldsymbol{\theta})$ or minimize $\Phi(\boldsymbol{\theta})$.
2. Estimate $\mathbf{V}^* = \tilde{\mathbf{V}}(\boldsymbol{\theta}^*)$ from Eq. (6). Here, too, the estimate is biased. See Section 7-13 for possible bias removal.

If the off-diagonal elements of $\mathbf{M}(\boldsymbol{\theta})$ are neglected, we have $\det \mathbf{M} = M_{11}M_{22}\cdots M_{mm}$, and $\log \det \mathbf{M} = \sum_{a=1}^{m} \log M_{aa}$. In that case, Eq. (9) reduces to Eq. (4-8-4).

The cases dealt with in this and the preceding sections may be regarded as the solving of weighted least squares problems with unknown weights. Formulas (4-8-4) and (9) give maximum likelihood estimates in the case of a normal distribution. One is tempted, however, to recommend their use even where the form of the distribution is unknown, provided assumptions (a) and (b) in Section 4-7 are valid (see Section 4-18). The use of Eq. (9) is illustrated by means of practical problems in Sections 5-23 and 9-7.

4-10. Independent Variables Subject to Error

Suppose our model is in reduced form, but the independent variables are also subject to error. It will be recalled that in this case the model equations take the form

$$\hat{\mathbf{y}}_\mu = \mathbf{f}(\hat{\mathbf{x}}_\mu, \boldsymbol{\theta}) \tag{4-10-1}$$

We now have residuals of two kinds:

$$\begin{aligned} \mathbf{e}_{x\mu}(\hat{\mathbf{x}}_\mu) &\equiv \mathbf{x}_\mu - \hat{\mathbf{x}}_\mu \\ \mathbf{e}_{y\mu}(\boldsymbol{\theta}, \hat{\mathbf{x}}_\mu) &\equiv \mathbf{y}_\mu - \mathbf{f}(\hat{\mathbf{x}}_\mu, \boldsymbol{\theta}) \end{aligned} \tag{4-10-2}$$

We adjoin the s-dimensional $\mathbf{e}_{x\mu}$ and m-dimensional $\mathbf{e}_{y\mu}$ into a single $(s+m)$-dimensional vector $\mathbf{e}_\mu$ of residuals

$$\mathbf{e}_\mu(\boldsymbol{\theta}, \hat{\mathbf{x}}_\mu) \equiv \begin{pmatrix} \mathbf{e}_{x\mu} \\ \mathbf{e}_{y\mu} \end{pmatrix} \tag{4-10-3}$$

If the $\mathbf{e}_\mu$ are assumed normally distributed with zero means and covariance matrices $\mathbf{V}$, the likelihood function is given by

$$\begin{aligned} \log L(\boldsymbol{\theta}, \hat{\mathbf{X}}_\mu, \mathbf{V}) = &-[(s+m)n/2] \log 2\pi - (n/2) \log \det \mathbf{V} \\ &- \tfrac{1}{2} \sum_{\mu=1}^{n} \mathbf{e}_\mu^{\mathrm{T}}(\boldsymbol{\theta}, \hat{\mathbf{x}}_\mu) \mathbf{V}^{-1} \mathbf{e}_\mu(\boldsymbol{\theta}, \hat{\mathbf{x}}_\mu) \end{aligned} \tag{4-10-4}$$

If $\mathbf{V}$ is known in its entirety, the function to be minimized is

$$\Phi(\boldsymbol{\theta}, \hat{\mathbf{X}}) \equiv \tfrac{1}{2} \sum_{\mu=1}^{n} \mathbf{e}_\mu^{\mathrm{T}}(\boldsymbol{\theta}, \hat{\mathbf{x}}_\mu) \mathbf{V}^{-1} \mathbf{e}_\mu(\boldsymbol{\theta}, \hat{\mathbf{x}}_\mu) \tag{4-10-5}$$

Unfortunately, when $\mathbf{V}$ is entirely unknown, it cannot be estimated by the method of maximum likelihood. To see why this is so, we partition the matrix $\mathbf{V}$ as follows:

$$\mathbf{V} = \begin{bmatrix} \mathbf{V}_{xx} & \mathbf{V}_{xy} \\ \mathbf{V}_{xy}^{\mathrm{T}} & \mathbf{V}_{yy} \end{bmatrix} \tag{4-10-6}$$

where

$$\mathbf{V}_{xx} \equiv E(\mathbf{e}_{x\mu} \mathbf{e}_{x\mu}^{\mathrm{T}}), \qquad \mathbf{V}_{xy} \equiv E(\mathbf{e}_{x\mu} \mathbf{e}_{y\mu}^{\mathrm{T}}), \qquad \mathbf{V}_{yy} \equiv E(\mathbf{e}_{y\mu} \mathbf{e}_{y\mu}^{\mathrm{T}}) \tag{4-10-7}$$

Let us set $\mathbf{V}_{xy} = \mathbf{0}$, and $\mathbf{V}_{xx} = \varepsilon \mathbf{I}$, where ε is a very small positive number, and let us set $\hat{\mathbf{x}}_\mu = \mathbf{x}_\mu$ $(\mu = 1, 2, \ldots, n)$, i.e., $\mathbf{e}_{x\mu} = \mathbf{0}$. Then Eq. (4) is reduced to

$$\begin{aligned} \log L = &-[(s+m)n/2] \log 2\pi - (n/2) \log \det \mathbf{V}_{yy} - (ns/2) \log \varepsilon \\ &- \tfrac{1}{2} \sum_{\mu=1}^{n} \mathbf{e}_{y\mu}^{\mathrm{T}} \mathbf{V}_{yy}^{-1} \mathbf{e}_{y\mu} \end{aligned} \tag{4-10-8}$$

Because of the term $-(ns/2)\log\varepsilon$, the quantity $\log L$ may be made arbitrarily large by choosing ε small enough. Thus, the likelihood function does not possess a maximum.

The above difficulty disappears when $\mathbf{V}_{xx}$ is known, say

$$\mathbf{V}_{xx} = \mathbf{P} \tag{4-10-9}$$

where $\mathbf{P}$ is a known positive definite matrix. If, in addition, we assume that the $\mathbf{x}$ and $\mathbf{y}$ errors are mutually uncorrelated, then the nonconstant part of Eq. (4) reduces to

$$\log L(\boldsymbol{\theta}, \hat{\mathbf{X}}, \mathbf{V}_{yy}) = -(n/2)\log\det\mathbf{V}_{yy} - \tfrac{1}{2}\sum_{\mu=1}^{n}(\mathbf{e}_{x\mu}^{\mathrm{T}}\mathbf{P}^{-1}\mathbf{e}_{x\mu} + \mathbf{e}_{y\mu}^{\mathrm{T}}\mathbf{V}_{yy}^{-1}\mathbf{e}_{y\mu}) \tag{4-10-10}$$

One verifies easily that the MLE for $\mathbf{V}_{yy}$ is

$$\tilde{\mathbf{V}}_{yy} = (1/n)\mathbf{M}_{yy} = (1/n)\sum_{\mu}\mathbf{e}_{y\mu}\mathbf{e}_{y\mu}^{\mathrm{T}} \tag{4-10-11}$$

so that the concentrated objective function to be minimized is

$$\Phi(\boldsymbol{\theta}, \hat{\mathbf{X}}) = (n/2)\log\det\mathbf{M}_{yy} + \tfrac{1}{2}\sum_{\mu=1}^{n}\mathbf{e}_{x\mu}^{\mathrm{T}}\mathbf{P}^{-1}\mathbf{e}_{x\mu} \tag{4-10-12}$$

The bias of the estimate Eq. (11) for $\mathbf{V}_{yy}$ can be considerable, and this estimate is not even consistent. A suitable correction factor is derived in Section 7-14.

Computationally it is often best to treat the problems discussed in this section as constrained minimization problems. That is, $\mathbf{f}(\hat{\mathbf{x}}_\mu, \boldsymbol{\theta})$ is not substituted for $\hat{\mathbf{y}}_\mu$ in the expression for Φ. The $\hat{\mathbf{y}}_\mu$ are retained as explicit unknowns, and Eq. (1) is treated as a set of equality constraints. In this form, the problem is amenable to solution by the method of Sections 4-11 and 6-6–6-8.

4-11. Exact Structural Models

Recall that the model takes the form

$$\mathbf{g}(\mathbf{u}_\mu, \hat{\mathbf{w}}_\mu, \boldsymbol{\theta}) = \mathbf{0} \tag{4-11-1}$$

The $\mathbf{u}_\mu$ have been measured precisely, the $\mathbf{w}_\mu$ are subject to measurement errors. The residuals are defined by Eq. (4-1-4). The likelihood function is given by Eq. (2-13-2). If the errors in each experiment are independently distributed as $N_r(\mathbf{0}, \mathbf{V})$ where r is the dimension of $\mathbf{w}_\mu$, then the likelihood takes the form

$$\log L(\hat{\mathbf{W}}, \mathbf{V}) = -(n/2)\log\det\mathbf{V} - \tfrac{1}{2}\sum_{\mu=1}^{n}\mathbf{e}_\mu^{\mathrm{T}}\mathbf{V}^{-1}\mathbf{e}_\mu \tag{4-11-2}$$

(constant terms have been dropped). The maximum likelihood estimate is found by determining the values of $\hat{\mathbf{W}}$ and $\boldsymbol{\theta}$ which maximize $\log L$ while satisfying the constraints [Eq. (1)]. We introduce an m-dimensional vector of Lagrange multipliers $\boldsymbol{\lambda}_\mu$ for each experiment to form the Lagrangian function

$$\Lambda(\mathbf{W}, \boldsymbol{\theta}, \mathbf{V}, \boldsymbol{\lambda}_1, \ldots, \boldsymbol{\lambda}_n) \equiv \log L + \sum_{\mu=1}^{n} \boldsymbol{\lambda}_\mu^{\mathrm{T}} \mathbf{g}(\mathbf{u}_\mu, \hat{\mathbf{w}}_\mu, \boldsymbol{\theta}) \qquad \text{(4-11-3)}$$

The solution to the estimation problem will be found at a stationary point of Λ. Numerical methods for finding the solution are described in Sections 6-6–6-8, and an example is worked out in detail in Section 6-13.

4-12. Data Requirements

In Sections 4-8 and 4-9, we saw that when unknown, the elements of the covariance matrix (or, equivalently, the weights for weighted least squares) could be estimated along with the model parameters. In the case of independent observations we found that we must minimize

$$\Phi_1(\boldsymbol{\theta}) = \sum_{a=1}^{m} \sum_{\mu=1}^{n} v_a^{-1} e_{\mu a}^2(\boldsymbol{\theta}) \qquad \text{(4-12-1)}$$

when the v_a are known, and

$$\Phi_2(\boldsymbol{\theta}) = \sum_{a=1}^{m} \log \sum_{\mu=1}^{n} e_{\mu a}^2(\boldsymbol{\theta}) \qquad \text{(4-12-2)}$$

when the v_a are not known. Clearly, $\Phi_1(\boldsymbol{\theta}) \geqslant 0$ for all $\boldsymbol{\theta}$, and the equality holds if and only if

$$y_{\mu a} = f_a(\mathbf{x}_\mu, \boldsymbol{\theta}) \qquad \text{or} \qquad g_{\mu a}(\mathbf{z}_\mu, \boldsymbol{\theta}) = 0 \qquad \text{(4-12-3)}$$

for all μ and a. Thus, we have a total of mn equations to be satisfied, and meaningful estimation can occur as soon as mn at least equals l, the number of parameters to be estimated. On the other hand, suppose we can find values of $\boldsymbol{\theta}$ which satisfy Eq. (3) exactly just for one specific value of a. This could occur if $l_a \geqslant n$, where l_a is the number of parameters appearing in the ath equation. But in this case the ath term in Eq. (2) is $-\infty$. For meaningful estimation, then, we must have $n > \max_a (l_a)$. In particular, if all l parameters appear in every equation, we must have $n > l$. The situation where $\mathbf{V}$ is not assumed to be diagonal is similar, but we have an additional restriction when Eq. (4-9-9) is used. The $m \times m$ matrix $\mathbf{M}(\boldsymbol{\theta})$ is the sum of n matrices $\mathbf{e}_\mu \mathbf{e}_\mu^{\mathrm{T}}$, each of rank one. Hence, the rank of $\mathbf{M}$ cannot exceed n, and for $\mathbf{M}$ to be nonsingular it is necessary that $n \geqslant m$. If $\mathbf{M}$ is singular, its determinant vanishes and Eq. (4-9-9) is meaningless.

To summarize, the number n of required experiments must satisfy:

1. $n > \max_a (l_a)$ if $\mathbf{V}$ is unknown. Also, $n \geqslant m$ if $\mathbf{V}$ is not known to be diagonal.
2. $n \geqslant l/m$ if $\mathbf{V}$ is known.‡

More observations are usually required when $\mathbf{V}$ is unknown than when it is known. This is not surprising.

4-13. Some Other Distributions

Perhaps the greatest virtue of the maximum likelihood method is its straightforward applicability to the formulation of a wide variety of estimation problems. Given a distribution for the errors, it is an easy matter to write down the expression for the likelihood function. When this function is continuous and smooth, its maximum can be found by means of some of the gradient methods of Chapter V as are applicable to the normal distribution problems. The situation is entirely different, however, in the case of a discontinuous distribution, such as the following: Suppose our measurement errors all follow uniform distributions. Let the range of $\varepsilon_{\mu a}$ be $\pm r_{\mu a}$. Any value $\boldsymbol{\theta}$ for which $|e_{\mu a}(\boldsymbol{\theta})| > r_{\mu a}$ for even one μ, a has likelihood zero. All values of $\boldsymbol{\theta}$ for which $|e_{\mu a}| \leqslant r_{\mu a}$ for all μ, a possess the same positive likelihood, and are all equally acceptable as maximum likelihood estimates. It may easily happen that no such values exist. The best procedure is to find the value of $\boldsymbol{\theta}$ for which

$$\Phi(\boldsymbol{\theta}) \equiv \max_{\mu,\, a} |e_{\mu a}(\boldsymbol{\theta})/r_{\mu a}| \tag{4-13-1}$$

attains its minimum value. If this minimum value turns out to be no greater than unity, we have found a maximum likelihood estimate. Otherwise, we know that no such estimate exists. What we have found, then, is a minimax weighted deviation estimate, as described in Section 4-17. When the range of $\varepsilon_{\mu a}$ is not known, but all errors are assumed to have the same range, then minimizing $\Phi(\boldsymbol{\theta}) \equiv \max |e_{\mu a}(\boldsymbol{\theta})|$ gives a maximum likelihood estimate, and the minimum value of Φ is an estimated value of all $r_{\mu a}$.

If the errors have the two-sided exponential distribution

$$p(\varepsilon_{\mu a}) = c_{\mu a} \exp(-k_{\mu a} |\varepsilon_{\mu a}|) \tag{4-13-2}$$

then, provided the $k_{\mu a}$ are known, the maximum likelihood estimate calls for minimizing the weighted sum of absolute values of the residuals

$$\Phi(\boldsymbol{\theta}) \equiv \sum_{\mu,\, a} k_{\mu a} |e_{\mu a}(\boldsymbol{\theta})| \tag{4-13-3}$$

‡ In some exceptional circumstances, a smaller number of experiments may suffice.

One verifies easily that the constants $c_{\mu a}$ and $k_{\mu a}$ are related as follows to the standard deviation $\sigma_{\mu a}$:

$$c_{\mu a} = 1/\sqrt{2}\,\sigma_{\mu a}, \qquad k_{\mu a} = \sqrt{2}/\sigma_{\mu a} \tag{4-13-4}$$

If we assume $\sigma_{\mu a} = \sigma_a$ for all μ, and if all errors are independent, the log likelihood is

$$\log L = -(nm/2)\log 2 - n\sum_{a=1}^{m} \log \sigma_a - \sqrt{2}\sum_{a=1}^{m}(1/\sigma_a)\sum_{\mu=1}^{n} |e_{\mu a}(\boldsymbol{\theta})| \tag{4-13-5}$$

Differentiating with respect to σ_a, equating to zero, and solving for σ_a gives the maximum likelihood estimate

$$\tilde{\sigma}_a = (\sqrt{2}/n)\sum_{\mu=1}^{n} |e_{\mu a}(\boldsymbol{\theta})| \tag{4-13-6}$$

Substituting back in Eq. (5) we eventually find that to estimate $\boldsymbol{\theta}$ when the σ_a are unknown, we must minimize

$$\Phi(\boldsymbol{\theta}) \equiv n\sum_{a=1}^{m} \log \sum_{\mu=1}^{n} |e_{\mu a}(\boldsymbol{\theta})| \tag{4-13-7}$$

The objective functions of Eqs. (3) and (7) may be brought into the realm of conventional mathematical programming problems by means of the following device: We define new variables $e_{\mu a}^{+}$ and $e_{\mu a}^{-}$ satisfying

$$e_{\mu a}(\boldsymbol{\theta}) = e_{\mu a}^{+} - e_{\mu a}^{-}, \qquad e_{\mu a}^{+} \geqslant 0, \qquad e_{\mu a}^{-} \geqslant 0 \tag{4-13-8}$$

Equation (3) is replaced with

$$\Phi(\boldsymbol{\theta}, e_{\mu a}^{+}, e_{\mu a}^{-}) = \sum_{\mu, a} k_{\mu a}(e_{\mu a}^{+} + e_{\mu a}^{-}) \tag{4-13-9}$$

Clearly, if $e_{\mu a}(\boldsymbol{\theta})$ is positive then $e_{\mu a}(\boldsymbol{\theta}) = e_{\mu a}^{+}$ and $e_{\mu a}^{-} = 0$, and vice versa. The theory of mathematical programming leads us to expect that the number of nonzero variables in the solution will equal mn, i.e., the number of equality constraints in Eq. (8). Among these will be the l parameters θ, leaving only $mn - l$ nonzero residuals. This means that the fitted equations will pass exactly through at least l of the observed data points. Hence this estimation method is relatively insensitive to the presence of a few observations with very large errors; these are simply ignored.

The mathematical programming formulation of problems given by Eqs. (1) and (3) are discussed by Kelley (1958) and Wagner (1959), who deal specifically with the linear programming problems which arise when the model equations are linear in the parameters.

C. Bayesian Estimation

4-14. Definition

In the estimation methods discussed so far we have made no use of the prior information, which in Sections 2-16–2-19 we have treated as an integral part of the problem. As we have seen, the posterior pdf $p^*(\boldsymbol{\phi})$ is given (Eq. 2-19-1) by Bayes' theorem as

$$p^*(\boldsymbol{\phi}) = cL(\boldsymbol{\phi})p_0(\boldsymbol{\phi}) \tag{4-14-1}$$

where $L(\boldsymbol{\phi})$ is the likelihood function, and $p_0(\boldsymbol{\phi})$ the prior pdf, which summarizes the prior information. Estimates which make use of the prior information are usually based on the posterior distribution, and are therefore known as *Bayesian* estimates.

If $p^*(\boldsymbol{\phi})$ is to be a pdf, we must have $\int p^*(\boldsymbol{\phi})\, d\boldsymbol{\phi} = 1$, and hence c must be $1/I$, where

$$I \equiv \int L(\boldsymbol{\phi})p_0(\boldsymbol{\phi})\, d\boldsymbol{\phi} \tag{4-14-2}$$

We refer to the function $p^*(\boldsymbol{\phi})$ as a *proper* or *improper* posterior distribution if the integral I does or does not exist, respectively. In the latter case, we let $c = 1$. The following are sufficient but by no means necessary conditions for the existence of I:

1. $L(\boldsymbol{\phi})$ is bounded and $p_0(\boldsymbol{\phi})$ is normal.
2. $L(\boldsymbol{\phi})$ and $p_0(\boldsymbol{\phi})$ are bounded, and $p_0(\boldsymbol{\phi})$ vanishes everywhere except in a bounded region of $\boldsymbol{\phi}$ space.

To select an estimate for the parameters $\boldsymbol{\phi}$, we pick some typical values of the posterior distribution, such as the mean, median, or mode. Such values are referred to as *location parameters* of the distribution since they locate the region in $\boldsymbol{\phi}$ space where most realizations of the random variable occur. Some of these location parameters exist even if the posterior distribution is improper, while others may not exist even for proper distributions. Among those parameters which exist for a given problem, the choice is somewhat arbitrary. In the sequel we shall describe two distinct approaches toward making this choice.

At this point it is well to summarize some of the benefits that accrue from Bayesian estimation:

1. One is sure to obtain estimates which are physically meaningful. It is guaranteed that estimates for parameters known to be positive are indeed positive.

2. The model equations may be degenerate relative to some of the parameters. For instance, the model for the falling sphere [Eq. (2-14-5)] contains only two independent combinations of five distinct parameters. Non-Bayesian methods can be used to estimate at most two of these parameters, and exact knowledge of the others is required. But if inexact prior information is available on at least three of the parameters, then the posterior density, being a nondegenerate function of all five, can be used to estimate all five.

4-15. Mode of the Posterior Distribution

The natural extension of the maximum likelihood method to Bayesian estimation problems consists of looking for the *mode* of the posterior distribution. That is, we accept as our estimate the value of $\boldsymbol{\phi}$ for which $p^*(\boldsymbol{\phi})$ is maximum. This method, to which we refer as MPD (maximum of posterior distribution), offers the following advantages:

1. The estimate coincides with the maximum likelihood estimate in case of a uniform prior distribution, since then $p^*(\boldsymbol{\phi})$ is proportional to $L(\boldsymbol{\phi})$. The estimates coincide even if $p_0(\boldsymbol{\phi})$ is uniform only within a bounded region and zero elsewhere, provided only the maximum of $L(\boldsymbol{\phi})$ occurs within this region. The practitioner who accepts the MLE when no prior information is given would naturally wish his estimates to be affected only slightly when a slight amount of prior information becomes available. MPD satisfies this requirement.

2. We know from a theorem by von Mises (1919), that if $p_0(\boldsymbol{\phi})$ is continuous and does not vanish at the maximum of $L(\boldsymbol{\phi})$, then the MPD converges to the MLE as the number of experiments is increased indefinitely. The MPD shares the consistency and asymptotic efficiency of the MLE.

3. The MPD can be obtained whether or not $p^*(\boldsymbol{\phi})$ is a proper distribution.

4. It is usually much easier to compute the MPD than other Bayesian estimates.

In computing MPD estimates, we distinguish two cases:

(a) The prior distribution does not vanish anywhere. In this case, we maximize

$$\Phi(\boldsymbol{\phi}) = \log L(\boldsymbol{\phi}) + \log p_0(\boldsymbol{\phi}) \tag{4-15-1}$$

The same techniques as are used for MLE can be applied here. In particular, if $L(\boldsymbol{\phi})$ is one of the normal cases discussed in Section 4-8 and Section 4-9, and if p_0 does not depend on the elements of $\mathbf{V}$, then those may be eliminated as before, and the concentrated likelihood may replace the likelihood in

Eq. (1). Care must be taken, however, to retain any constants multiplying the concentrated likelihood. For instance, if L is given by Eq. (4-7-6), we may use Eq. (4-9-8) to replace Eq. (1) by

$$\Phi(\boldsymbol{\theta}) = (n/2) \log \det \mathbf{M}(\boldsymbol{\theta}) - \log p_0(\boldsymbol{\theta}) \tag{4-15-2}$$

which is to be minimized. For numerical examples, see Sections 5-22 and 8-7. Note that in the presence of $\log p_0(\boldsymbol{\theta})$, the factor $n/2$ may not be dropped. In the case of single equation least squares with unknown σ, the term $(n/2) \log \det \mathbf{M}$ takes the form $(n/2) \log \sum_{\mu=1}^{n} \mathbf{e}_\mu^{\,2}(\boldsymbol{\theta})$.

(b) The prior distribution vanishes outside the region defined by a set of constraints

$$\mathbf{h}(\boldsymbol{\phi}) \geqslant \mathbf{0} \tag{4-15-3}$$

In this case we have a typical nonlinear programming problem; find the maximum of Eq. (1) subject to all the applicable constraints. Methods of dealing with this problem are described in Chapter VI.

4-16. Minimum Risk Estimates

So far our motive has been to find values of $\boldsymbol{\theta}$ which are most likely to be close to the true values. Sometimes, however, the estimated value is required for a specific purpose, e.g., for designing a plant, and we are interested in finding the value of $\boldsymbol{\theta}$ which is best for this particular purpose. In many situations, what is "best" is determined by economic considerations, and the choice of the best estimate can be made by means of decision theory.

In decision theory, a cost is assigned to any loss suffered because of an error in the estimate. That is, to the act of using the parameter value $\boldsymbol{\phi}^*$ when the true value is $\hat{\boldsymbol{\phi}}$ we assign the cost $c(\boldsymbol{\phi}^*, \hat{\boldsymbol{\phi}})$. Since $\hat{\boldsymbol{\phi}}$ is unknown, the actual cost $c(\boldsymbol{\phi}^*, \hat{\boldsymbol{\phi}})$ cannot be computed. However, if we are willing to say that $\boldsymbol{\phi}$ is distributed according to the posterior distribution, then we can compute the *risk*, defined as the expected value of the cost of assigning the value $\boldsymbol{\phi}^*$ to $\boldsymbol{\phi}$

$$R(\boldsymbol{\phi}^*) \equiv Ec(\boldsymbol{\phi}^*, \boldsymbol{\phi}) = \int c(\boldsymbol{\phi}^*, \boldsymbol{\phi}) p^*(\boldsymbol{\phi})\, d\boldsymbol{\phi} \tag{4-16-1}$$

The *minimum risk estimate* (MRE) is defined as the value of $\boldsymbol{\phi}^*$ which minimizes $R(\boldsymbol{\phi}^*)$. Here $p^*(\boldsymbol{\phi})$ must be a proper pdf. The following is a simple example:

A manufacturer conducts experiments to measure the tensile strength θ of an alloy. He intends to use the alloy to manufacture a component whose size, and hence cost, will be inversely proportional to θ. Let θ^* be the estimate to be used for θ. Then the cost of the component will be \$$a/\theta^*$ (any additional fixed cost is irrelevant to the present discussion). The component will fail if the true value $\hat{\theta}$ is less than θ^*. However, if the component does fail, the manufacturer will have to pay a fine of \$$K$. His total cost will be

$$c(\theta^*, \hat{\theta}) = \begin{cases} a/\theta^* & (\theta^* \leqslant \hat{\theta}) \\ K + a/\theta^* & (\theta^* > \hat{\theta}) \end{cases} \tag{4-16-2}$$

Assuming that the posterior density $p^*(\theta)$ summarizes all available information on θ, then the risk or expected cost is

$$R(\theta^*) = \int_{-\infty}^{\infty} c(\theta^*, \theta)p^*(\theta)\, d\theta = a/\theta^* + K \int_{-\infty}^{\theta^*} p^*(\theta)\, d\theta \tag{4-16-3}$$

To find the minimum risk estimate, we differentiate

$$dR/d\theta^* = -a/(\theta^*)^2 + Kp^*(\theta^*) = 0 \tag{4-16-4}$$

Hence, one should use the value θ^* which satisfies the equation

$$(\theta^*)^2 p^*(\theta^*) = a/K \tag{4-16-5}$$

In a sense, the MRE is not really an estimate. The value of θ^* which satisfies Eq. (5) cannot be considered the most likely to be true; it is merely the value which in the given economic situation involves the least risk.

Attempts to use decision-theory-like methods in pure (i.e., economics-free) estimation problems usually start with the assumption of quadratic cost functions taking the form

$$c(\boldsymbol{\phi}^*, \hat{\boldsymbol{\phi}}) = (\boldsymbol{\phi}^* - \hat{\boldsymbol{\phi}})^{\mathrm{T}}\mathbf{P}(\boldsymbol{\phi}^* - \hat{\boldsymbol{\phi}}) \tag{4-16-6}$$

where $\mathbf{P}$ is a given positive definite weighting matrix. This essentially defines the cost as a weighted sum of squares of the estimation errors. Substituting Eq. (6) in Eq. (1) yields

$$R(\boldsymbol{\phi}^*) = \int (\boldsymbol{\phi}^* - \boldsymbol{\phi})^{\mathrm{T}}\mathbf{P}(\boldsymbol{\phi}^* - \boldsymbol{\phi})p^*(\boldsymbol{\phi})\, d\boldsymbol{\phi} \tag{4-16-7}$$

and

$$\partial R/\partial \boldsymbol{\phi}^* = 2\mathbf{P} \int (\boldsymbol{\phi}^* - \boldsymbol{\phi})p^*(\boldsymbol{\phi})\, d\boldsymbol{\phi} \tag{4-16-8}$$

Where $R(\phi^*)$ attains its minimum, $\partial R/\partial \boldsymbol{\phi}^*$ vanishes, whence (assuming $\mathbf{P}$ nonsingular)

$$\int \boldsymbol{\phi}^* p^*(\boldsymbol{\phi})\, d\boldsymbol{\phi} = \int \boldsymbol{\phi} p^*(\boldsymbol{\phi})\, d\boldsymbol{\phi} \tag{4-16-9}$$

Since $\boldsymbol{\phi}^*$ is a constant and $\int p^*(\boldsymbol{\phi})\, d\boldsymbol{\phi} = 1$, Eq. (9) reduces to

$$\boldsymbol{\phi}^* = \int \boldsymbol{\phi} p^*(\boldsymbol{\phi})\, d\boldsymbol{\phi} \tag{4-16-10}$$

We conclude, then, that the MRE for a quadratic cost function is the *mean* of the posterior distribution. More explicitly, Eq. (10) can be written as

$$\boldsymbol{\phi}^* = \int \boldsymbol{\phi} L(\boldsymbol{\phi}) p_0(\boldsymbol{\phi})\, d\boldsymbol{\phi} \Big/ \int L(\boldsymbol{\phi}) p_0(\boldsymbol{\phi})\, d\boldsymbol{\phi} \qquad (4\text{-}16\text{-}11)$$

Fortunately, the estimate Eq. (11) does not depend on the weights **P**. Hence, one need not worry about what values should be assigned to them.

There are many practical disadvantages associated with this MRE:

(a) The estimate does not exist if $p^*(\boldsymbol{\phi})$ is an improper distribution. Consider the case where our model takes the form

$$\hat{y}_\mu = \alpha \exp(-\theta x_\mu) \qquad (4\text{-}16\text{-}12)$$

α being a known constant. Assuming a normal distribution with standard deviation σ, we have

$$L(\theta) = (2\pi)^{-n/2} \sigma^{-n} \exp\left\{ -(1/2\sigma^2) \sum_{\mu=1}^{n} [y_\mu - \alpha \exp(-\theta x_\mu)]^2 \right\} \qquad (4\text{-}16\text{-}13)$$

Suppose all x_μ are positive. As θ increases beyond bound, $L(\theta)$ becomes proportional to $\exp[-(1/2\sigma^2) \sum_{\mu=1}^{n} y_\mu^{\ 2}] \neq 0$. Hence, if $p_0(\theta)$ is uniform for all values of θ, the integrals in Eq. (11) diverge. What is even worse, however, is the fact that if we assume $p_0(\theta) = 0$ outside the region $0 \leqslant \theta \leqslant A$, the integrals in Eq. (11) exist, but their ratio tends to infinity with A. Thus, the estimate Eq. (11) is not robust under seemingly unimportant changes in $p_0(\theta)$. After all, the value chosen for A is arbitrary, and the estimate should not depend strongly on this choice. The source of the difficulty here is the sensitivity of the MRE to the tails of the assumed distribution.

(b) Even when the integrals in (11) exist, their evaluation may be impractical. If $\boldsymbol{\phi}$ is an l-dimensional vector, then $l + 1$ integrals must be evaluated: one in the denominator, and one for each component of $\boldsymbol{\phi}$ in the numerator. Each one of these integrations must be carried out over an l-dimensional space, each dimension possibly extending from $-\infty$ to $+\infty$. No satisfactory methods for performing such integrations (unless $l = 1$) are available. In addition, any reasonable approach to this integration problem requires finding, as a first step, the location of the mode of the posterior distribution. Thus, computation of the MPD is a prerequisite to the computation of the MRE.

(c) The MRE is not invariant under reparametrization, whereas the MPD is.

(d) The MRE does not generally converge to the MLE as $p_0(\boldsymbol{\phi})$ approaches the uniform distribution.

In conclusion, the MRE can be recommended only where called for by true economic decision making purposes. For a further discussion of minimum risk estimates, not confined to quadratic cost functions, the reader is referred to Chapters 2 and 3 of Deutsch (1965) and Chapter 6 of Raiffa and Schlaifer (1961).

D. Other Methods

4-17. Minimax Deviation

The parameters are determined in such a way as to minimize the maximum deviation of the model from the data. This is particularly useful for design purposes (see Section 2-5), or for obtaining maximum likelihood estimates with uniform error distributions (Section 4-13). Such estimates are sometimes called *Chebyshev estimates.*

Let ζ denote the magnitude of the largest residual. Then the following conditions are satisfied

$$e_{\mu a}(\boldsymbol{\theta}) \leqslant \zeta, \qquad -e_{\mu a}(\boldsymbol{\theta}) \leqslant \zeta \qquad (\mu = 1, 2, \ldots, n;\ a = 1, 2, \ldots, m) \tag{4-17-1}$$

or, equivalently

$$e_{\mu a}^2(\boldsymbol{\theta}) \leqslant \zeta^2 \qquad (\mu = 1, 2, \ldots, n;\ a = 1, 2, \ldots, m) \tag{4-17-2}$$

Our problem may then be formulated as follows:

Find the values of the parameters $\boldsymbol{\theta}$ and ζ which minimize the objective function

$$\Phi(\boldsymbol{\theta}, \zeta) \equiv \zeta \tag{4-17-3}$$

while satisfying Eq. (1) or Eq. (2). This is a classical nonlinear programming problem.

It is possible to attach different weights to different residuals, i.e., by replacing Eq. (2) with

$$\alpha_{\mu a}\, \mathbf{e}_{\mu a}^2 \leqslant \zeta^2 \qquad (\mu = 1, 2, \ldots, n;\ a = 1, 2, \ldots, m) \tag{4-17-4}$$

with $\alpha_{\mu a}$ given positive numbers.

Numerical procedures for solving this problem are discussed in Section 6-5. If the model equations are linear, the algorithm of Bartels and Golub (1968) may be used.

4-18. Pseudomaximum Likelihood

In this method we employ the maximum likelihood equations derived on the assumption of a normal distribution, regardless of whether the distribution is or is not in fact normal. Since in practice we often assume normality even when we have no basis for doing so, pseudomaximum likelihood is perhaps the most widely used method. We may regard any use of nonlinear least squares or weighted least squares as an application of this method. Equations (4-8-4) and (4-9-9) are the most important extensions of pseudomaximum likelihood beyond the weighted least squares concept.

4-19. Linearizing Transformations

Consider the model equations $\mathbf{y} = \mathbf{f}(\mathbf{x}, \boldsymbol{\theta})$. Suppose we were able to effect a transformation of variables $\tilde{\mathbf{y}} = \boldsymbol{\tau}(\mathbf{y})$ in such a way that the function $\boldsymbol{\tau}[\mathbf{f}(\mathbf{x}, \boldsymbol{\theta})]$ is linear in $\boldsymbol{\theta}$. Then we apply the method of multiple linear regression to estimate $\boldsymbol{\theta}$. The advantage gained derives from the fact that this estimate may be obtained by direct calculations, whereas nonlinear estimation procedures require complicated iterative schemes. Understandably, this method was very popular before nonlinear estimation codes for electronic computers became available.

We illustrate by means of a simple example:

Let

$$y = x_1 \exp(-\theta x_2) \tag{4-19-1}$$

be our model equation, with θ to be estimated. Letting $\tilde{y} \equiv \log y$ transforms the equation into

$$\tilde{y} = \log x_1 - \theta x_2 \tag{4-19-2}$$

This is linear in θ, which can be estimated, say, by minimizing

$$\Phi(\theta) \equiv \sum_{\mu=1}^{n} (\log y_\mu - \log x_{\mu 1} + \theta x_{\mu 2})^2.$$

The method can be applied equally well when the transformed equations are linear not in the l parameters $\boldsymbol{\theta}$, but in a set of l independent functions of them, say $\boldsymbol{\pi}(\boldsymbol{\theta})$. Then we can obtain linear regression estimates $\boldsymbol{\pi}^*$ of the $\boldsymbol{\pi}$, and estimates of $\boldsymbol{\theta}$ by solving the equations $\boldsymbol{\pi}(\boldsymbol{\theta}^*) = \boldsymbol{\pi}^*$. To illustrate by means of a trivial example, let

$$y = \theta_1 \exp(-\theta_2 x) \tag{4-19-3}$$

with θ_1 and θ_2 to be estimated. As before, let $\tilde{y} \equiv \log y$ so that

$$\tilde{y} = \log \theta_1 - \theta_2 x \tag{4-19-4}$$

Letting $\pi_1 \equiv \log \theta_1$, $\pi_2 \equiv \theta_2$, we have

$$\tilde{y} = \pi_1 - \pi_2 x \tag{4-19-5}$$

which is linear in π_1 and π_2. Estimates for $\boldsymbol{\theta}$ may be obtained from those of $\boldsymbol{\pi}$ by means of

$$\theta_1{}^* = \exp(\pi_1{}^*); \qquad \theta_2{}^* = \pi_2{}^*$$

Other examples, arising in chemical reaction kinetics, are

(a)

$$y = x_1 \exp[-\theta_1 x_3 \exp(-\theta_2/x_2)] \tag{4-19-6}$$

which, under

$$\tilde{y} \equiv \log[-\log(y/x_1)], \qquad \pi_1 \equiv \log \theta_1, \qquad \pi_2 \equiv \theta_2$$

transforms into

$$\tilde{y} = \pi_1 - (1/x_2)\pi_2 + \log x_3 \tag{4-19-7}$$

and

(b)

$$y = \theta_1 x_1/(1 + \theta_2 x_2 + \theta_3 x_3 + \theta_4 x_4)^2 \tag{4-19-8}$$

which under

$$\tilde{y} \equiv (x_1/y)^{1/2}, \quad \pi_1 \equiv 1/\theta_1{}^{1/2}, \quad \pi_2 \equiv \theta_2/\theta_1{}^{1/2}, \quad \pi_3 \equiv \theta_3/\theta_1{}^{1/2}, \quad \pi_4 \equiv \theta_4/\theta_1{}^{1/2}$$

becomes

$$\tilde{y} = \pi_1 + x_2 \pi_2 + x_3 \pi_3 + x_4 \pi_4 \tag{4-19-9}$$

The main objection that can be raised to this method (other than its limited applicability) is that the statistical distribution of the errors on the calculated values of $\tilde{\mathbf{y}}_\mu \equiv \boldsymbol{\tau}(\mathbf{y}_\mu)$ is not the same as that of the errors in $\mathbf{y}_\mu$. Therefore it may be appropriate to apply the least squares criterion to the residuals in $\mathbf{y}$ but not in $\tilde{\mathbf{y}}$. We can overcome this problem in part by recognizing that if $\mathbf{V}_\mu$ is the covariance matrix of the errors in $\mathbf{y}_\mu$, then (provided these errors are small and the transformation $\boldsymbol{\tau}$ and its derivatives are continuous and bounded) the covariance matrix of $\tilde{\mathbf{y}}_\mu$ is approximately

$$\tilde{\mathbf{V}}_\mu = (\partial\boldsymbol{\tau}/\partial\mathbf{y}_\mu)\mathbf{V}_\mu\,(\partial\boldsymbol{\tau}/\partial\mathbf{y}_\mu)^{\mathrm{T}} \tag{4-19-10}$$

Hence, in place of minimizing $\sum_{\mu=1}^{n}(\mathbf{y}_\mu - \mathbf{f}_\mu)^{\mathrm{T}}\mathbf{V}_\mu^{-1}(\mathbf{y}_\mu - \mathbf{f}_\mu)$ we should minimize $\sum_{\mu=1}^{n}[\tilde{\mathbf{y}}_\mu - \boldsymbol{\tau}(\mathbf{f}_\mu)]^{\mathrm{T}}\tilde{\mathbf{V}}_\mu^{-1}[\tilde{\mathbf{y}}_\mu - \boldsymbol{\tau}(\mathbf{f}_\mu)]$. In the single equation case, a variance σ^2 of y_μ is translated, approximately, into a variance $\tilde{\sigma}^2 = (\partial\tilde{y}_\mu/\partial y_\mu)^2\sigma^2$ of $\tilde{y}_\mu$.

The transformation $\boldsymbol{\tau}$ usually introduces a bias; i.e., if the errors in $\mathbf{y}_\mu$ have zero means, those in $\tilde{\mathbf{y}}_\mu$ do not. This bias can usually be neglected under the above assumptions.

4-20. Minimum Chi-Square Method

The minimum chi-square method (Cramer, 1946; Rao, 1957) is used to put statistical estimation problems (i.e., estimating the parameters in a probability distribution) in a least squares form. Suppose we have observed N realizations x_i of a random variable ξ, and suppose ξ is supposed to have a pdf $p(\xi \mid \boldsymbol{\theta})$. If we divide the entire range of ξ into n disjoint intervals $[a_j, b_j]$ then the expected number of observations x_i falling in each interval is

$$E_j(\boldsymbol{\theta}) \equiv N \Pr(a_j < \xi \leqslant b_j) = N \int_{a_j}^{b_j} p(x \mid \boldsymbol{\theta}) \, dx$$

Let N_j be the number of x_i actually observed in this interval. The minimum chi-square method consists in finding the value of $\boldsymbol{\theta}$ which minimizes

$$\Phi(\boldsymbol{\theta}) \equiv \sum_j [N_j - E_j(\boldsymbol{\theta})]^2 / E_j(\boldsymbol{\theta}) \tag{4-20-1}$$

In the modified minimum chi-square method, $1/N_j$ is used in place of $1/E_j$ as the weighting factor

$$\Phi(\boldsymbol{\theta}) \equiv \sum_j [N_j - E_j(\boldsymbol{\theta})]^2 / N_j \tag{4-20-2}$$

We must choose our intervals in such a way that $N_j \neq 0$ for all j. The modified form Eq. (2) is easier to use, since the denominators are constants.

Both estimates are consistent and asymptotically efficient (if each N_j goes to infinity). These properties are identical to those of the maximum likelihood method, and the latter may be preferred on account of its greater simplicity. The minimum chi-square method, however, enables us to examine the fit of the proposed distribution to the data throughout the range of the distribution, by comparing the observed occurrences N_j with the predicted values $E_j(\boldsymbol{\theta}^*)$. When the number of observations is small, the loss of information due to the grouping of the data may be considerable, and the method cannot be recommended.

4-21. Problems

1. Work out the statistical assumptions on the distribution of errors which correspond to the weighting schemes of Eqs. (4-3-3)–(4-3-6).

2. Verify the relation Eq. (4-4-15) for the two parameter case (i.e., $\mathbf{V}$ is

2×2) with $\mathbf{U} = \mathbf{I}$. Explain why the latter assumption entails no loss of generality.

3. Derive Eqs. (4-4-17)–(4-4-18).

4. Show that if the model equations are linear in the parameters, the covariance matrix is known, and the distribution of errors is normal, then the MLE is efficient.

5. Show that if the errors in each experiment are normally and independently distributed with covariance matrix $\mathbf{V} = \tau\mathbf{Q}$, where $\mathbf{Q}$ is a known matrix and τ is an unknown constant, then the MLE can be found by minimizing the concentrated objective function

$$\Phi(\boldsymbol{\theta}) = (nm/2) \log \operatorname{Tr}[\mathbf{Q}^{-1}\mathbf{M}(\boldsymbol{\theta})] \qquad (4\text{-}21\text{-}1)$$

and τ can be estimated from

$$\tilde{\tau}(\boldsymbol{\theta}) = (1/nm) \operatorname{Tr}[\mathbf{Q}^{-1}\mathbf{M}(\boldsymbol{\theta})] \qquad (4\text{-}21\text{-}2)$$

6. Consider the exact structural model $\hat{y}_\mu - \theta\hat{x}_\mu = 0$ ($\mu = 1, 2, \ldots, n$). The measurement errors of x_μ and y_μ are normal and independent with known variances σ_x^2 and σ_y^2, respectively. Show that the MLE for θ is

$$\theta^* = \{\alpha S_{yy} - S_{xx} + [(\alpha S_{yy} - S_{xx})^2 + 4\alpha S_{xy}^2]^{1/2}\}/2\alpha S_{xy}$$

where

$$\alpha = \sigma_x^2/\sigma_y^2, \quad S_{xx} = \textstyle\sum_{\mu=1}^n x_\mu^2, \quad S_{yy} = \sum_{\mu=1}^n y_\mu^2, \quad S_{xy} = \sum_{\mu=1}^n x_\mu y_\mu$$

Show that this estimate converges to the usual least squares estimate S_{xy}/S_{xx} as $\sigma_x \to 0$. See also Barnett (1967) for a slightly more general case.

7. Consider a sequence of disjoint time intervals of lengths $t_1, t_2, \ldots, t_n$. Let the number of phone calls passing through an exchange in the μth interval be N_μ. Under the assumption of a Poisson process, the probability that $N_\mu = k$ ($k = 0, 1, 2, \ldots$) is $[(\lambda t_\mu)^k/k!] \exp(-\lambda t_\mu)$, where λ is the average arrival rate of calls. Find the maximum likelihood and minimum chi-square estimates for λ.

8. Consider the following model: Certain "state" variables $\mathbf{s}$ are functions of the independent variables $\mathbf{x}$

$$\mathbf{s}_\mu = \mathbf{f}(\mathbf{x}_\mu, \boldsymbol{\theta}) + \boldsymbol{\varepsilon}_\mu$$

and the dependent variables $\mathbf{y}_\mu$ are linear functions of the state variables ($\mathbf{B}$ is a known matrix)

$$\mathbf{y}_\mu = \mathbf{B}\mathbf{s}_\mu + \boldsymbol{\delta}_\mu$$

Assume that $\boldsymbol{\varepsilon}_\mu$ and $\boldsymbol{\delta}_\mu$ are random variables independently distributed as $N_r(\mathbf{0}, \mathbf{P})$ and $N_m(\mathbf{0}, \mathbf{Q})$ respectively, that $\mathbf{Q}$ is a known matrix, and that $m > r$. Show that the MLE of $\boldsymbol{\theta}$ can be obtained as follows:

i. Compute the multiple linear regression estimates of the "true values" of the state variables

$$\hat{\mathbf{s}}_\mu = (\mathbf{B}^\mathrm{T}\mathbf{Q}^{-1}\mathbf{B})^{-1}\mathbf{B}^\mathrm{T}\mathbf{Q}^{-1}\mathbf{y}_\mu \qquad (\mu = 1, 2, \ldots, n)$$

ii. Use the computed $\hat{\mathbf{s}}_\mu$ as though they were actually measured values of $\mathbf{s}_\mu$, and apply the appropriate MLE estimate from Sections 4-7–4-9.

Note: For an application, see Section 8-8.

9. Suppose each experiment consists of measuring the same quantity y several (m) times. Show how the results of the previous problem can be applied to the present case. Note that under a reasonable set of assumptions one has $\mathbf{Q} = \sigma^2\mathbf{I}$, and one does not have to know the value of σ^2 in order to apply the method.

10. Let $\mathbf{S}$ be a matrix such that $\mathbf{S}^\mathrm{T}\mathbf{S} = \mathbf{V}^{-1}$. Define $\tilde{\mathbf{B}} = \mathbf{SB}$ and $\tilde{\mathbf{Y}} = \mathbf{SY}$. Show that this transformation reduces Eq. (4-4-4) to an unweighted sum of squares.

Chapter

V

Computation of the Estimates I: Unconstrained Problems

5-1. Introduction

Most parameter estimation methods require that we find values $\boldsymbol{\phi}^*$ of the parameters $\boldsymbol{\phi}$ for which some objective function $\Phi(\boldsymbol{\phi})$ attains its maximum or minimum. Typical objective functions to be minimized are the sum of squares, weighted sum of squares, and risk function. Those to be maximized are the likelihood, concentrated likelihood, and posterior density functions. All these were described in Chapter IV.

In most practical applications, any unknown distribution parameters $\boldsymbol{\psi}$ are eliminated from the objective function by methods such as those described in Sections 4-8 and 4-9. Therefore, we shall write our objective function as $\Phi(\boldsymbol{\theta})$ rather than $\Phi(\boldsymbol{\phi})$ and we regard $\boldsymbol{\theta}$ as a vector with l components. Some of the methods of solution we shall discuss are easily extended to the case when distribution parameters remain present in the objective function.

Sometimes the unknown parameters are free to assume any values whatsoever, and we speak of *unconstrained optimization*.‡ In other cases, only values satisfying certain inequalities and/or equations are admissible. The problem is then to find $\boldsymbol{\theta}$ such that $\Phi(\boldsymbol{\theta})$ is maximum (or minimum) subject to:

$$\mathbf{h}(\boldsymbol{\theta}) \geqslant \mathbf{0} \tag{5-1-1}$$

$$\mathbf{g}(\boldsymbol{\theta}) = \mathbf{0} \tag{5-1-2}$$

where $\mathbf{h}$ and $\mathbf{g}$ are vectors (possibly vacuous) of given functions.

We sacrifice nothing by restricting our attention to minimization, for maximizing a function can always be accomplished by minimizing its negative. Minimizing a function subject to Eq. (1) and Eq. (2) constitutes the problem of *nonlinear programming*. In spite of extensive treatment in the literature [see Daniel (1971), Wilde and Beightler (1967), Abadie (1967a), Künzi and

‡ Minimization or maximization, as appropriate.

Krelle (1966), Hadley (1964), ...] no single method has emerged which is best for the solution of all nonlinear programming problems. One cannot even hope that a "best" method will ever be found, since problems vary so much in size and nature. For parameter estimation problems we must seek methods which are particularly suitable to the special nature of these problems which may be characterized as follows:

1. A relatively small number of unknowns, rarely exceeding a dozen or so.
2. A highly nonlinear (though continuous and differentiable) objective function, whose computation is often very time consuming.
3. A relatively small number (sometimes zero) of inequality constraints. Those are usually of a very simple nature, e.g., upper and lower bounds.
4. No equality constraints, except in the case of exact structural models (where, incidentally, the number of unknowns is large). These will be treated separately in Sections 6-6–6-8.

Since the constraints play a relatively minor role in most estimation problems, we shall first discuss some methods for unconstrained optimization. Additional methods can be found in Kowalik and Osborne (1968). In Section 6-1 we shall show how constrained problems can be converted into unconstrained ones, making the previously described methods applicable.

5-2. Iterative Scheme

The methods we shall discuss are *iterative* in nature. We start with a given point $\boldsymbol{\theta}_1$ known as the *initial guess*, and proceed to generate a sequence of points $\boldsymbol{\theta}_2, \boldsymbol{\theta}_3, \ldots$ which we hope converges to the point $\boldsymbol{\theta}^*$ at which $\Phi(\boldsymbol{\theta})$ is minimum. The computation of $\boldsymbol{\theta}_{i+1}$ is called the ith *iteration*, and the point $\boldsymbol{\theta}_i$ the ith *iterate*. In practice, one *terminates* the sequence after a finite number N of iterations, and one accepts $\boldsymbol{\theta}_N$ as an approximation to $\boldsymbol{\theta}^*$. The vector

$$\boldsymbol{\sigma}_i \equiv \boldsymbol{\theta}_{i+1} - \boldsymbol{\theta}_i \tag{5-2-1}$$

is called the ith *step*. We wish each step to bring us closer to the minimum. Since we do not know where the minimum is, we cannot test for this condition directly. In a sense, however, we may consider the ith step to have "improved" our situation (by bringing us closer to the minimum in Φ space, if not in $\boldsymbol{\theta}$ space) if

$$\Phi_{i+1} < \Phi_i \tag{5-2-2}$$

where

$$\Phi_j \equiv \Phi(\boldsymbol{\theta}_j) \qquad (j = 1, 2, \ldots) \tag{5-2-3}$$

jth step

We call the ith step *acceptable* if Eq. (2) holds. An iterative method is *acceptable* if all the steps it produces are acceptable. We shall only consider acceptable methods.

All the methods we shall discuss adhere to the following scheme:

1. Set $i = 1$. An initial guess $\mathbf{\theta}_1$ must be provided externally.
2. Determine a vector $\mathbf{v}_i$ in the direction of the proposed ith step.
3. Determine a scalar ρ_i such that the step

$$\mathbf{\sigma}_i = \rho_i \mathbf{v}_i \qquad \rho_i = \frac{|\sigma_i|}{|\mathbf{v}_i|} \qquad (5\text{-}2\text{-}4)$$

is acceptable. That is, we take

$$\mathbf{\theta}_{i+1} = \mathbf{\theta}_i + \rho_i \mathbf{v}_i \qquad (5\text{-}2\text{-}5)$$

and require that ρ_i be chosen so that Eq. (2) holds.

4. Test whether the termination criterion (see Section 5-15) is met. If not, increase i by one and return to step 2. If yes, accept $\mathbf{\theta}_{i+1}$ as the value of $\mathbf{\theta}^*$.

The various methods to be described below differ only in the manner of choosing $\mathbf{v}_i$ and ρ_i. We refer to these quantities as *step direction* and *step size* respectively. Since $\mathbf{v}_i$ is not required to be a unit vector, ρ_i is only proportional, but not necessarily equal, to the step length in the usual sense.

5-3. Acceptability

Consider the ith iteration of a minimization procedure. Suppose we strike out from $\mathbf{\theta}_i$ along some direction $\mathbf{v}$, generating the ray

$$\mathbf{\theta}(\rho) \equiv \mathbf{\theta}_i + \rho \mathbf{v} \qquad (\rho \geqslant 0) \qquad (5\text{-}3\text{-}1)$$

Along this ray, the objective function varies as ρ is changed, thus becoming a function of ρ alone. We designate this function

$$\Psi_{i\mathbf{v}}(\rho) \equiv \Phi(\mathbf{\theta}(\rho)) = \Phi(\mathbf{\theta}_i + \rho \mathbf{v}) \qquad (5\text{-}3\text{-}2)$$

Its derivative is given by

$$d\Psi_{i\mathbf{v}}/d\rho = (\partial\Phi/\partial\mathbf{\theta})^{\mathsf{T}}(\partial\mathbf{\theta}/\partial\rho) = (\partial\Phi/\partial\mathbf{\theta})^{\mathsf{T}}\mathbf{v} \qquad (5\text{-}3\text{-}3)$$

The *gradient vector* of $\Phi(\mathbf{\theta})$ is $\partial\Phi/\partial\mathbf{\theta}$, which we designate as $\mathbf{q}(\mathbf{\theta})$. The αth component of $\mathbf{q}$ is the quantity $\partial\Phi/\partial\theta_\alpha$. Denoting by $\mathbf{q}_i$ the gradient vector evaluated at $\mathbf{\theta} = \mathbf{\theta}_i$, we have

$$\Psi'_{i\mathbf{v}} \equiv d\Psi_{i\mathbf{v}}/d\rho)_{\rho=0} = \mathbf{q}_i^{\mathsf{T}}\mathbf{v} \qquad (5\text{-}3\text{-}4)$$

In the sequel we assume $\mathbf{q}_i \neq \mathbf{0}$.

The quantity $\Psi'_{i\mathbf{v}}$ is called the *directional derivative* of Φ relative to $\mathbf{v}$ at $\boldsymbol{\theta}_i$. If $\Psi'_{i\mathbf{v}}$ is negative, then $\Phi(\boldsymbol{\theta})$ decreases in value when one starts moving away from $\boldsymbol{\theta}_i$ in the direction of $\mathbf{v}$. Therefore, if ρ is a sufficiently small positive number, the step $\rho\mathbf{v}$ is acceptable. On the other hand, if $\Psi'_{i\mathbf{v}} \geqslant 0$, there may not exist any positive value of ρ for which $\rho\mathbf{v}$ is an acceptable step. We call $\mathbf{v}$ an *acceptable direction* if $\Psi'_{i\mathbf{v}} < 0$.

We can now prove the following:

Theorem A direction $\mathbf{v}$ is acceptable if and only if there exists a positive definite matrix $\mathbf{R}$ such that

$$\mathbf{v} = -\mathbf{R}\mathbf{q}_i \tag{5-3-5}$$

Proof

1. Let $\mathbf{R}$ be a positive definite matrix, and let $\mathbf{v}$ be given by Eq. (5). Then, from Eq. (4) and the definition of positive definiteness

$$\Psi'_{i\mathbf{v}} = \mathbf{q}_i^{\mathrm{T}}\mathbf{v} = -\mathbf{q}_i^{\mathrm{T}}\mathbf{R}\mathbf{q}_i < 0 \tag{5-3-6}$$

2. Suppose $\mathbf{q}_i^{\mathrm{T}}\mathbf{v} < 0$. Select

$$\mathbf{R} = [\mathbf{I} - (\mathbf{q}_i\mathbf{q}_i^{\mathrm{T}}/\mathbf{q}_i^{\mathrm{T}}\mathbf{q}_i) - (\mathbf{v}\mathbf{v}^{\mathrm{T}}/\mathbf{v}^{\mathrm{T}}\mathbf{q}_i)]$$

Then Eq. (5) holds, and $\mathbf{R}$ is positive definite (we leave the details as an exercise for the reader).

The requirement $\Psi'_{i\mathbf{v}} = \mathbf{q}_i^{\mathrm{T}}\mathbf{v} < 0$ says that the direction $\mathbf{v}$ leads downhill if it forms a greater than 90° angle with the gradient $\mathbf{q}_i$. The theorem states that this condition can be insured if the direction is determined by operating on the negative gradient with a positive definite matrix according to Eq. (5). A minimization method in which the directions are obtained in this manner is called an *acceptable gradient method* (it is simply a *gradient method* if $\mathbf{R}$ is not required to be positive definite). The basic equation of the *i*th iteration in any gradient method is

$$\boldsymbol{\theta}_{i+1} = \boldsymbol{\theta}_i - \rho_i\mathbf{R}_i\mathbf{q}_i \tag{5-3-7}$$

Various gradient methods differ in the manner of choosing the $\mathbf{R}_i$ and ρ_i.

In devising or choosing an optimization method one attempts to minimize the total computation time required for convergence to the minimum. This time is composed primarily of the following two factors

1. Function and derivative evaluations.
2. Algebraic manipulations such as matrix inversions or eigenvalue determinations. It is usually possible to trade off these factors against each other. A method employing more laborious algebraic procedures may require fewer iterations, and hence fewer function evaluations. This is likely to pay off if

the objective function is a complicated one. In parameter estimation problems, the objective function is synthesized from the model equations and from the data obtained in many experiments. Its computation is usually time consuming. We do not hesitate therefore to recommend methods which are sophisticated algebraically, as long as they are efficient in terms of the number of required function and derivative evaluations.

5-4. Convergence

One would like to be able to prove that the method one has selected converges to the true minimum of the objective function. Unfortunately convergence proofs usually require that certain assumptions be made concerning the nature of the objective function, and the validity of these assumptions is difficult to verify on any given problem. Even more significantly, the existence of a convergence proof is no guarantee of reasonable performance in practice. A method may converge in theory, yet take an excessive number of iterations, or require computations to be carried out with an unreasonable number of significant digits. For this reason, our discussion of convergence theorems will be brief.

Let Φ_i denote the value of $\Phi(\boldsymbol{\theta}_i)$. If at each iteration we select an acceptable point, then the sequence $\{\Phi_i\} \equiv \{\Phi_0, \Phi_1, \Phi_2, \ldots\}$ is monotone decreasing. If the values of the objective function possess a lower bound, then this sequence must converge to a limit Φ_∞. If the sequence $\{\boldsymbol{\theta}_i\}$ is bounded (i.e., there exists a number M such that $\boldsymbol{\theta}_i^{\mathrm{T}}\boldsymbol{\theta}_i < M$ for all i) then it has at least one limit point. It follows from the continuity of Φ that $\Phi(\boldsymbol{\theta}_\infty) = \Phi_\infty$, where $\boldsymbol{\theta}_\infty$ is any limit point of $\{\boldsymbol{\theta}_i\}$. Because of this, the sequence $\{\boldsymbol{\theta}_i\}$ can have more than one limit point only by remarkable coincidence. In all practical cases, the sequence $\{\boldsymbol{\theta}_i\}$ is either unbounded, or converges to a point $\boldsymbol{\theta}_\infty$. The rate of convergence, however, may be so slow that the sequence appears nonconvergent.

A *stationary point* of the objective function is one at which $\mathbf{q}(\boldsymbol{\theta}) = \mathbf{0}$. If $\boldsymbol{\theta}_i$ is stationary, i.e., $\mathbf{q}_i = \mathbf{0}$, then Eq. (5-3-7) shows that all $\boldsymbol{\theta}_j$ $(j \geqslant i)$ coincide with $\boldsymbol{\theta}_i$. Therefore, the most that we can hope to prove about any gradient method is that it converges to a stationary point. Convergence to the true minimum can be guaranteed only if it can be shown that the objective function has no other stationary points. In practice, convergence to a local maximum or saddle point requires an improbable coincidence. One usually reaches at least a local minimum.

Convergence proofs require that the ρ_i be chosen sufficiently large, and the matrices $\mathbf{R}_i$ sufficiently positive definite. The following theorem is typical. Its proof is given in Appendix F:

Theorem Let $\mathscr{D}$ denote the set of all $\boldsymbol{\theta}$ such that $\Phi(\boldsymbol{\theta}) \leqslant \Phi_1$. Suppose the following conditions are satisfied:

1. Φ has continuous first and bounded second derivatives in $\mathscr{D}$.
2. Let μ_i be the smallest nonnegative value of ρ at which $\Psi_{i\mathbf{v}_i}(\rho)$ attains a local maximum, where $\mathbf{v}_i = -\mathbf{R}_i \mathbf{q}_i$. Let α be a positive number less than one, and ρ_0 a positive number. We choose each ρ_i so that either $\alpha\mu_i \leqslant \rho_i \leqslant \mu_i$ or $\min(\rho_0, \alpha\mu_i) \leqslant \rho_i \leqslant \mu_i$.
3. Let β and γ be constants satisfying $\gamma > \beta > 0$. We choose each $\mathbf{R}_i$ so that all its eigenvalues lie between β and γ.

Then all the limit points of $\{\boldsymbol{\theta}_i\}$ are stationary points of Φ.

These conditions are sufficient, but not necessary. In the algorithms that we shall describe the $\mathbf{R}_i$ are usually chosen so as to satisfy condition 3. There does not seem to be any need, in practice, to trouble oneself with satisfying condition 2 precisely. Condition 1 is almost always satisfied in principle, but the limited accuracy of numerical calculations sometimes causes trouble.

5-5. Steepest Descent

The simplest gradient method employs $\mathbf{R}_i = \mathbf{I}$, so that $\mathbf{v}_i = -\mathbf{q}_i$ in all iterations. The direction $-\mathbf{q}_i$ is the one in which the objective function decreases most rapidly, at least initially. Hence this method is called *steepest descent.* Unfortunately, as discussed more fully in subsequent sections, this method is often very inefficient, requiring a large number of steps which tend to zigzag in a so-called hemstitching pattern. The method is not recommended for practical applications, and is discussed here only for reference.

5-6. Newton's Method

The Hessian matrix $\mathbf{H}(\boldsymbol{\theta})$ of the function $\Phi(\boldsymbol{\theta})$ is the matrix of second partial derivatives, i.e.,

$$H_{\alpha\beta}(\boldsymbol{\theta}) \equiv \partial^2\Phi/\partial\theta_\alpha\,\partial\theta_\beta \tag{5-6-1}$$

Let $\mathbf{H}_i$ be the Hessian matrix of Φ evaluated at $\boldsymbol{\theta} = \boldsymbol{\theta}_i$. We define the function

$$Q_i(\boldsymbol{\theta}) = \Phi_i + \mathbf{q}_i^{\mathrm{T}}(\boldsymbol{\theta} - \boldsymbol{\theta}_i) + \tfrac{1}{2}(\boldsymbol{\theta} - \boldsymbol{\theta}_i)^{\mathrm{T}}\mathbf{H}_i(\boldsymbol{\theta} - \boldsymbol{\theta}_i) \tag{5-6-2}$$

which consists of the terms up to second order in the Taylor series expansion of Φ around the point $\boldsymbol{\theta}_i$. In a sense, $Q_i(\boldsymbol{\theta})$ matches the behavior of $\Phi(\boldsymbol{\theta})$ at $\boldsymbol{\theta} = \boldsymbol{\theta}_i$ more closely than does any other second order surface.

Suppose we wish to find the point at which $Q_i(\boldsymbol{\theta})$ is stationary. We equate to zero the gradient of Q_i

$$\partial Q_i/\partial \boldsymbol{\theta} = \mathbf{q}_i + \mathbf{H}_i(\boldsymbol{\theta} - \boldsymbol{\theta}_i) = \mathbf{0} \qquad (5\text{-}6\text{-}3)$$

which, if $\mathbf{H}_i$ is nonsingular, has the solution

$$\boldsymbol{\theta}_{i+1} = \boldsymbol{\theta}_i - \mathbf{H}_i^{-1}\mathbf{q}_i \qquad (5\text{-}6\text{-}4)$$

Equation (4) defines the ith iteration of the Newton (also known as Newton–Raphson) method. It conforms to the general formula of gradient methods Eq. (5-3-7) with $\rho_i = 1$ and $\mathbf{R}_i = \mathbf{H}_i^{-1}$.

If $\Phi(\boldsymbol{\theta})$ coincides with $Q_i(\boldsymbol{\theta})$, i.e., if Φ is a quadratic function, then $\boldsymbol{\theta}_{i+1}$ is a stationary point of Φ. This is a minimum if $\mathbf{H}_i$ is positive definite. In that case $\mathbf{R}_i$ is positive definite, the method is acceptable, and it converges in a single iteration. If $\mathbf{H}_i$ is negative definite, $\boldsymbol{\theta}_{i+1}$ is a maximum, and if $\mathbf{H}_i$ is indefinite, $\boldsymbol{\theta}_{i+1}$ is a saddle point. In both cases the method is not acceptable.

When Φ is not quadratic, $\boldsymbol{\theta}_{i+1}$ does not generally coincide with the stationary point, and the method does not converge in a single iteration. The method is acceptable, however, as long as $\mathbf{H}_i$ is positive definite, as it should be at least in some neighborhood of the minimum. In this neighborhood, convergence is quadratic. This means that the number of correct digits in $\boldsymbol{\theta}$ is approximately doubled by each iteration, until further improvement is barred by the rounding errors in the calculations. Outside this neighborhood, convergence cannot be guaranteed. It is worth noting that the step $\mathbf{v} = -\mathbf{Rq}$ ($\mathbf{R}$ positive definite) solves the problem of minimizing the function $\mathbf{q}^{\mathrm{T}}\mathbf{v} + \frac{1}{2}\mathbf{v}^{\mathrm{T}}\mathbf{R}^{-1}\mathbf{v}$. The closer $\mathbf{R}$ is to $\mathbf{H}^{-1}$, the closer is this modified problem to the original one.

Returning to the case of a quadratic function Φ with a positive definite Hessian $\mathbf{H}$, we find that the Newton step $-\mathbf{H}^{-1}\mathbf{q}$ is the only one that takes us to the minimum in a single iteration. Any other step $-\mathbf{Rq}$ with $\mathbf{R} \neq \mathbf{H}^{-1}$ will miss the minimum. If we define the efficiency of a method as the decrease in function value obtained by a single step of the method, divided by the maximum possible decrease, then the efficiency of the Newton method is 1. It was shown by Greenstadt (1967) that the efficiency $e(\mathbf{R})$ of a method using the direction $-\mathbf{Rq}$ is bounded by

$$4\gamma/(1+\gamma)^2 \leqslant e(\mathbf{R}) \leqslant 1 \qquad (5\text{-}6\text{-}5)$$

where γ is the condition number, i.e., the ratio of largest to smallest eigenvalues of the matrix $\mathbf{R}^{1/2}\mathbf{H}\mathbf{R}^{1/2}$. With $\mathbf{R} = \mathbf{I}$ (the method of steepest descent), $\gamma = \gamma_{\mathbf{H}}$ is the condition number of $\mathbf{H}$ itself. It is quite common to find cases where $\gamma > 10^5$, so that the Newton method may be 25,000 times more efficient than steepest descent!

It is worth remarking that Marquardt's method finds the step $\mathbf{v}$ which minimizes the quadratic approximation to Φ given by

$$Q_i(\mathbf{v}) \equiv \Phi_i + \mathbf{v}^{\mathrm{T}}\mathbf{q}_i + \tfrac{1}{2}\mathbf{v}^{\mathrm{T}}\mathbf{A}_i\mathbf{v} \tag{5-8-7}$$

subject to the restriction that

$$\mathbf{v}^{\mathrm{T}}\mathbf{B}_i^{\,2}\mathbf{v} = c \tag{5-8-8}$$

That is, the step $\mathbf{v}$ takes us to that point on the ellipsoid defined by Eq. (8) at which the function $Q_i(\mathbf{v})$ attains its minimum.

To prove this, we form the Lagrangian

$$\Lambda(\mathbf{v}) = \Phi_i + \mathbf{v}^{\mathrm{T}}\mathbf{q}_i + \tfrac{1}{2}\mathbf{v}^{\mathrm{T}}\mathbf{A}_i\mathbf{v} + \tfrac{1}{2}\lambda_i(\mathbf{v}^{\mathrm{T}}\mathbf{B}_i^{\,2}\mathbf{v} - c) \tag{5-8-9}$$

We differentiate with respect to $\mathbf{v}$, equate to zero

$$\mathbf{q}_i + \mathbf{A}_i\mathbf{v} + \lambda_i\mathbf{B}_i^{\,2}\mathbf{v} = \mathbf{0} \tag{5-8-10}$$

and solve for $\mathbf{v}$

$$\mathbf{v} = -(\mathbf{A}_i + \lambda_i\mathbf{B}_i^{\,2})^{-1}\mathbf{q}_i \tag{5-8-11}$$

in agreement with Eq. (1). The particular ellipsoid chosen depends on λ_i, since by substituting Eq. (11) into Eq. (8) we find

$$c = \mathbf{q}_i^{\mathrm{T}}(\mathbf{A}_i + \lambda_i\mathbf{B}_i^{\,2})^{-1}\mathbf{B}_i^{\,2}(\mathbf{A}_i + \lambda_i\mathbf{B}_i^{\,2})^{-1}\mathbf{q}_i \tag{5-8-12}$$

The larger λ_i is, the smaller is c, and the smaller is the ellipsoid to which we are confined. The algorithm starts with an ellipsoid of a certain size, determined through Eq. (12) by the initial choice for λ_i. If the corresponding step $\mathbf{v}$ fails to decrease the objective function, this is an indication that the chosen ellipsoid is much larger than the region within which the quadratic approximation Eq. (7) is valid. By increasing λ, we shrink the ellipsoid and try again.

The Marquardt method has proven very reliable in practice. In the problem of Section 5-21 with difficult initial guess, and in several cases of the problem of Section 5-23, the Marquardt method proved faster than directional discrimination methods. On the other hand, in a series of ten other test problems (Bard, 1970), the reverse held true. The need for further testing is evident.

5-9. The Gauss Method

In most parameter estimation problems, the unknown parameters appear only indirectly in the objective function. The latter depends explicitly on the model equations, which in turn depend on the parameters. To compute derivatives of the objective function, we first differentiate it with respect to the model equations, and then differentiate those with respect to the parameters. The Gauss (1809) method, originally applied to least squares problems, consists of simply omitting the second derivatives of the model equations when the Hessian is being computed.

We illustrate by means of the simplest example, that of single equation nonlinear least squares. Here we minimize

$$\Phi(\boldsymbol{\theta}) = \sum_{\mu=1}^{n} [y_\mu - f(\mathbf{x}_\mu, \boldsymbol{\theta})]^2 = \sum_{\mu=1}^{n} (y_\mu - f_\mu)^2 = \sum_{\mu=1}^{n} e_\mu{}^2 \tag{5-9-1}$$

whence

$$q_\alpha = \partial\Phi/\partial\theta_\alpha = 2\sum_{\mu=1}^{n} e_\mu\, \partial e_\mu/\partial\theta_\alpha = -2\sum_{\mu=1}^{n} e_\mu\, \partial f_\mu/\partial\theta_\alpha \tag{5-9-2}$$

and

$$H_{\alpha\beta} = \partial^2\Phi/\partial\theta_\alpha\,\partial\theta_\beta = -2\sum_{\mu=1}^{n} e_\mu\, \partial^2 f_\mu/\partial\theta_\alpha\,\partial\theta_\beta + 2\sum_{\mu=1}^{n} (\partial f_\mu/\partial\theta_\alpha)(\partial f_\mu/\partial\theta_\beta) \tag{5-9-3}$$

In the Gauss method, we neglect the first term, and use **N** in place of **H**, where **N** is defined by

$$N_{\alpha\beta} = 2\sum_{\mu=1}^{n} (\partial f_\mu/\partial\theta_\alpha)(\partial f_\mu/\partial\theta_\beta) \tag{5-9-4}$$

A numerical example appears in Section 5-21.

In the preceding discussion we have derived **N** as an approximation to **H**, and the Gauss method as an approximation to the Newton method. There is an alternative interpretation: suppose the model equations are replaced by their tangents; that is, the nonlinear (in $\boldsymbol{\theta}$) model is approximated by one that is linear. If we now try to solve the corresponding linear least squares problem we find the solution to be precisely $\tilde{\boldsymbol{\theta}} = \boldsymbol{\theta}_i - \mathbf{N}^{-1}\mathbf{q}$. Now $\tilde{\boldsymbol{\theta}}$ is usually not the correct solution to the nonlinear problem. Yet, if we accept $\boldsymbol{\theta}_{i+1} = \tilde{\boldsymbol{\theta}}$, we may regard the Gauss method as solving a sequence of linear problems. This interpretation is pursued further in Section 5-10.

The term neglected in Eq. (3) contained the residual e_μ as a factor. Since the residuals are, hopefully, small, this provides some justification for regarding **N** as a good approximation to **H**, particularly near the minimum. The same justification applies to all of the more general cases in which the objective function depends on the parameters only through the elements of the moment matrix of the residuals $\mathbf{M}(\boldsymbol{\theta}) = \sum_\mu \mathbf{e}_\mu(\boldsymbol{\theta})\mathbf{e}_\mu{}^{\mathrm{T}}(\boldsymbol{\theta})$. This includes most of the least squares and maximum likelihood estimates for normal distributions that were discussed in Chapter IV, e.g., Eqs. (4-7-7), (4-7-9), (4-8-4), (4-9-9), and (4-21-1). In all these cases we have

$$\Phi(\boldsymbol{\theta}) = \Psi(\mathbf{M}(\boldsymbol{\theta})) \tag{5-9-5}$$

where Ψ is a suitable function.

Differentiation of Eq. (4-7-5) yields:

$$\partial M_{ab}/\partial\theta_\alpha = \sum_\mu (e_{\mu a}\, \partial e_{\mu b}/\partial\theta_\alpha + e_{\mu b}\, \partial e_{\mu a}/\partial\theta_\alpha) \tag{5-9-6}$$

$$\partial e_{\mu b}/\partial\theta_\alpha = \begin{cases} -\partial f_{\mu b}/\partial\theta_\alpha & \text{for reduced models} \\ \partial g_{\mu b}/\partial\theta_\alpha & \text{for inexact structural models} \end{cases} \tag{5-9-7}$$

Therefore, from Eq. (5) and because of the symmetry of **M**

$$\begin{aligned} q_\alpha = \partial\Phi/\partial\theta_\alpha &= \sum_{a,b} (\partial\Psi/\partial M_{ab})(\partial M_{ab}/\partial\theta_\alpha) \\ &= \sum_{\mu,a,b} (\partial\Psi/\partial M_{ab})(e_{\mu a}\, \partial e_{\mu b}/\partial\theta_\alpha + e_{\mu b}\, \partial e_{\mu a}/\partial\theta_\alpha) \\ &= 2\sum_{\mu,a,b} (\partial\Psi/\partial M_{ab}) e_{\mu a}(\partial e_{\mu b}/\partial\theta_\alpha) \end{aligned} \tag{5-9-8}$$

One can further work out that

$$\begin{aligned} H_{\alpha\beta} = \partial^2\Phi/\partial\theta_\alpha\, \partial\theta_\beta &= 2\sum_{\mu,a,b} (\partial\Psi/\partial M_{ab})(\partial e_{\mu a}/\partial\theta_\alpha)(\partial e_{\mu b}/\partial\theta_\beta) \\ &+ 2\sum_{\mu,a,b} (\partial\Psi/\partial M_{ab}) e_{\mu a}(\partial^2 e_{\mu b}/\partial\theta_\alpha\, \partial\theta_\beta) \\ &+ 4\sum_{\mu,a,b}\sum_{\eta,c,d} (\partial^2\Psi/\partial M_{ab}\, \partial M_{cd}) e_{\mu a}(\partial e_{\mu b}/\partial\theta_\alpha) e_{\eta c}(\partial e_{\eta d}/\partial\theta_\beta) \end{aligned} \tag{5-9-9}$$

As we see, the second derivatives of the model equations are always multiplied by residuals $e_{\mu a}$, and the terms involving them are dropped in the Gauss method. We note also that the terms involving $\partial^2\Psi/\partial M\, \partial M$ contain residuals and we drop these too. This leaves us with the approximate Hessian

$$N_{\alpha\beta} \equiv 2\sum_{\mu,a,b} (\partial\Psi/\partial M_{ab})(\partial e_{\mu a}/\partial\theta_\alpha)(\partial e_{\mu b}/\partial\theta_\beta)$$

or, in matrix notation

$$\mathbf{N} = 2\sum_{\mu=1}^{n} \mathbf{B}_\mu^{\mathrm{T}} \boldsymbol{\Gamma} \mathbf{B}_\mu \tag{5-9-10}$$

where $\mathbf{B}_\mu$ and $\boldsymbol{\Gamma}$ are matrices defined by

$$B_{\mu,a\alpha} \equiv -\partial e_{\mu a}/\partial\theta_\alpha = \partial f_{\mu a}/\partial\theta_\alpha, \qquad \Gamma_{ab} \equiv \partial\Psi/\partial M_{ab} \tag{5-9-11}$$

Using the same notation, the gradient Eq. (8) is given by

$$\mathbf{q} = -2\sum_{\mu=1}^{n} \mathbf{B}_\mu^{\mathrm{T}} \boldsymbol{\Gamma} \mathbf{e}_\mu \tag{5-9-12}$$

In Table 5-1 we give formulas for $\boldsymbol{\Gamma}$ in the cases of the normal distributions discussed in Sections 4-7–4-9. It is significant that in all these cases $\boldsymbol{\Gamma}$ turns out to be positive definite (or at least semidefinite). It follows from Eq. (10) that

Table 5-1
Some Log Likelihood Functions and Their Derivatives

Objective function given by	Description	Objective function $\Psi(\mathbf{M})$	$\mathbf{\Gamma} = \frac{\partial \Psi}{\partial \mathbf{M}}$
Eq. (4-7-7)	Normal distribution, known covariance matrix, weighted least squares.	$\frac{1}{2}\,\mathrm{Tr}(\mathbf{V}^{-1}\mathbf{M})$	$\frac{1}{2}\mathbf{V}^{-1}$
Eq. (4-21-1)	Normal distribution, covariance matrix known except for multiplicative factor.	$\frac{nm}{2} \log \mathrm{Tr}(\mathbf{Q}^{-1}\mathbf{M})$	$\frac{nm}{2\,\mathrm{Tr}(\mathbf{Q}^{-1}\mathbf{M})}\mathbf{Q}^{-1}$
Eq. (4-9-9)	Normal distribution, unknown covariance matrix.	$\frac{n}{2} \log \det \mathbf{M}$	$\frac{n}{2}\mathbf{M}^{-1}$

$\mathbf{N}$ and $\mathbf{R} = \mathbf{N}^{-1}$ are also positive definite. In particular, this is true in the single equation least squares case, Eq. (4). Application of the Gauss method to an objective function of the form Eq. (4-9-9) is illustrated in Section 5-23.

If the objective function is a posterior density as in Eq. (4-15-1) or Eq. (4-15-2), the Gauss method may still be used to approximate the Hessian of the log likelihood, to which the true Hessian of the log prior density must be added. The latter is frequently easy to compute. For instance, if $p_0(\boldsymbol{\theta})$ is normal with covariance matrix $\mathbf{V}$, then the Hessian of $\log p_0$ is simply $-\mathbf{V}^{-1}$. See Section 5-22 for a numerical example.

5-10. The Gauss Method as a Sequence of Linear Regression Problems

The essence of the Gauss method is to use in the ith iteration a step whose direction is given by

$$\mathbf{v}_i = -\mathbf{N}_i^{-1}\mathbf{q}_i \tag{5-10-1}$$

Equivalently, $\mathbf{v}_i$ is the solution of the set of simultaneous linear equations

$$\mathbf{N}_i \mathbf{v}_i = -\mathbf{q}_i \tag{5-10-2}$$

which, in view of Eq. (5-9-10) and Eq. (5-9-12), may be written out as

$$\sum_{\mu=1}^{n} \mathbf{B}_\mu{}^{\mathrm{T}} \mathbf{\Gamma} \mathbf{B}_\mu \mathbf{v} = \sum_{\mu=1}^{n} \mathbf{B}_\mu{}^{\mathrm{T}} \mathbf{\Gamma} \mathbf{e}_\mu \tag{5-10-3}$$

where the subscript i has been dropped for convenience.

Referring back to Section 4-4, we find that if we want to determine by multiple linear regression the coefficients $\mathbf{v}$ in the expressions

$$\mathbf{e}_\mu = \mathbf{B}_\mu \mathbf{v} \qquad (\mu = 1, 2, \ldots, n) \tag{5-10-4}$$

and if we assume the covariance matrix

$$E(\mathbf{e}_\mu - \mathbf{B}_\mu \mathbf{v})(\mathbf{e}_\eta - \mathbf{B}_\eta \mathbf{v})^{\mathrm{T}} = \delta_{\mu\eta} \mathbf{\Gamma}^{-1} \tag{5-10-5}$$

we are led to minimizing the function

$$\Phi^{(i)}(\mathbf{v}) \equiv \sum_{\mu=1}^{n} (\mathbf{e}_\mu - \mathbf{B}_\mu \mathbf{v})^{\mathrm{T}} \mathbf{\Gamma} (\mathbf{e}_\mu - \mathbf{B}_\mu \mathbf{v}) \tag{5-10-6}$$

The normal equations corresponding to $\Phi^{(i)}(\mathbf{v})$ are given precisely by Eq. (3), with all quantities evaluated for $\boldsymbol{\theta} = \boldsymbol{\theta}_i$. Thus, each iteration of the Gauss method may be regarded as the solution of a multiple linear regression problem.

The above remark applies only to the determination of the direction $\mathbf{v}_i$ but not of the length ρ_i. The solution of the linear problem is $\boldsymbol{\theta}_i + \mathbf{v}_i$, i.e., $\rho_i = 1$. This step may prove unacceptable in the nonlinear problem, and we use an interpolation–extrapolation scheme as described in Section 5-14 to determine a better value of ρ_i. Such schemes were originated by Box (1957), Hartley (1961), McGhee (1963).

The linear regression problem represented by the objective function Eq. (6) and its normal equations Eq. (3) can be generated by inspection, without reference to Table 5-1 and the derivations of Section 5-9. All we have to do is replace the model equations by their linear approximations around the current value $\boldsymbol{\theta}_i$

$$\mathbf{f}_\mu(\boldsymbol{\theta}) = \mathbf{f}_\mu(\boldsymbol{\theta}_i) + (\partial \mathbf{f}_\mu / \partial \boldsymbol{\theta})(\boldsymbol{\theta} - \boldsymbol{\theta}_i) = \mathbf{f}_\mu(\boldsymbol{\theta}_i) + \mathbf{B}_\mu \mathbf{v}_i \tag{5-10-7}$$

so that $\mathbf{e}_\mu = \mathbf{y}_\mu - \mathbf{f}_\mu(\boldsymbol{\theta}_i) = \mathbf{B}_\mu \mathbf{v}_i$ as in Eq. (4). These are the linearized model equations. For the weighting matrix $\mathbf{\Gamma}$ we take $\mathbf{V}^{-1}$ when known, or its current estimate (e.g., $n\mathbf{M}^{-1}$) when not known.

It goes without saying that various tricks that are useful for solving linear regression problems are applicable here too. For instance, it is well known that the condition of the normal equations is usually improved if one "subtracts out the means." This strategy applies provided the model has a constant term. Suppose, for instance, that for a single-equation model, Eq. (4) has the form

$$e_\mu = v_1 + \sum_{\alpha=2}^{l} b_{\mu\alpha} v_\alpha \tag{5-10-8}$$

Let $\bar{b}_\alpha$ be the average value of $b_{\mu\alpha}$, i.e.,

$$\bar{b}_\alpha \equiv (1/n) \sum_{\mu=1}^{n} b_{\mu\alpha} \tag{5-10-9}$$

Then Eq. (8) may be rewritten as

$$e_\mu = \tilde{v}_1 + \sum_{\alpha=2}^{l} \tilde{b}_{\mu\alpha} v_\alpha \tag{5-10-10}$$

where

$$\tilde{v}_1 \equiv v_1 + \sum_{\alpha=2}^{l} \bar{b}_\alpha v_\alpha \tag{5-10-11}$$

$$\tilde{b}_{\mu\alpha} \equiv b_{\mu\alpha} - \bar{b}_\alpha \tag{5-10-12}$$

We now use model Eq. (4) to calculate $\tilde{v}_1, v_2, \ldots, v_l$, and from these and Eq. (11), we can compute v_1.

It is a remarkable property of the normal equations in regression problems that they always have a solution, even when **N** is singular. In fact, when **N** is singular there are infinitely many solutions. Among these, the one of minimum length is given by

$$\mathbf{v}_i = -\mathbf{N}_i^{+}\mathbf{q}_i \tag{5-10-13}$$

Other solutions may be obtained, for instance, by means of stepwise regression (one continues pivoting until no nonzero pivots are left. See Section A-3.) In this solution, the number of nonzero components in $\mathbf{v}_i$ does not exceed the rank of $\mathbf{N}_i$. We prefer the pseudoinverse solution because of its minimum length property. However, presence of rounding errors makes use of the exact pseudoinverse undesirable, and it is best to make $\mathbf{N}_i$ nonsingular by the directional discrimination or Marquardt methods.

5-11. The Implementation of the Gauss Method

There are several ways in which the direction $\mathbf{v}_i$ given by Eq. (5-10-1) may be computed. Any method suitable for the solution of multiple linear regression problems is also useful here. In a sense, however, linear problems place more stringent requirements on methods of solution: We expect to obtain the correct answer in a single step, and must therefore compute $\mathbf{N}^{-1}$ very precisely. A nonlinear problem, on the other hand, requires several iterations; slight errors in each iteration can be tolerated, as long as the chosen directions are acceptable. In other words, $\mathbf{N}_i^{-1}$ need in principle only be positive definite for nonlinear problems. However, substantial errors in the computation of $\mathbf{N}_i^{-1}$ may greatly increase the number of iterations required.

In the presence of a prior density (as when we seek the mode of the posterior distribution) or a penalty function (when inequality constraints apply, see Section 6-1), appropriate terms must be added to both $\mathbf{N}_i$ and $\mathbf{q}_i$. The terms added to **N** are usually positive definite, and when **N** is ill-conditioned, an improvement in its condition may result. The linear regression structure is seemingly lost, but at the end of this section we indicate how it can sometimes be recovered.

Numerical techniques for computing the direction $\mathbf{v}_i$ fall into two classes: First, methods for solving the normal equations, without taking account of their particular structure. These methods are obviously applicable whether or not the equations possess the linear regression structure. Second, methods that rely on the linear regression structure, and are sometimes inapplicable in the presence of a prior distribution. We may also classify methods into those which simply compute $\mathbf{v}_i = -\mathbf{N}_i^{-1}\mathbf{q}_i$, and those which allow the inverse to be adjusted in favorable ways.

(a) Normal Equation Methods. The simplest method simply solves the Eq. (5-10-2) for $\mathbf{v}_i$, using standard simultaneous equation techniques. The fastest method is the Cholesky decomposition (see Section A-5), but is not recommended unless $\mathbf{N}_i$ is known to be positive definite and fairly well-conditioned.‡ In general, we recommend one of the directional discrimination methods (Section 5-7) or the Marquardt method (Section 5-8), all of which allow us to improve the condition of $\mathbf{N}_i$.

(b) Regression Methods. The method of Jennrich and Sampson (1968) consists of applying the stepwise regression technique (see Appendix A-3) to the regression problem of Section 5-10. The normal equations are formed, but components of $\mathbf{v}_i$ which cannot significantly reduce the value of the objective function are set to zero. This is a directional discrimination method, the directions coinciding with the coordinate axes. We hazard the guess that backward stepwise regression would be more efficient than forward regression in most nonlinear problems.

Methods which do not require formation of the normal equations are capable of greater numerical accuracy, and are particularly suitable when precise solutions to highly ill-conditioned linear regression problems are required. The main disadvantage of these methods is the need to keep in computer storage all the $\mathbf{B}_\mu$ matrices and $\mathbf{e}_\mu$ vectors (to form the normal equations, these may be generated and discarded one at a time).

We mention here two of these methods: (1) Golub (1965) generates the Cholesky decomposition of $\mathbf{N}$ directly from the $\mathbf{B}_\mu$, using Householder transformations;§ and (2) modified Gram–Schmidt orthogonalization has been

‡ The method can be adapted to the singular or near singular case, as shown by Healy (1968), but this adaptation has performed poorly in some test cases we tried. This is because although the Cholesky method gives an accurate solution to the normal equations even when they are nearly singular, the step direction thus generated is so far from the negative gradient as to be almost unacceptable.

§ Golub's method can be adapted to operate sequentially on small segments of the matrix **B**, thereby overcoming the computer storage problem. The resulting algorithm is, however, rather unattractive.

found by Longley (1967) to be considerably more accurate than solution of normal equations. Golub (1969) reports it to be slightly more accurate than the Householder procedure.

We present here the details of orthogonalization: Let us adopt the following notation, similar to the one used in Section 4-4

$$\mathbf{B} \equiv \begin{bmatrix} \mathbf{B}_1 \\ \mathbf{B}_2 \\ \vdots \\ \mathbf{B}_n \end{bmatrix}, \qquad \mathbf{\Pi} \equiv \begin{bmatrix} \mathbf{\Gamma} & \mathbf{0} & \cdots & \mathbf{0} \\ \mathbf{0} & \mathbf{\Gamma} & & \\ \vdots & & & \\ \mathbf{0} & & \cdots & \mathbf{\Gamma} \end{bmatrix}, \qquad \mathrm{E} \equiv \begin{bmatrix} \mathbf{e}_1 \\ \mathbf{e}_2 \\ \vdots \\ \mathbf{e}_n \end{bmatrix} \tag{5-11-1}$$

$\mathbf{B}$ has mn rows and l columns; we denote the latter as $\mathbf{b}_1, \mathbf{b}_2, \ldots, \mathbf{b}_l$. $\mathbf{\Pi}$ is $mn \times mn$, and E is a column vector with mn elements. The normal Eq. (5-10-3) may be rewritten as

$$\mathbf{B}^{\mathrm{T}}\mathbf{\Pi}\mathbf{B}\mathbf{v} = \mathbf{B}^{\mathrm{T}}\mathbf{\Pi}\mathrm{E} \tag{5-11-2}$$

Suppose the $\mathbf{b}_\alpha$ are linearly independent. Then we can find a set of l vectors $\mathbf{p}_1, \mathbf{p}_2, \ldots, \mathbf{p}_l$ which are orthonormal relative to $\mathbf{\Pi}$, i.e.,

$$\mathbf{p}_i{}^{\mathrm{T}}\mathbf{\Pi}\mathbf{p}_j = \delta_{ij} \qquad (i, j = 1, 2, \ldots, l) \tag{5-11-3}$$

and which form a basis for the $\mathbf{b}_\alpha$. This means that the $\mathbf{b}_\alpha$ are independent linear combinations of the $\mathbf{p}_i$, i.e., there exists a nonsingular $l \times l$ matrix $\mathbf{A}$ such that

$$\mathbf{b}_\alpha = \sum_{i=1}^{l} A_{i\alpha}\mathbf{p}_i \qquad (\alpha = 1, 2, \ldots, l) \tag{5-11-4}$$

Let $\mathbf{P}$ be the matrix whose columns are the $\mathbf{p}_i$. Then Eq. (3) and Eq. (4) are equivalent to:

$$\mathbf{P}^{\mathrm{T}}\mathbf{\Pi}\mathbf{P} = \mathbf{I} \tag{5-11-5}$$

$$\mathbf{B} = \mathbf{P}\mathbf{A} \tag{5-11-6}$$

In the sequel, whenever we use the term "orthogonal," we mean orthogonal relative to $\mathbf{\Pi}$. The vector E can be decomposed into a component which is a linear combination of the $\mathbf{p}_i$, and a component D which is orthogonal to all of them. This can be stated concisely in the following equations

$$\mathrm{E} = \mathrm{D} + \sum_{i=1}^{l} t_i\mathbf{p}_i = \mathrm{D} + \mathbf{P}\mathbf{t} \tag{5-11-7}$$

where D satisfies $\mathbf{P}^{\mathrm{T}}\mathbf{\Pi}\mathrm{D} = \mathbf{0}$ and $\mathbf{t}$ is an l-vector of coefficients. We verify easily that $\mathbf{P}^{\mathrm{T}}\mathbf{\Pi}\mathrm{E} = \mathbf{P}^{\mathrm{T}}\mathbf{\Pi}\mathrm{D} + \mathbf{P}^{\mathrm{T}}\mathbf{\Pi}\mathbf{P}\mathbf{t}$, i.e.,

$$\mathbf{t} = \mathbf{P}^{\mathrm{T}}\mathbf{\Pi}\mathrm{E} \tag{5-11-8}$$

Employing Eqs. (5)–(8), we can write the solution to Eq. (2) as

$$\mathbf{v} = (\mathbf{B}^{\mathrm{T}}\mathbf{\Pi}\mathbf{B})^{-1}\mathbf{B}^{\mathrm{T}}\mathbf{\Pi}\mathrm{E} = (\mathbf{A}^{\mathrm{T}}\mathbf{P}^{\mathrm{T}}\mathbf{\Pi}\mathbf{P}\mathbf{A})^{-1}\mathbf{A}^{\mathrm{T}}\mathbf{P}^{\mathrm{T}}\mathbf{\Pi}\mathrm{E} = (\mathbf{A}^{\mathrm{T}}\mathbf{A})^{-1}\mathbf{A}^{\mathrm{T}}\mathbf{t} = \mathbf{A}^{-1}\mathbf{t} \tag{5-11-9}$$

The computation of $\mathbf{v}$ is particularly easy if $\mathbf{A}$ is an easily inverted matrix, e.g., of upper triangular form. The following procedure generates such a matrix. It is known as modified Gram–Schmidt orthogonalization.

1. Form the matrix $\mathbf{C} \equiv [\mathbf{B}, \mathbf{E}]$. This matrix, which has mn rows and $l+1$ columns, will be transformed as described below. We let $\mathbf{c}_i$ denote the ith column of $\mathbf{C}$.
2. Set $k = 1$.
3. Let $S_{kk} = (\mathbf{c}_k{}^T \mathbf{\Pi} \mathbf{c}_k)^{1/2}$
4. Replace $\mathbf{c}_k$ with $\mathbf{c}_k / S_{kk}$. Note that now $\mathbf{c}_k$ is normalized in the sense that $\mathbf{c}_k{}^T \mathbf{\Pi} \mathbf{c}_k = 1$.
5. Let $S_{ki} = \mathbf{c}_i{}^T \mathbf{\Pi} \mathbf{c}_k$ for $i = k+1, k+2, \ldots, l+1$.
6. Replace $\mathbf{c}_i$ with $\mathbf{c}_i - S_{ki}\mathbf{c}_k$ for $i = k+1, k+2, \ldots, l+1$. Note that thereby the $\mathbf{c}_i$ $(i > k)$ are rendered orthogonal to $\mathbf{c}_k$, without losing their previously established orthogonality to the $\mathbf{c}_j$ $(j < k)$.
7. If $k = l$, terminate. Otherwise, replace k by $k+1$ and return to step 3.

It is clear from the remarks accompanying steps 4 and 6 that the first l columns of the final $\mathbf{C}$ tableau form an orthonormal (relative to $\mathbf{\Pi}$) basis for $\mathbf{B}$, and that the last column of $\mathbf{C}$ contains the component of $\mathbf{E}$ orthogonal to the vectors in that basis. In other words, we now have

$$\mathbf{C} = [\mathbf{P}; \mathbf{D}] \tag{5-11-10}$$

It is easily verified by reference to our algorithm that

$$\mathbf{p}_\alpha = (1/S_{\alpha\alpha})\left(\mathbf{b}_\alpha - \sum_{i=1}^{\alpha-1} S_{i\alpha}\mathbf{p}_i\right) \qquad (\alpha = 1, 2, \ldots, l) \tag{5-11-11}$$

and

$$\mathbf{D} = \mathbf{E} - \sum_{i=1}^{l} S_{i,\,l+1}\,\mathbf{p}_i \tag{5-11-12}$$

Hence

$$\mathbf{b}_\alpha = \sum_{i=1}^{\alpha-1} S_{i\alpha}\mathbf{p}_i + S_{\alpha\alpha}\mathbf{p}_\alpha \tag{5-11-13}$$

and

$$\mathbf{E} = \mathbf{D} + \sum_{i=1}^{l} S_{i,\,l+1}\,\mathbf{p}_i \tag{5-11-14}$$

Comparing Eq. (13) to Eq. (4) yields:

$$\begin{aligned} A_{i\alpha} &= S_{i\alpha} \qquad (i = 1, 2, \ldots, \alpha - 1) \\ A_{\alpha\alpha} &= S_{\alpha\alpha} \\ A_{i\alpha} &= 0 \qquad (i = \alpha+1, \alpha+2, \ldots, l) \end{aligned} \tag{5-11-15}$$

Similarly, from Eq. (14) and Eq. (7) it appears that

$$t_i = S_{i,\,l+1} \qquad (i = 1, 2, \ldots, l) \tag{5-11-16}$$

The matrix **A** and the vector **t** are thus fully determined. The system of equations $\mathbf{Av} = \mathbf{t}$ can now be solved for **v** by successive substitutions:

$$v_l = t_l/A_{ll}$$
$$v_\alpha = \left(t_\alpha - \sum_{\beta=\alpha+1}^{l} A_{\alpha\beta} v_\beta\right) \bigg/ A_{\alpha\alpha} \qquad (\alpha = l-1, l-2, \ldots, 1) \quad (5\text{-}11\text{-}17)$$

It may happen that the **b** are not linearly independent. Suppose the rank of **B** (i.e., the number of linearly independent $\mathbf{b}_\alpha$) is $l_1 < l$. Then in $l - l_1$ of the iterations in the orthogonalization procedure it will turn out that $\mathbf{c}_k = \mathbf{0}$, hence $S_{kk} = 0$ and steps 4–6 cannot be carried through. The simplest solution is to leave **C** unchanged and set $s_{ki} = 0$ $(i = k+1,\ k+2,\ \ldots,\ l+1)$. It follows then that $l - l_1$ rows of **A** and the corresponding elements of **t** will be zero, and the corresponding v_α will be indeterminate according to Eq. (17). To these v_α we may assign arbitrary values, e.g., zero, and the remaining v_α are computed using Eq. (17).

The following is a simple numerical example. Let

$$\mathbf{B} = \begin{bmatrix} 2 & 12 \\ 6 & 19 \\ 15 & 8 \\ 16 & 10 \end{bmatrix}, \qquad \mathbf{E} = \begin{bmatrix} 14 \\ 4 \\ 9 \\ 13 \end{bmatrix}, \qquad \mathbf{\Pi} = \begin{bmatrix} 2 & 1 & 0 & 0 \\ 1 & 1 & 0 & 0 \\ 0 & 0 & 2 & 1 \\ 0 & 0 & 1 & 1 \end{bmatrix}$$

Hence follow the steps of the algorithm:

1.
$$\mathbf{C} = \begin{bmatrix} 2 & 12 & 14 \\ 6 & 19 & 4 \\ 15 & 8 & 9 \\ 16 & 10 & 13 \end{bmatrix}$$

2. $k = 1$
3. $S_{11} = 35.41186$
4. Replace the first column of **C** with $[0.056478,\ 0.169435,\ 0.423587,\ 0.451826]^{\mathrm{T}}$.
5. $S_{12} = 26.82717$, $S_{13} = 27.92849$
6. The second and third columns of **C** are replaced by

$$\begin{bmatrix} 10.48485 & 12.42265 \\ 14.45455 & -0.73206 \\ -3.36364 & -2.83014 \\ -2.12121 & 0.38118 \end{bmatrix}$$

7. $k = 2$
3. $S_{22} = 27.80833$
4. Replace the second column of **C** with $[0.37704,\ 0.51979,\ -0.12096,\ -0.07628]^{\mathrm{T}}$.
5. $S_{23} = 15.99368$

6. Replace the third column of $\mathbf{C}$ with $[6.39239, \ -9.04545, \ -0.89558, 1.60117]^T$.

Now we have $\mathbf{A} = \begin{bmatrix} 35.41186 & 26.82717 \\ 0 & 27.80832 \end{bmatrix}$, $\mathbf{t} = \begin{bmatrix} 27.92849 \\ 15.99368 \end{bmatrix}$, so that $\mathbf{v} = \begin{bmatrix} 0.35296 \\ 0.57514 \end{bmatrix}$. It is easily verified that this satisfies the normal equations

$$\begin{bmatrix} 1254 & 950 \\ 950 & 1493 \end{bmatrix} \mathbf{v} = \begin{bmatrix} 989 \\ 1194 \end{bmatrix}$$

In some cases, the normal equations take the form

$$(\mathbf{B}^T \mathbf{\Pi} \mathbf{B} + \mathbf{Q})\mathbf{v} = \mathbf{B}^T \mathbf{\Pi} \mathbf{E} + \boldsymbol{\phi} \tag{5-11-18}$$

where $\mathbf{Q}$ is a given positive definite matrix and $\boldsymbol{\phi}$ is a given vector. For instance:

1. In the Marquardt method, $\mathbf{Q}$ consists of the diagonal elements of $\mathbf{B}^T\mathbf{\Pi}\mathbf{B}$ multiplied by a scalar λ, and $\boldsymbol{\phi} = \mathbf{0}$.
2. If $\boldsymbol{\theta}$ has a normal prior distribution with covariance $\mathbf{V}_0$ and mean $\boldsymbol{\theta}_0$, then $\mathbf{Q} = \mathbf{V}_0^{-1}$ and $\boldsymbol{\phi} = \mathbf{V}_0^{-1}(\boldsymbol{\theta}_0 - \boldsymbol{\theta}_i)$.
3. If $\boldsymbol{\theta}$ is subject to inequality constraints and the penalty function method is used, then Eq. (6-1-9) supplies $\mathbf{Q}$ and Eq. (6-1-7) supplies $-\boldsymbol{\phi}$, both to be summed over all constraints.

By appending to the *nm* model equations a fictitious l additional equations one can reduce Eq. (18) to the normal regression form Eq. (2). Let $\mathbf{S}$ be a matrix such that $\mathbf{S}\mathbf{S}^T = \mathbf{Q}$ (e.g., the Cholesky decomposition). Define

$$\tilde{\mathbf{B}} \equiv \begin{bmatrix} \mathbf{B} \\ \mathbf{S}^T \end{bmatrix}, \qquad \tilde{\mathbf{E}} \equiv \begin{bmatrix} \mathbf{E} \\ \mathbf{S}^{-1}\boldsymbol{\phi} \end{bmatrix}, \qquad \tilde{\mathbf{\Pi}} \equiv \begin{bmatrix} \mathbf{\Pi} & \mathbf{0} \\ \mathbf{0} & \mathbf{I}_l \end{bmatrix}$$

One verifies easily that $\tilde{\mathbf{B}}^T\tilde{\mathbf{\Pi}}\tilde{\mathbf{B}} = \mathbf{B}^T\mathbf{\Pi}\mathbf{B} + \mathbf{Q}$, and $\tilde{\mathbf{B}}^T\tilde{\mathbf{\Pi}}\tilde{\mathbf{E}} = \mathbf{B}^T\mathbf{\Pi}\mathbf{E} + \boldsymbol{\phi}$. Therefore, performing a linear regression with $\tilde{\mathbf{B}}$, $\tilde{\mathbf{E}}$, and $\tilde{\mathbf{\Pi}}$ replacing $\mathbf{B}$, $\mathbf{E}$, and $\mathbf{\Pi}$ is equivalent to solving Eq. (18).

5-12. Variable Metric Methods

The Gauss method in its various forms is undoubtedly the best available for the solution of those problems to which it applies. When the objective function is not one of those shown in Table 5-1, however, the method may not be applicable. One of the so-called *variable metric* methods is recommended in such cases. The term variable metric methods was coined by Davidon (1959) to designate schemes in which the matrix $\mathbf{R}$ is systematically adjusted from iteration to iteration in such a way as to make it behave like $\mathbf{H}^{-1}$. These methods may be viewed as sophisticated finite difference schemes for computing the second derivatives of Φ. The specific scheme proposed by Davidon has been modified slightly by Fletcher and Powell (1963), and has been widely used in this form, gaining a reputation of being the most efficient general unconstrained optimization method available. This particular implementation was admittedly arbitrary, and subsequent papers have come forth with alternative implementations, e.g., Broyden (1967), Greenstadt (1970), Fiacco and McCormick (1968), Davidon (1968), Pearson (1969), Bard (1970), and Fletcher (1970).

Following a general introduction to these methods, we shall describe in detail the ROC and IROC variations which we have found (Bard, 1970) somewhat more efficient than others. This will be followed by a brief description of the Davidon–Fletcher–Powell method, which is well documented in the literature.

The main idea behind the variable metric methods is the following: From the definitions of the gradient $\mathbf{q}$ and the Hessian $\mathbf{H}$ we have

$$\mathbf{H}_i = (\partial \mathbf{q}/\partial \boldsymbol{\theta})_{\boldsymbol{\theta}=\boldsymbol{\theta}_i} \tag{5-12-1}$$

Therefore, to a first-order approximation

$$\mathbf{H}_i \boldsymbol{\sigma}_i = \boldsymbol{\eta}_i \tag{5-12-2}$$

where $\boldsymbol{\sigma}_i = \boldsymbol{\theta}_{i+1} - \boldsymbol{\theta}_i$, and $\boldsymbol{\eta}_i = \mathbf{q}_{i+1} - \mathbf{q}_i$. This means that

$$\boldsymbol{\sigma}_i = \mathbf{H}_i^{-1} \boldsymbol{\eta}_i \tag{5-12-3}$$

Suppose that before the ith iteration we have a matrix $\mathbf{A}_i$ which is an approximation to $\mathbf{H}_i^{-1}$. We wish to add to it a correction $\Delta\mathbf{A}_i$ in such a way that the resulting matrix $\mathbf{A}_{i+1}$ satisfies Eq. (3) when replacing $\mathbf{H}_i^{-1}$. That is, with

$$\mathbf{A}_{i+1} \equiv \mathbf{A}_i + \Delta\mathbf{A}_i \tag{5-12-4}$$

we require that

$$\boldsymbol{\sigma}_i = \mathbf{A}_{i+1} \boldsymbol{\eta}_i = \mathbf{A}_i \boldsymbol{\eta}_i + \Delta\mathbf{A}_i \boldsymbol{\eta}_i \tag{5-12-5}$$

Hence

$$\Delta\mathbf{A}_i \boldsymbol{\eta}_i = \mathbf{p}_i \tag{5-12-6}$$

where

$$\mathbf{p}_i = \boldsymbol{\sigma}_i - \mathbf{A}_i \boldsymbol{\eta}_i \tag{5-12-7}$$

Eq. (6) does not determine $\Delta\mathbf{A}_i$ uniquely, since it contains only l conditions for the $l(l+1)/2$ independent elements of the symmetric matrix $\Delta\mathbf{A}_i$.

The simplest possible matrix $\Delta\mathbf{A}_i$ is of rank one, i.e., it has the form

$$\Delta\mathbf{A}_i = \mathbf{r}_i\mathbf{r}_i^{\mathrm{T}} \tag{5-12-8}$$

where $\mathbf{r}_i$ is some vector. Substituting in Eq. (6) we obtain

$$\mathbf{r}_i\mathbf{r}_i^{\mathrm{T}}\boldsymbol{\eta}_i = \mathbf{p}_i \tag{5-12-9}$$

that is

$$\mathbf{r}_i = (1/\mathbf{r}_i^{\mathrm{T}}\boldsymbol{\eta}_i)\mathbf{p}_i = \alpha\mathbf{p}_i \tag{5-12-10}$$

where $\alpha \equiv (\mathbf{r}_i^{\mathrm{T}}\boldsymbol{\eta}_i)^{-1}$ is an unknown constant. Substituting in Eq. (9) we find

$$\alpha^2\mathbf{p}_i(\mathbf{p}_i^{\mathrm{T}}\boldsymbol{\eta}_i) = \mathbf{p}_i \tag{5-12-11}$$

Therefore

$$\alpha^2 = 1/\mathbf{p}_i^{\mathrm{T}}\boldsymbol{\eta}_i \tag{5-12-12}$$

Finally

$$\Delta\mathbf{A}_i = \mathbf{r}_i\mathbf{r}_i^{\mathrm{T}} = \alpha^2\mathbf{p}_i\mathbf{p}_i^{\mathrm{T}} = (1/\mathbf{p}_i^{\mathrm{T}}\boldsymbol{\eta}_i)\mathbf{p}_i\mathbf{p}_i^{\mathrm{T}} \tag{5-12-13}$$

Eq. (13) defines the *Rank One Correction* method (ROC). Broyden (1967), Davidon (1968), and Fiacco and McCormick (1968), have all proven the following:

Theorem Suppose $\Phi(\boldsymbol{\theta})$ is a quadratic function with a constant nonsingular Hessian matrix $\mathbf{H}$. Let $\boldsymbol{\theta}_1, \boldsymbol{\theta}_2, \ldots, \boldsymbol{\theta}_{l+1}$ be a set of points such that the vectors $\boldsymbol{\sigma}_i \equiv \boldsymbol{\theta}_{i+1} - \boldsymbol{\theta}_i$ $(i = 1, 2, \ldots, l)$ are linearly independent. Let $\mathbf{A}_1$ be an arbitrary symmetric matrix, and let $\mathbf{A}_i$ $(i = 2, 3, \ldots, l+1)$ be defined recursively by means of Eq. (4) and Eq. (13). Then, provided $\mathbf{p}_i^{\mathrm{T}}\boldsymbol{\eta}_i \neq 0$ for $i = 1, 2, \ldots, l$, we have

$$\mathbf{A}_{l+1} = \mathbf{H}^{-1} \tag{5-12-14}$$

The theorem says that if Φ is quadratic, the ROC method produces the exact inverse Hessian in l steps. Once the inverse Hessian is known, a single Newton step converges to the minimum. When Φ is not quadratic, one expects $\mathbf{A}_i$ $(i \geqslant l)$ to represent an approximation to $\mathbf{H}^{-1}$ evaluated somewhere in the region of the last l iterates. This should be particularly true near the minimum, where successive iterates lie close together. We expect the matrices $\mathbf{A}_i$ to converge to the value of $\mathbf{H}^{-1}$ at the minimum. Though no rigorous proof of this proposition has been found as yet, numerical tests have confirmed its validity.

Although the theorem holds in principle for arbitrary $\mathbf{A}_1$, numerical stability of the calculations [see Bard (1968)] requires that the elements of $\mathbf{A}_1$ have the right order of magnitude. A good choice is a diagonal matrix with

$$A_{1\alpha\alpha} = -\theta_{1\alpha}/q_{1\alpha} \tag{5-12-15}$$

Since $\mathbf{A}_i$ is an approximation to $\mathbf{H}_i^{-1}$, we would like to take $\mathbf{A}_i$ for $\mathbf{R}_i$. There is no guarantee, however, that $\mathbf{A}_i$ is positive definite. This may be remedied by means of a slightly modified Greenstadt procedure:

Let Eq. (A-5-21) be the scaled decomposition of $\mathbf{A}_i$, that is

$$\mathbf{A}_i = \sum_{j=1}^{l} \pi_j \mathbf{f}_j \mathbf{f}_j^{\mathrm{T}} \tag{5-12-16}$$

Let us define

$$\gamma_j = \max[\beta, \min(|\pi_j|, \gamma)] \tag{5-12-17}$$

where β and γ are small and large positive constant, respectively. Then

$$\mathbf{R}_i \equiv \sum_{j=1}^{l} \gamma_j \mathbf{f}_j \mathbf{f}_j^{\mathrm{T}} \tag{5-12-18}$$

is positive definite. It coincides with $\mathbf{A}_i$ when the latter is positive definite with all eigenvalues between β and γ.

Marquardt type corrections to the diagonal elements could, in principle, be also used for rendering $\mathbf{A}_i$ positive definite. This type of correction does not appear to work very well when applied to a matrix that is an approximation to the inverse, rather than to the Hessian itself. It happens, however, that a procedure entirely analogous to the ROC method can be used to construct an approximation to the Hessian directly. We call this method inverse rank one correction (IROC) (Bard, 1970). In this case we wish to satisfy $(\mathbf{A}_i + \Delta\mathbf{A}_i)\boldsymbol{\sigma}_i = \boldsymbol{\eta}_i$ [see Eq. (2)], and we are led to

$$\mathbf{A}_i = (1/\mathbf{s}_i^{\mathrm{T}}\boldsymbol{\sigma}_i)\mathbf{s}_i\mathbf{s}_i^{\mathrm{T}} \tag{5-12-19}$$

where

$$\mathbf{s}_i \equiv \boldsymbol{\eta}_i - \mathbf{A}_i\boldsymbol{\sigma}_i \tag{5-12-20}$$

We initialize $\mathbf{A}_1$ as the inverse of the matrix given by Eq. (15). The matrices $\mathbf{A}_i$ converge to $\mathbf{H}$ in the quadratic case. Since $\mathbf{A}_i$ is an approximation to $\mathbf{H}$, we can use the Cholesky decomposition with the Marquardt method to compute $\mathbf{v}_i$ efficiently, as described in Section 5-8.

In the Davidon–Fletcher–Powell method (DFP) (Davidon, 1959; Fletcher and Powell, 1963), the matrix $\Delta\mathbf{A}_i$ is of rank two, instead of rank one. The simplest such choice satisfying Eq. (6) is

$$\Delta\mathbf{A}_i = (1/\boldsymbol{\sigma}_i^{\mathrm{T}}\boldsymbol{\eta}_i)\boldsymbol{\sigma}_i\boldsymbol{\sigma}_i^{\mathrm{T}} - (1/\boldsymbol{\eta}_i^{\mathrm{T}}\mathbf{A}_i\boldsymbol{\eta}_i)\mathbf{A}_i\boldsymbol{\eta}_i\boldsymbol{\eta}_i^{\mathrm{T}}\mathbf{A}_i \qquad (5\text{-}12\text{-}21)$$

Suppose we choose $\boldsymbol{\sigma}_i = -\mu_i\mathbf{A}_i\mathbf{q}_i$, where μ_i is a positive value of ρ at which $\Phi(\boldsymbol{\theta}_i - \rho\mathbf{A}_i\mathbf{q}_i)$ attains a minimum. Fletcher and Powell have shown that under these conditions $\mathbf{A}_{i+1} = \mathbf{A}_i + \Delta\mathbf{A}_i$ is positive definite provided $\mathbf{A}_i$ was so. Therefore using $\mathbf{R}_i = \mathbf{A}_i$ always produces an acceptable step.

5-13. Step Size

In the preceeding sections we were concerned primarily with choosing the direction of the step taken in the ith iteration, that is, with the choice of $\mathbf{R}_i$. We shall now turn our attention to the determination of step size, i.e., to the choice of ρ_i. The methods that have been used fall into three categories:

1. $\rho_i = 1$. Required by Newton's method (to guarantee quadratic convergence near the minimum) and by Marquardt's method. In the latter case, the step size is determined indirectly through the choice of λ_i.
2. $\rho_i = \mu_i$. I.e., we proceed along the chosen direction to the point at which Φ ceases to decrease, as required by the Davidon–Fletcher–Powell method. Suitable methods of searching for μ_i are given by Fletcher and Powell (1963), Bard (1970), Goldfarb and Lapidus (1968), and others.
3. Interpolation–extrapolation, employed in conjunction with the Gauss and ROC methods. Here one expends a certain amount of effort on finding a good, acceptable value of ρ_i, without bothering to locate μ_i precisely.

It is true, on the whole, that the closer ρ_i is to μ_i, the smaller is the total number of iterations required. On the other hand, the more precisely we wish to determine the value of μ_i, the larger is the number of times that we must evaluate the objective function in each iteration. The difference between cases 2 and 3 is that in the former the best balance is struck when μ_i is determined with much greater precision than is required in the latter. In the succeeding section we suggest a simple algorithm for determining ρ_i in 3. This has worked with a reasonable degree of success, but there is no evidence that it is the most efficient possible. There is no end to the degree of ingenuity that may be expended on devising such algorithms.

In all cases the search for ρ_i proceeds without computation of derivatives. It would be wasteful to compute at each point $l + 1$ functions (Φ and the l components of its gradient) in order to conduct a one-dimensional search. The gradient is required only at the main iterates $\boldsymbol{\theta}_1, \boldsymbol{\theta}_2, \ldots$.

In the algorithm of the succeeding section it is assumed that at each iteration we are given an upper bound $\rho_{i,\max}$ on the feasible values of ρ_i. When inequality constraints need to be satisfied (see Section 6-1) $\rho_{i,\max}$ is the smallest positive value of ρ for which $\boldsymbol{\theta}_i - \rho \mathbf{R}_i \mathbf{q}_i$ lies on the boundary of the feasible region. If no inequality constraints apply, $\rho_{i,\max}$ can be chosen as an arbitrarily large number. We are also given a lower bound $\rho_{i,\min}$ (see Section 5-15). If no acceptable $\rho > \rho_{i,\min}$ can be found, the search is terminated.

5-14. Interpolation–Extrapolation

Assuming that we have chosen an acceptable direction, there always exists a number η_i such that if $0 < \rho < \eta_i$, then $\Psi_i(\rho) \equiv \Phi(\boldsymbol{\theta}_i - \rho \mathbf{R}_i \mathbf{q}_i) < \Phi_i$. The basic idea of the *interpolation* method is that if we have initially picked a value $\rho = \rho^{(0)}$ such that $\Psi_i(\rho^{(0)}) \geqslant \Phi_i$, we next try a smaller value of ρ, and keep repeating the process until an acceptable value is found. The idea behind *extrapolation* is that if our initial choice $\rho = \rho^{(0)}$ turned out acceptable, it pays to try at least one other value of ρ to see whether we cannot do even better. In both cases, the new trial value of ρ is chosen so as to minimize a quadratic approximation to $\Psi_i(\rho)$. We know that $\Psi_i(0) = \Phi_i$, and $d\Psi_i/d\rho)_{\rho=0} = -\mathbf{q}_i^{\mathrm{T}} \mathbf{R}_i \mathbf{q}_i$ (see Eq. (5-3-6).

Suppose we have computed $\Psi_i(\rho^{(0)})$. Let us define $\alpha \equiv \Phi_i$, $\beta \equiv \Psi_i(\rho^{(0)})$, $\gamma \equiv -\mathbf{q}_i^{\mathrm{T}} \mathbf{R}_i \mathbf{q}_i$, and let us try to find a quadratic function $a + b\rho + c\rho^2$ whose values match those of $\Psi_i(\rho)$ at $\rho = 0$ and $\rho = \rho^{(0)}$, and whose slope matches that of $\Psi_i(\rho)$ at $\rho = 0$. We have then:

$$a = \alpha \tag{5-14-1}$$

$$b = \gamma \tag{5-14-2}$$

$$a + b\rho^{(0)} + c\rho^{(0)2} = \beta \tag{5-14-3}$$

Whence

$$c = (\beta - \alpha - \gamma\rho^{(0)})/\rho^{(0)2} \tag{5-14-4}$$

The quadratic $a + b\rho + c\rho^2$ has a stationary point at

$$\rho^* = -b/2c = \gamma\rho^{(0)2}/2(\gamma\rho^{(0)} + \alpha - \beta) \tag{5-14-5}$$

The initial value of $\rho^{(0)}$ for each iteration is determined cautiously or optimistically depending on whether or not the previous iteration did or did not require interpolations. A detailed implementation of these ideas is given in the flowcharts of Fig. 5-2.

Entry: Given Φ_i, $\boldsymbol{\theta}_i$, $\mathbf{R}_i$, $\mathbf{q}_i$, $\rho_{i,\max}$, $\rho_{i,\min}$, J (an integer set $= 1$ in the first iteration), a ($= 0.5$ if penalty functions are used, $= 1$ otherwise).

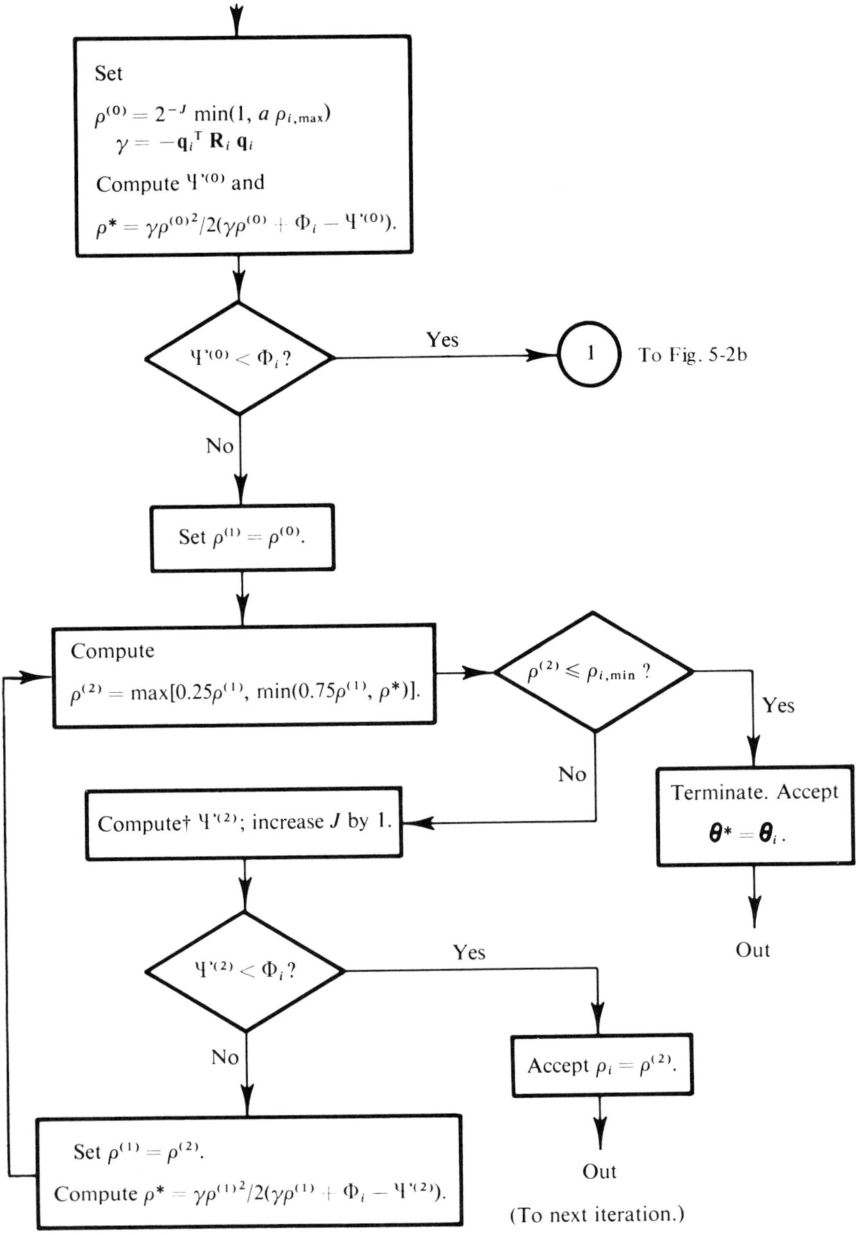

Notation:

$$\Psi'^{(j)} = \Psi'_i(\rho^{(j)}) = \Phi(\boldsymbol{\theta}_i - \rho^{(j)}\mathbf{R}_i\mathbf{q}_i)$$

Fig. 5-2a. Determination of step length, interpolation. † If the computation of $\Psi'^{(2)}$ is impossible (due, e.g., to an excessively large argument in an exponential) then increase J by 2, halve $\rho^{(2)}$, and try again.

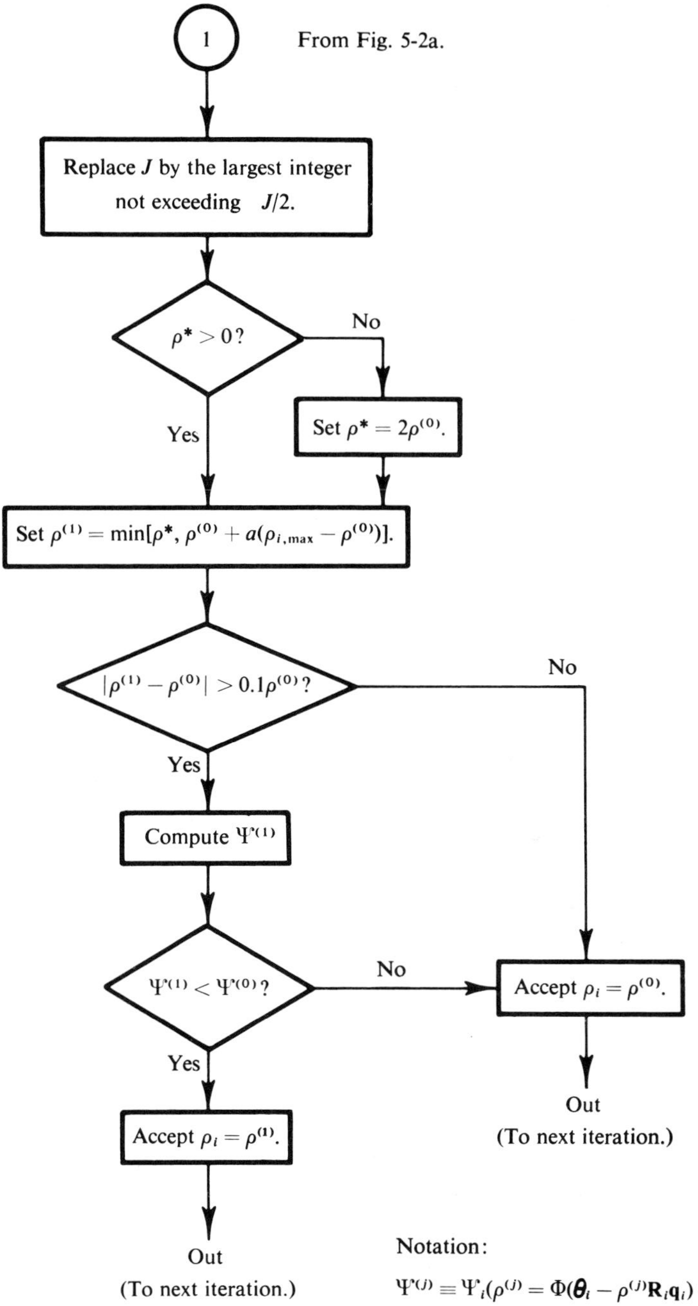

Fig. 5-2b. Determination of step length, extrapolation.

5-15. Termination

It is necessary to devise a criterion for stopping the iterative search for the minimum of $\Phi(\boldsymbol{\theta})$. As was stated before, all one can hope for is convergence to a stationary point of Φ. It may seem natural, therefore, to adopt the vanishing of the gradient as the termination criterion. Unfortunately, rounding errors and poor scaling often make the goal of a vanishing gradient unattainable even approximately. In many cases, the computer comes up with parameter values very close to the minimum, yet the gradient is still sizable. In addition, if perchance the algorithm fails to converge at all, a termination rule based entirely on the gradient leaves the program to iterate endlessly.

A more practical criterion dictates stopping as soon as further iterations fail to change the parameter values significantly. That is, given a set of small numbers ε_α $(\alpha = 1, 2, \ldots, l)$, we accept $\boldsymbol{\theta}_{i+1}$ as the solution $\boldsymbol{\theta}^*$ provided

$$|\theta_{i+1,\alpha} - \theta_{i,\alpha}| \leqslant \varepsilon_\alpha \qquad (\alpha = 1, 2, \ldots, l) \tag{5-15-1}$$

Where $\theta_{i,\alpha}$ is the αth component of $\boldsymbol{\theta}_i$. The numbers ε_α may either be prescribed in advance, or they may be computed by the program. In the latter case, following Marquardt (1963), we recommend

$$\varepsilon_\alpha = 10^{-4}(\theta_{i,\alpha} + 10^{-3}) \tag{5-15-2}$$

where the additive term 10^{-3} is designed to avoid embarrassment if θ_α happens to be nearly zero. This criterion has worked very well in practice. It tends to be on the conservative side, sometimes allowing a few more iterations than are strictly necessary. The rationale for the criterion is that, convergence or no convergence, it does not pay to keep iterating if the parameter values cease changing.

Suppose in the ith iteration a step direction $\mathbf{v}_i$ has been determined. Then Eq. (1) is satisfied if for each α, $\rho|v_{i,\alpha}| \leqslant \varepsilon_\alpha$, i.e., if $\rho \leqslant \min_\alpha[\varepsilon_\alpha/|v_{i,\alpha}|]$. Hence the minimum admissible ρ for the ith iteration is $\rho_{i,\min} = \min_\alpha[\varepsilon_\alpha/|v_{i,\alpha}|]$. As shown in Fig. 5-2*a*, termination occurs if the algorithm is forced to choose $\rho_i \leqslant \rho_{i,\min}$.

The above criterion does not offer an ironclad guarantee that the process will terminate in a finite number of steps. If the objective function is known to have a finite minimum, then termination can be guaranteed if we stop whenever $\Phi_{i-1} - \Phi_i < \varepsilon$ for some small prespecified positive number ε. That is, we stop as soon as no significant progress is made in reducing the value of the objective function. It may be safer, however, to require $\Phi_{i-l} - \Phi_i < \varepsilon$, i.e., to continue unless no significant progress has been made over a number of iterations. The variable metric methods are particularly liable to stall over a number of iterations, then to make sudden progress.

Finally, an upper bound may be placed on the number of iterations allowed. This should be coupled with a restart procedure to permit continuation with possibly a different algorithm.

Once the iterative process is terminated at $\boldsymbol{\theta} = \boldsymbol{\theta}^*$, one would like to know whether or not one has arrived at a minimum. We assume that we know the gradient $\mathbf{q}^* = \mathbf{q}(\boldsymbol{\theta}^*)$ and at least some approximation $\mathbf{H}^*$ to the Hessian $\mathbf{H}(\boldsymbol{\theta}^*)$. If we cut a cross section of the Φ surface along the θ_α axis, we have a curve whose approximate equation near $\boldsymbol{\theta}^*$ is given by

$$\Psi(\theta_\alpha) = \Phi^* + q_\alpha^*(\theta_\alpha - \theta_\alpha^*) + \tfrac{1}{2}H_{\alpha\alpha}^*(\theta_\alpha - \theta_\alpha^*)^2 \tag{5-15-3}$$

which has a stationary point at

$$\theta_\alpha = \theta_\alpha^* - q_\alpha^*/H_{\alpha\alpha}^* \tag{5-15-4}$$

The quantity $\delta_\alpha = |q_\alpha^*/H_{\alpha\alpha}^*|$ is therefore a measure of the error in the determination of θ^*. If each δ_α is small on the scale by which θ_α is measured, then it is likely that $\boldsymbol{\theta}^*$ is very close to a stationary point of Φ. Though not foolproof, this test works in most cases. Its reliability is improved if applied in the coordinate system of the canonical variables (see Section 7-3).

If $\mathbf{H}^*$ is indeed the Hessian of Φ at $\boldsymbol{\theta}^*$, then we may easily determine whether $\boldsymbol{\theta}^*$ (already known to be a stationary point) is really a minimum. All that is required is that $\mathbf{H}^*$ be positive definite (see Section 3-5), i.e., that all its eigenvalues be positive. When using the Gauss method, our approximation $\mathbf{N}^*$ is constructed so as to be automatically positive definite, regardless of whether or not $\mathbf{H}(\boldsymbol{\theta}^*)$ is so. In these cases, then, $\mathbf{N}^*$ contains no information pertaining to the nature of the point $\boldsymbol{\theta}^*$. Our only recourse is to explore directly the behavior of Φ around $\boldsymbol{\theta}^*$. In the ROC method, on the other hand, the matrix $\mathbf{R}_i$ from the last iteration may be a true approximation to $[\mathbf{H}(\boldsymbol{\theta}^*)]^{-1}$. We cannot prove that $\mathbf{R}_i$ is or is not positive definite when $\boldsymbol{\theta}^*$ is or is not a minimum; however, if $\mathbf{R}_i$ is not positive definite we suspect that $\boldsymbol{\theta}^*$ is not a minimum, and vice versa.

If one has reason to doubt that $\boldsymbol{\theta}^*$ is a minimum, one should restart the iterative procedure from a point close but not identical to $\boldsymbol{\theta}^*$. If convergence to the same $\boldsymbol{\theta}^*$ is obtained, this is likely to be at least a local minimum.

5-16. Remarks on Convergence

Suppose our process has converged to a point $\boldsymbol{\theta}^*$ at which Φ turns out to be not stationary. We assume that we have used a gradient method that is at least nominally acceptable. The directional derivative $-\mathbf{q}_i^{\mathrm{T}}\mathbf{R}_i\mathbf{q}_i$ of Φ in the direction of the vector $-\mathbf{R}_i\mathbf{q}_i$ is supposed to be negative, yet even a very small step in this direction fails to produce a decrease in Φ. The cause is probably one of the following two:

(a) The gradient is calculated with insufficient accuracy. If an incorrect vector $\mathbf{p}$ is used in place of $\mathbf{q}_i$ in the calculation of the direction, the true directional derivative is $-\mathbf{q}_i^{\mathrm{T}}\mathbf{R}_i\mathbf{p}$, which need not be negative even when $\mathbf{R}_i$ is positive definite. What is needed, then, is increased accuracy in the computation of the derivatives. This is why the derivatives should be computed from their analytic formulas. If this is impossible, one must revert to finite difference approximations. The question of finite differences is discussed in Section 5-18.

When the model equations are thought to be too complicated for analytic differentiation by hand, one may use the computer to perform this task. Computer programs to perform analytic differentiation are available. In fact, the FORMAC system (Bond *et al.*, 1964) is utilized by Eisenpress *et al.* (1966) to supply first and second analytic derivatives of nonlinear model equations in an implementation of the Newton-Greenstadt algorithm (see Eisenpress and Greenstadt (1966)) for parameter estimation of econometric models. For a survey of other analytic differentiation by computer schemes see Sammet (1966).

(b) The matrix $\mathbf{R}_i$ is not sufficiently positive definite, due to accumulation of rounding errors. This cannot happen with either the Gauss or ROC methods, as long as we use the discrimination or Marquardt trick to insure positive definiteness. It can, however, happen in ill-conditioned cases as a result of rounding errors when conventional matrix inversion or simultaneous equations solution methods are used. These are to be avoided.

We do not possess any methods that guarantee convergence to the global minimum. If we have computed a solution which we suspect is not the global minimum, we can restart the calculations from a radically different initial guess, and repeat the process until we are satisfied. Such procedures are rarely required in well-posed parameter estimation problems, that is, in problems where: (1) the errors in the data are not excessive, (2) the model fits the data well (with the proper values of the parameters), (3) the true parameter values are not outside the permitted range, and (4) the data were obtained from properly designed experiments (see Section 7-18 and Chapter X). When these conditions do not hold, almost anything can happen: the objective function may possess multiple minima, or may slope down asymptotically as certain parameters increase beyond bound. These things do occasionally happen even in well-posed problems, but not very frequently.

The reader should realize that the state of the art of nonlinear optimization is such that one cannot as yet write a computer program that will produce the correct answer to every parameter estimation problem in a single computer run. All too often, the first run produces unacceptable results. By studying these results one can perhaps obtain better starting guesses; one can choose to impose bounds or a prior distribution on the variables, or to relax previously imposed bounds; one can search for errors in the coding of the model equations or their derivatives. By careful coaxing, the computer may be made to yield acceptable results in subsequent runs. An interactive computer system can be particularly useful for this purpose.

5-17. Derivative Free Methods

We have dwelt at length on gradient methods because these have proven to be fastest and most reliable for a large number of problems.‡ This is not surprising; precise knowledge of the objective function gradient at any point immediately puts at our disposal the totality of downhill directions at that point. To test whether a given direction belongs to this class, all we need to do is verify that it forms an obtuse angle with the gradient. As we have remarked in the previous section, we lose this crucial ability as soon as the true gradient is replaced by an approximation.

Nevertheless, the burden of differentiating the model equations may at times prove too onerous, and while precise derivatives may be crucial in some problems, other (perhaps most) problems can be solved without them. We discuss below some of the methods of doing so.

5-18. Finite Differences

The most obvious, and in our experience most successful method for avoiding analytic differentiation of the objective function is to use a gradient method, with finite difference approximations supplying the required derivatives.

The simplest finite difference approximation to the gradient is given by the *one-sided difference* method

$$q_\alpha \approx \frac{\Phi(\theta_1, \theta_2, \ldots, \theta_\alpha + \delta\theta_\alpha, \ldots, \theta_l) - \Phi(\theta_1, \theta_2, \ldots, \theta_\alpha, \ldots, \theta_l)}{\delta\theta_\alpha} \qquad (\alpha = 1, 2, \ldots, l) \qquad (5\text{-}18\text{-}1)$$

‡ Colville (1968) reports this to be the case even with some highly constrained nonlinear programming problems.

Two sources of error contribute to the inaccuracy of q_α: (1) the rounding error arising when two closely spaced values of Φ are subtracted from each other, and (2) the *truncation error* due to the inexact nature of Eq. (1),which is accurate only in the limit as $\delta\theta_\alpha \to 0$.

The rounding error increases as $\delta\theta_\alpha$ decreases. We shall henceforth write $q_\alpha = [\Phi(\boldsymbol{\theta} + \delta\theta_\alpha) - \Phi(\boldsymbol{\theta})]/\delta\theta_\alpha$ as shorthand for Eq. (1). Let ε be the relative error in the computed values of Φ (at best $\varepsilon = 2^{-b}$ where b is the number of binary digits carried by the computer in use). The actual error in Φ has magnitude $\varepsilon|\Phi|$, and the error in $\Phi(\boldsymbol{\theta} + \delta\theta_\alpha) - \Phi(\boldsymbol{\theta})$ can be as high as $2\varepsilon|\Phi|$, although the root-mean-square error is only $\sqrt{2}\,\varepsilon|\Phi|$. The maximum rounding error in Eq. (1) is, therefore

$$\delta_{R,\alpha} \equiv 2\varepsilon|\Phi|/|\delta\theta_\alpha| \tag{5-18-2}$$

On the other hand, we have the Taylor series expansion

$$\Phi(\boldsymbol{\theta} + \delta\theta_\alpha) = \Phi(\boldsymbol{\theta}) + q_\alpha\,\delta\theta_\alpha + \tfrac{1}{2}H_{\alpha\alpha}\,\delta\theta_\alpha{}^2 + \cdots \tag{5-18-3}$$

The truncation error in Eq. (1) is, therefore, approximately

$$\delta_{T,\alpha} \equiv \tfrac{1}{2}|H_{\alpha\alpha}|\,|\delta\theta_\alpha| \tag{5-18-4}$$

The maximum total error is approximately

$$\delta_\alpha \equiv \delta_{R,\alpha} + \delta_{T,\alpha} = 2\varepsilon|\Phi/\delta\theta_\alpha| + \tfrac{1}{2}|H_{\alpha\alpha}\,\delta\theta_\alpha| \tag{5-18-5}$$

This has a minimum at

$$|\delta\theta_\alpha| = (4\varepsilon|\Phi/H_{\alpha\alpha}|)^{1/2} \tag{5-18-6}$$

If we are interested in the mean square error instead, we would minimize

$$2\varepsilon^2\Phi^2/\delta\theta_\alpha{}^2 + H_{\alpha\alpha}^2\,\delta\theta_\alpha{}^2/4$$

so that

$$|\delta\theta_\alpha| = (2\sqrt{2}\,\varepsilon|\Phi/H_{\alpha\alpha}|)^{1/2} \tag{5-18-7}$$

Eq. (6) or Eq. (7) can be used as a basis for estimating the step sizes $\delta\theta_\alpha$ ($\alpha = 1, 2, \ldots, l$) required for computing the differences. The same equations could have been obtained by requiring that the two error sources contribute equally. In the case of Eq. (6) the total maximum error turns out to be $2(\varepsilon|\Phi H_{\alpha\alpha}|)^{1/2}$, whereas the root-mean-square error attendant upon Eq. (7) is $2^{1/4}(\varepsilon|\Phi H_{\alpha\alpha}|)^{1/2}$.

To apply these formulas we need estimates of the Hessian $\mathbf{H}$. These are available in the Gauss and variable metric methods. In the latter, we usually have $\mathbf{A} \approx \mathbf{H}^{-1}$ rather than $\mathbf{H}$. However, if we start out with a diagonal $\mathbf{A}_1$ we can form $\mathbf{A}_1^{-1}$ easily. Then the easily verifiable formula

$$\mathbf{A}_{i+1}^{-1} = \mathbf{A}_i^{-1} - (1/\boldsymbol{\sigma}_i{}^{\mathrm{T}}\mathbf{A}_i^{-1}\mathbf{p}_i)\mathbf{A}_i^{-1}\mathbf{p}_i\mathbf{p}_i{}^{\mathrm{T}}\mathbf{A}_i^{-1} \tag{5-18-8}$$

enables us in the ROC method to compute successively the matrices $\mathbf{A}_i^{-1}$ which are approximations to $\mathbf{H}$. A similar procedure for the Davidon–Fletcher–Powell method is given by Stewart (1967).

In the Gauss method we need the derivatives of the individual model equations, rather than of the objective function directly. In the absence of anything better, we would still recommend using Eq. (6) or Eq. (7) for choosing $\delta\theta_\alpha$. In place of Eq. (1), however, we would apply similar equations to the model equations for each experiment in turn.

These formulas should not be used blindly. Gross errors in the estimated $\mathbf{H}$ may lead to absurd values of $\delta\theta_\alpha$. Lower and upper limits should be imposed on the $\delta\theta_\alpha$, e.g., $10^{-5}|\theta_\alpha| \leqslant |\delta\theta_\alpha| \leqslant 10^{-2}|\theta_\alpha|$. A smaller lower bound would be appropriate if the calculations are performed in double precision.

Eq. (1) represents the crudest possible estimate of q_α. A better estimate is given by the *central difference* scheme

$$q_\alpha \approx \frac{\Phi(\theta_1, \theta_2, \ldots, \theta_\alpha + \delta\theta_\alpha, \ldots, \theta_l) - \Phi(\theta_1, \theta_2, \ldots, \theta_\alpha - \delta\theta_\alpha, \ldots, \theta_l)}{2\,\delta\theta_\alpha} \quad (\alpha = 1, 2, \ldots, l) \qquad \text{(5-18-9)}$$

Unfortunately, this scheme requires computation of two additional function values for each gradient component, instead of the one required by the one-sided difference. The truncation error of Eq. (9) is $\Phi_\alpha^{(4)}\,\delta\theta_\alpha{}^3/24$, where $\Phi_\alpha^{(4)} \equiv \partial^4\Phi/\partial\theta_\alpha{}^4$. The step with least maximum error has length $|\delta\theta_\alpha| = (16\varepsilon|\Phi|/|\Phi_\alpha^{(4)}|)^{1/4}$, and the attendant error has magnitude $\frac{4}{3}[\varepsilon^3|\Phi|^3\,|\Phi_\alpha^{(4)}|)^{1/4}$. These formulas are not very useful, since we rarely know $\Phi_\alpha^{(4)}$. However, it is safe to say that somewhat larger values of $\delta\theta_\alpha$ may be used here than with the one-sided difference scheme.

For economy of calculation we suggest that one-sided differences be used for several iterations, until no further progress can be made. Then one may switch to central differences if one feels that the solution has not been attained.

5-19. Direct Search Methods

The term *direct search* was coined by Hooke and Jeeves (Hooke and Jeeves, 1961). It has come to be applied to methods which (like Hooke and Jeeves') search for the minimum without explicit evaluation of derivatives, analytic or numerical. The idea of direct search methods is appealing, and they have performed well in certain cases [see, e.g., the survey by Box (Box, 1966)]. Our own experience, however, has been disappointing; gradient methods, even using finite difference approximations, have outperformed direct search methods on all but the most trivial parameter estimation problems, both in reliability and speed of convergence. For this reason we shall mention a few of the more promising or popular methods, but not describe any of them in detail.

The methods that performed best in the Box (1966) survey were those due to Powell (1964, 1965). The first one minimizes an arbitrary function; it has been amended by Zangwill (1967b), who also describes a method of his own. The second Powell method (Powell, 1965) is designed specifically for minimizing a sum of squares, but can be adapted easily to other problems which admit the Gauss approximation. This algorithm is related to the Gauss method, with finite differences taken along the search directions (instead of along the coordinate directions, as would be the case with the usual finite difference version of the Gauss method). The weakness of the method derives from the fact that the differences are taken only in a single direction per iteration, so that one's estimated derivatives in all other directions are perennially out of date. This effect worsens as the dimension of the parameter vector increases.

Other methods that have found considerable use in solving optimization problems are those of Hooke and Jeeves (1961), Rosenbrock (1960) [see also Rosenbrock and Storey (1966)], Buzzi Ferraris (1968), Brent (1971), and the Simplex method of Nelder and Mead (1965). The latter method was adapted to least squares problems by Spendley (1969). A review of direct search methods appears in Fletcher (1965).

5-20. The Initial Guess

All the optimization methods that we have described require that one supply an initial guess $\boldsymbol{\theta}_1$ for the values of the parameters. The choice of a good initial guess can spell the difference between success and failure in locating the optimum, or between rapid and slow convergence to the solution. Unfortunately, while we can prescribe algorithms for proceeding from the initial guess, we must rely heavily on intuition and prior knowledge in selecting the initial guess. Nevertheless, we can provide some suggestions which may be helpful in many cases. A comprehensive discussion of such methods can be found in Kittrell *et al.* (1965).

At the outset we must caution the reader not to exaggerate the importance of finding a good initial guess. In many cases the proper solution has been obtained starting from the first initial guess that came to mind. In these cases, at the possible expense of a few additional minutes of computer time, one has saved oneself a considerable amount of trouble. We suggest, therefore, that (unless computer time is exceptionally scarce or expensive) one attempt to estimate the parameters "by brute force". Only if this strategy fails should one resort to more delicate techniques.

The most obvious method for making the initial guesses is by the use of prior information. Estimates calculated from previous experiments, known values from similar systems, values computed from theoretical considerations: all these form ideal initial guesses.

On the opposite end of the spectrum stand problems in which our only information concerning the parameter values is given in the form of upper and lower bounds on their values. If we do not even have such bounds, we can transform our variables into bounded ones; e.g., a positive variable θ can be replaced by the bounded variable $\phi = 1/(1 + \theta)$, or a completely free variable θ may be replaced by $\phi = \arctan \theta$.

Once we have all our parameters confined to a rectangular region in $\boldsymbol{\theta}$ space, we can conduct a grid search: compute the value of the objective function at every point on a regular rectangular grid, and choose the point with the best value as the initial guess. The main difficulty with this approach is "the curse of dimensionality"; in a grid with k levels in each one of the l dimensions, the total number of points at which the objective function must be evaluated is k^l. This is a prohibitively large number for all but the smallest values of k and l.

An alternative to the grid search is random search. Here a number of points within the feasible region are chosen at random, and the one giving the best value of the objective function is used as the initial guess. It is true that among a hundred points there is a good chance of finding one that is within 1% of the solution. However, this 1% applies to the volume of the feasible region. If there are l parameters, the relative accuracy of each parameter is only $0.01^{1/l}$, or 31.6% of the permitted range when $l = 4$. The random search method does not overcome the curve of dimensionality, but it does offer some advantages over grid search. One may bias the sampling so as to favor certain regions of parameter space (this can be regarded as sampling from a prior distribution), and one may use a sequential termination criterion: stop sampling as soon as a function value significantly better than the average has been found. Sometimes, a transformation of variables is called for prior to commencement of the search. For instance, if even the order of magnitude of a parameter is unknown, it should be replaced by its logarithm.

It is not always necessary to provide initial guesses for all the parameters in a model. If some of the parameters enter the model equations linearly, and an initial guess is provided for the other parameters, then the linear parameters can be estimated by linear multiple regression. Suppose, for instance, that the model has the form $\hat{y}_\mu = \theta_1 \exp(-\theta_2 x)$. If we have the initial guess $\theta_2 = 6$, and let $z_\mu \equiv \exp(-6x_\mu)$, then an initial guess for θ_1 can be found by solving the linear least squares problem $\min \sum_\mu (y_\mu - \theta_1 z_\mu)^2$. Special versions of the Gauss method to deal with partly linear models have been devised (Lawton and Sylvestre, 1971; Golub and Pereyra, 1972).

The most fruitful approach to finding an initial guess is to substitute a simpler problem for the original estimation problem. The answers to the simpler problem can be used as initial guesses for the original problem. There is no systematic way of applying this idea to all problems, but the following is a partial list of what may be attempted.

(a) Linearization. We try, by means of transformation of variables, to change the model equations into ones that are linear in the parameters (see Section 4-19). The linear problem can be solved by multiple linear regression with no need for an initial guess.

(b) Multistage Estimation. By breaking up the data into groups, we may estimate certain auxiliary parameters for each group; then we estimate the original parameters as functions of the auxiliary parameters. For example, the rate of a chemical reaction is given by the expression

$$y = kx \exp(-E/T) \qquad \text{(5-20-1)}$$

where y is the rate, x the concentration, T the temperature, and k and E the parameters to be estimated. The rate y is measured as a function of x at several values of T, say $T_1, T_2, \ldots, T_q$. Suppose we use the data taken at T_i to estimate the coefficient K_i in the equation

$$y = K_i x \qquad (i = 1, 2, \ldots, q) \qquad \text{(5-20-2)}$$

The estimated K_i can then be used as data for estimating $\log k$ and E in the linearized model

$$\log K_i = \log k - E/T_i \qquad (i = 1, 2, \ldots, q) \qquad \text{(5-20-3)}$$

Of course, in this case we could have linearized Eq. (1) directly; however, in kinetics models involving simultaneous reactions, the original equations cannot be linearized, whereas the multistage procedure still applies.

(c) Model Simplification. It is frequently possible to approach the final model through a sequence of simpler ones, in which various effects are neglected and the corresponding parameters suppressed. After the parameters have been estimated for a simple model, analysis of the residuals (see Section 7-13) can provide an indication as to what terms should be added to the model next. This method serves not only to obtain initial guesses for a given model, but also (and perhaps more importantly) to synthesize a final model where none is given. Examples for such syntheses are given by Box and Youle (1955), Peterson (1962), Box and Hunter (1962), Hunter and Mezaki (1964), and Kittrell, Hunter, and Mezaki (1966).

(d) Simpler Estimation Method. We replace the proper objective function by one which is easier to minimize. For instance: (1) We linearize the model as under (a) above. (2) In a multi-equation model, we use one of the equations to obtain preliminary estimates of the parameters. It is true, however, that sometimes it is easiest to obtain estimates when all equations are used simultaneously, since otherwise the information relating to the values of some of the parameters is lost. To give a trivial example, let the three model equations be

$$y_1 = (\theta_1 + \theta_2)x, \qquad y_2 = (\theta_2 + \theta_3)x, \qquad y_3 = (\theta_3 + \theta_1)x$$

Clearly, the three parameters can be estimated independently only if all three equations are used. (3) In dynamic models we may use easy to apply data integration or differentiation methods to obtain initial guesses for the integration of equations method (see Section 8-1).

5-21. A Single-Equation Least Squares Problem‡

Let y be the fraction remaining at time x_1 of a chemical compound A undergoing the first order reaction

$$A \rightarrow B \tag{5-21-1}$$

The variable y satisfies the differential equation

$$dy/dx_1 = -ky \tag{5-21-2}$$

where k is the rate constant. The solution to this equation with the initial condition $y = 1$ at $x_1 = 0$ is

$$y = \exp(-kx_1) \tag{5-21-3}$$

The rate constant k depends on the absolute temperature x_2 as follows

$$k = \theta_1 \exp(-\theta_2/x_2) \tag{5-21-4}$$

‡ The numerical results quoted in the discussion of this and subsequent problems were obtained as output from calculations performed in single precision floating point arithmetic on an IBM System/360 computer. The results were converted to decimal from a binary representation inside the computer. Therefore, the results of performing the same calculations on a decimal desk calculator (or, for that matter, on any other computer, or even using a different program on the same computer) would differ slightly from those presented here. In a long iterative procedure such differences can build up to such an extent that a different number of iterations may be required to reach substantially the same end result.

where θ_1 is the so-called frequency constant, and θ_2 is the activation energy (expressed in suitable units). Our model equation takes the form

$$y = f(x_1, x_2, \theta_1, \theta_2) = \exp[-\theta_1 x_1 \exp(-\theta_2/x_2)] \qquad (5\text{-}21\text{-}5)$$

Data for a set of fifteen observations on **x** and y are given in Table 5-2.

Table 5-2
Data for Least Squares Problem

Experiment number, μ	Time, $x_{\mu 1}$(hr)	Temperature, $x_{\mu 2}$(°K)	Fraction A remaining, y_μ
1	0.1	100	0.980
2	0.2	100	0.983
3	0.3	100	0.955
4	0.4	100	0.979
5	0.5	100	0.993
6	0.05	200	0.626
7	0.1	200	0.544
8	0.15	200	0.455
9	0.2	200	0.225
10	0.25	200	0.167
11	0.02	300	0.566
12	0.04	300	0.317
13	0.06	300	0.034
14	0.08	300	0.016
15	0.1	300	0.066

Our aim is to estimate θ_1 and θ_2. As far as we know, the errors in the $\mathbf{x}_\mu$ are negligible, whereas those of the y_μ are all independent and with equal standard deviations. The least squares criterion is, therefore, appropriate. We seek to minimize

$$\Phi(\boldsymbol{\theta}) = \sum_{\mu=1}^{15} e_\mu^{\,2}(\boldsymbol{\theta}) = \sum_{\mu=1}^{15} [y_\mu - f_\mu(\boldsymbol{\theta})]^2 \qquad (5\text{-}21\text{-}6)$$

Following Eq. (5-9-2), we find that the gradient of Φ is given by:

$$q_1 \equiv \partial\Phi/\partial\theta_1 = -2\sum_{\mu=1}^{15} e_\mu\, \partial f_\mu/\partial\theta_1 = 2\sum_{\mu=1}^{15} e_\mu f_\mu \exp(-\theta_2/x_{\mu 2}) x_{\mu 1} \qquad (5\text{-}21\text{-}7)$$

$$q_2 \equiv \partial\Phi/\partial\theta_2 = -2\sum_{\mu=1}^{15} e_\mu\, \partial f_\mu/\partial\theta_2 = -2\sum_{\mu=1}^{15} e_\mu (\theta_1 x_{\mu 1}/x_{\mu 2}) f_\mu \exp(-\theta_2/x_{\mu 2}) \qquad (5\text{-}21\text{-}8)$$

The approximate Hessian **N** is given by Eq. (5-9-4)

$$N_{\alpha\beta} = 2\sum_{\mu=1}^{15} (\partial f_\mu/\partial\theta_\alpha)(\partial f_\mu/\partial\theta_\beta) \qquad (\alpha, \beta = 1, 2) \tag{5-21-9}$$

Let our initial guess be

$$\boldsymbol{\theta}_1 = \begin{bmatrix}\theta_{1,1}\\ \theta_{1,2}\end{bmatrix} = \begin{bmatrix}750\\ 1200\end{bmatrix}$$

Table 5-3 gives the values of the f_μ, e_μ, and $\partial f_\mu/\partial\boldsymbol{\theta}$ for $\boldsymbol{\theta} = \boldsymbol{\theta}_1$. From Table 5-3 entries and Eqs. (6)–(9) one easily calculates

$$\Phi_1 = 1.090441$$

$$\mathbf{q}_1 = \begin{bmatrix}-0.002230450\\ 0.006863795\end{bmatrix}, \qquad \mathbf{N}_1 = \begin{bmatrix}0.2689478 & -0.7730614\\ -0.7730614 & 2.310325\end{bmatrix} \times 10^{-5}$$

Table 5-3
Least Squares Problem Functions at $\boldsymbol{\theta}^T = [750, 1200]$

μ	f_μ	$e_\mu = y_\mu - f_\mu$	$10^6 \times \partial f_\mu/\partial\theta_1$	$10^5 \times \partial f_\mu/\partial\theta_2$
1	0.9995393	−0.0195393	−0.6141379	0.4606032
2	0.9990788	−0.0160788	−1.227710	0.9207821
3	0.9986185	−0.0436185	−1.840716	1.380537
4	0.9981585	−0.0191585	−2.453158	1.839868
5	0.9976986	−0.0046986	−3.065035	2.298776
6	0.9112362	−0.2852362	−112.9364	42.35113
7	0.8303515	−0.2863515	−205.8234	77.18373
8	0.7566463	−0.3016463	−281.3304	105.4990
9	0.6894834	−0.4644834	−341.8115	128.1793
10	0.6282821	−0.4612821	−389.3387	146.0020
11	0.7597739	−0.1937739	−278.3146	69.57867
12	0.5772563	−0.2602563	−422.9124	105.7281
13	0.4385841	−0.4045841	−481.9767	120.4942
14	0.3332248	−0.3172248	−488.2577	122.0644
15	0.2531757	−0.1871757	−463.7071	115.9268

To determine the first step direction $\mathbf{v}_1$ we must compute $\mathbf{v}_1 = -\mathbf{N}^{-1}\mathbf{q}_1$, i.e., solve the set of simultaneous equations $\mathbf{N}_1\mathbf{v}_1 = -\mathbf{q}_1$. In our two-dimensional problem this can be done trivially on a desk calculator. However, for purposes of illustration, we shall apply the Greenstadt method. For this, we need to compute the inverse scaled decomposition of **N** (we omit the subscript 1). We follow the steps outlined in Section A-5:

1. $N_{11}^{1/2} = 0.001639963, \qquad N_{22}^{1/2} = 0.004806584.$

Therefore

$$\mathbf{N} = \mathbf{BCB} = \begin{bmatrix} 0.001639963 & 0 \\ 0 & 0.004806584 \end{bmatrix} \begin{bmatrix} 1 & -0.980716 \\ -0.980716 & 1 \end{bmatrix}$$
$$\times \begin{bmatrix} 0.001639963 & 0 \\ 0 & 0.004806584 \end{bmatrix}$$

2. The matrix $\mathbf{C}$ has the form $\begin{bmatrix} 1 & -a \\ -a & 1 \end{bmatrix}$. The reader may verify that such a matrix has eigenvalues $1 + a$ and $1 - a$ with corresponding eigenvectors $[1/\sqrt{2}, -1/\sqrt{2}]$ and $[1/\sqrt{2}, 1/\sqrt{2}]$ respectively. Hence the eigenvalue decomposition of $\mathbf{C}$ is given by

$$\mathbf{C} = \mathbf{U\Pi U}^{\mathrm{T}} = \begin{bmatrix} 1/\sqrt{2} & 1/\sqrt{2} \\ -1/\sqrt{2} & 1/\sqrt{2} \end{bmatrix} \begin{bmatrix} 1.980716 & 0 \\ 0 & 0.019284 \end{bmatrix} \begin{bmatrix} 1/\sqrt{2} & -1/\sqrt{2} \\ 1/\sqrt{2} & 1/\sqrt{2} \end{bmatrix}$$

3. The inverse decomposition of $\mathbf{N}$ is given by $\mathbf{N}^{-1} = \mathbf{G\Pi}^{-1}\mathbf{G}^{\mathrm{T}}$ where

$$\mathbf{G} = \mathbf{B}^{-1}\mathbf{U} = \begin{bmatrix} (1/\sqrt{2})1/0.001639963 & (1/\sqrt{2})1/0.001639963 \\ -(1/\sqrt{2})1/0.004806584 & (1/\sqrt{2})1/0.004806584 \end{bmatrix}$$
$$= \begin{bmatrix} 431.1723 & 431.1723 \\ -147.1122 & 147.1122 \end{bmatrix}$$
$$\mathbf{\Pi}^{-1} = \begin{bmatrix} 1/1.980716 & 0 \\ 0 & 1/0.019284 \end{bmatrix} = \begin{bmatrix} 0.504868 & 0 \\ 0 & 51.8573 \end{bmatrix}$$

Hence

$$\mathbf{N}^{-1} = \begin{bmatrix} 431.1723 & 431.1723 \\ -147.1122 & 147.1122 \end{bmatrix} \begin{bmatrix} 0.504868 & 0 \\ 0 & 51.8573 \end{bmatrix} \begin{bmatrix} 431.1723 & -147.1122 \\ 431.1723 & 147.1122 \end{bmatrix}$$

The ratio of eigenvalues was $1.980716/0.019284 \approx 100$, indicating that $\mathbf{N}$ is mildly ill-conditioned. We can tolerate, however, eigenvalue ratios of up to 10^4 or 10^5, hence there is no need to adjust the value of the smaller eigenvalue.

4. To compute $-\mathbf{N}^{-1}\mathbf{q}$ we proceed as follows:

$$\begin{bmatrix} 431.1723 & -147.1122 \\ 431.1723 & 147.1122 \end{bmatrix} \begin{bmatrix} 0.002230450 \\ -0.006863795 \end{bmatrix} = \begin{bmatrix} 1.971456 \\ -0.04803973 \end{bmatrix}$$

$$\begin{bmatrix} 0.5048679 & 0 \\ 0 & 51.85727 \end{bmatrix} \begin{bmatrix} 1.971456 \\ -0.04803973 \end{bmatrix} = \begin{bmatrix} 0.9953249 \\ -2.491209 \end{bmatrix}$$

$$\mathbf{v}_1 = \begin{bmatrix} 431.1723 & 431.1723 \\ -147.1122 & 147.1122 \end{bmatrix} \begin{bmatrix} 0.9953249 \\ -2.491209 \end{bmatrix} = \begin{bmatrix} -644.9785 \\ -512.9099 \end{bmatrix}$$

Along the ray $\boldsymbol{\theta} = \boldsymbol{\theta}_1 + \rho \mathbf{v}_1$, the directional derivative at $\rho = 0$ is

$$\begin{aligned}\partial \Psi / \partial \rho = \mathbf{v}_1{}^{\mathrm{T}} \mathbf{q}_1 &= (-644.9785)(-0.002230450) + (-512.9099)(0.006863795) \\ &= -2.081916\end{aligned}$$

This quantity being negative confirms the fact that Φ decreases, at least initially, as one proceeds from $\boldsymbol{\theta}_1$ in this direction.

Trying initially $\rho^{(0)} = 1$ we arrive at

$$\boldsymbol{\theta}^{(0)} = \begin{bmatrix} 750 - 644.9785 \\ 1200 - 512.9099 \end{bmatrix} = \begin{bmatrix} 105.0215 \\ 687.0901 \end{bmatrix}$$

where $\Phi(\boldsymbol{\theta}^{(0)}) = 0.9133969 < \Phi_1$. This indicates that we are at an acceptable point. We try to find an even better point by fitting a parabola to $\Phi(\boldsymbol{\theta}_1 + \rho \mathbf{v}_1)$. The equation of the parabola is

$$\Psi(\rho) = a + b\rho + c\rho^2$$

And it must assume the following values:

$$\begin{aligned}\Psi(0) &= 1.090441 = \alpha \\ \Psi(1) &= 0.9133969 = \beta \\ d\Psi/d\rho)_{\rho=0} &= -2.081916 = \gamma\end{aligned}$$

Using Eq. (5-14-5) we find that $\Psi(\rho)$ has a stationary point at

$$\begin{aligned}\rho^* &= -2.081916 \times 1/2(-2.081916 \times 1 + 1.090441 - 0.9133969) \\ &= 0.5464714\end{aligned}$$

Trying $\boldsymbol{\theta}^{(1)} = \boldsymbol{\theta}_1 + \rho^* \mathbf{v}_1 = [397.5376,\ \ 919.7092]^{\mathrm{T}}$ we find $\Phi(\boldsymbol{\theta}^{(1)}) = 0.3345645$ which is a great improvement over both $\boldsymbol{\theta}_1$ and $\boldsymbol{\theta}^{(0)}$. We accept $\boldsymbol{\theta}^{(1)}$ then as $\boldsymbol{\theta}_2$, the starting point for the next iteration.

A computer program using the flowchart of Fig. 5-2 produced the sequence of iterations given in Table 5-4. No further reduction of $\Phi(\boldsymbol{\theta})$ was obtained

Table 5-4
Least Squares Problem, Good Initial Guess (Gauss Iterations)

i	$\Phi(\boldsymbol{\theta}_i)$	$\boldsymbol{\theta}_i$
1	1.090441	$[750,\ 1200]^{\mathrm{T}}$
2	0.3345645	$[397.5376,\ 919.7092]^{\mathrm{T}}$
3	0.05765885	$[646.0847,\ 938.5288]^{\mathrm{T}}$
4	0.04038005	$[810.6260,\ 965.7625]^{\mathrm{T}}$
5	0.03980731	$[818.3628,\ 962.1228]^{\mathrm{T}}$
6	0.03980599	$[813.4583,\ 960.9063]^{\mathrm{T}}$

after six iterations, so that we took as our estimate $\boldsymbol{\theta}^* = [813.4583, 960.9063]^T$ with $\Phi^* = 0.03980599$. At this point, the gradient was

$$\mathbf{q}^* = \begin{bmatrix} -0.218524 \\ 0.631308 \end{bmatrix} \times 10^{-6}$$

and the approximate Hessian

$$\mathbf{N}^* = \begin{bmatrix} 0.271890 & -0.957336 \\ -0.957336 & 3.50371 \end{bmatrix} \times 10^{-5}$$

Applying the test of Section 5-15 we find that

$$\delta_1 = |q_1^*/N_{11}^*| \approx 0.1, \qquad \delta_2 = |q_2^*/N_{22}^*| \approx 0.02$$

These values are negligible compared to θ_1^* and θ_2^*, so we may assume that we have converged to a stationary point. The final residuals corresponding to this solution are given in Table 5-5.

Table 5-5
Least Squares Problem, Final Residuals

μ	$e_\mu^* = y_\mu - f(\mathbf{x}_\mu, \boldsymbol{\theta}^*)$	μ	$e_\mu^* = y_\mu - f(\mathbf{x}_\mu, \boldsymbol{\theta}^*)$
1	−0.0145552	9	−0.0387225
2	−0.00613993	10	−0.0219878
3	−0.0287542	11	0.0497515
4	0.000602186	12	0.0504873
5	0.0199295	13	−0.103587
6	−0.0906165	14	−0.0550289
7	0.0304608	15	0.0293314
8	0.0869893		

In the preceding calculations we were fortunate enough to have started from a good initial guess; $\boldsymbol{\theta}_1 = [750, 1200]^T$ as compared to the final estimate $\boldsymbol{\theta}^* = [813.4583, 960.9063]^T$. Suppose now that we had started from the much poorer initial guess $\boldsymbol{\theta}_1 = [100, 2000]^T$. Proceeding as before we find

$$\Phi_1 = 5.299502$$

$$\mathbf{q}_1 = \begin{bmatrix} -0.0007098080 \\ 0.0002442936 \end{bmatrix} \tag{5-21-10}$$

$$\mathbf{N}_1 = \begin{bmatrix} 0.7036033 & -0.2354773 \\ -0.2354773 & 0.07896382 \end{bmatrix} \times 10^{-7}$$

$$\mathbf{v}_1 = -\mathbf{N}_1^{-1}\mathbf{q}_1 = \begin{bmatrix} -134608.0 \\ -432361.0 \end{bmatrix} \tag{5-21-11}$$

$$\boldsymbol{\theta}^{(0)} = \boldsymbol{\theta}_1 + \mathbf{v}_1 = \begin{bmatrix} -134508.0 \\ -430361.0 \end{bmatrix}$$

When we attempt to compute $\Phi(\boldsymbol{\theta}^{(0)})$ we note that the exponents occurring in the formulas for f_μ are so large that computer capacity is exceeded. We attempt to remedy things by halving ρ, but we have to repeat this process eight times before the exponentials are brought under control. We have then, with $\rho^{(0)} = 2^{-8} = 0.00390625$

$$\boldsymbol{\theta}^{(0)} = \boldsymbol{\theta}_1 + 0.00390625\mathbf{v}_1 = \begin{bmatrix} -425.8140 \\ 311.0039 \end{bmatrix}$$

with $\Psi^{(0)} = \Phi(\boldsymbol{\theta}^{(0)}) = 0.3366272 \times 10^{20}$. Since this exceeds Φ_1, $\boldsymbol{\theta}^{(0)}$ is not acceptable and we must interpolate. Here

$$\gamma = \mathbf{v}_1{}^{\mathrm{T}}\mathbf{q}_1 = -10.08228$$

Hence, from Eq. (5-14-5)

$$\rho^* = \frac{-10.08228 \times (0.00390625)^2}{2(-10.08228 \times 0.00390625 + 5.299502 - 0.3366272 \times 10^{20})}$$
$$\approx 5 \times 10^{-25}$$

Since this is less than $0.25\rho^{(0)}$, we follow Fig. 5-2A and set

$$\rho^{(2)} = 0.25\rho^{(0)} = 0.0009765625$$

leading to

$$\boldsymbol{\theta}^{(2)} = \begin{bmatrix} -31.45349 \\ 1577.782 \end{bmatrix}, \qquad \Psi^{(2)} = 5.471375$$

This is still unacceptable. Repetition of the interpolation procedure once more forces us to take

$$\rho^{(3)} = 0.25\rho^{(2)} = 0.000244141$$

and

$$\boldsymbol{\theta}^{(3)} = \begin{bmatrix} 67.13663 \\ 1894.446 \end{bmatrix}, \qquad \Psi^{(3)} = 5.301888$$

Once more we need to interpolate

$$\rho^* = \frac{-10.08228 \times (0.000244141)^2}{2(-10.08228 \times 0.000244141 + 5.299502 - 5.301888)}$$
$$= 0.0000619701$$

This time, $0.75\rho^{(3)} > \rho^* > 0.25\rho^{(3)}$, so we take $\rho^{(4)} = \rho^*$ and

$$\boldsymbol{\theta}^{(4)} = \begin{bmatrix} 91.65955 \\ 1973.211 \end{bmatrix}, \qquad \Psi^{(4)} = 5.299135$$

We have finally an acceptable point. We set $\boldsymbol{\theta}_2 = \boldsymbol{\theta}^{(4)}$ and proceed to the next iteration. The procedure converges, though rather slowly, in 25 iterations, which are plotted in Fig. 5-3. Also shown is the direction of the negative gradient at $\boldsymbol{\theta}_1$. Since the vector pointing from $\boldsymbol{\theta}_1$ to $\boldsymbol{\theta}^*$ lies between the negative gradient and the Gauss direction from $\boldsymbol{\theta}_1$ to $\boldsymbol{\theta}_2$, it appears that the Marquardt method may prove efficient in this case, and indeed it turns out to be so.

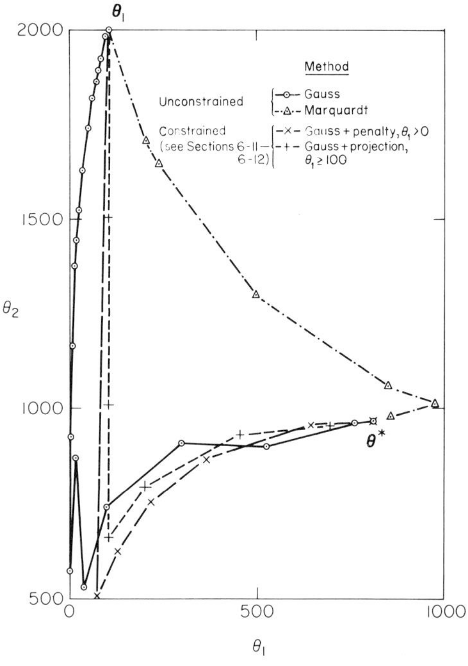

Fig. 5-3. Least squares problem

Returning to the first iteration, the Marquardt step would be given by

$$\mathbf{v}_1 = -10^7 \begin{bmatrix} 0.7036033(1 + \lambda_1) & -0.2354773 \\ -0.2354773 & 0.07896382(1 + \lambda_1) \end{bmatrix}^{-1} \times \begin{bmatrix} -0.0007098080 \\ 0.0002442936 \end{bmatrix}$$

Trying first $\lambda_1 = 0.01$, we obtain a step leading to a value of $\boldsymbol{\theta}$ for which Φ cannot be computed, and similarly for $\lambda_1 = 0.1$. With $\lambda_1 = 1$, however, we find

$$\mathbf{v}_1 = \begin{bmatrix} 3272.001 \\ -10589.988 \end{bmatrix}, \qquad \boldsymbol{\theta}^{(0)} = \begin{bmatrix} 3372.001 \\ -8589.988 \end{bmatrix}, \qquad \Phi^{(0)} = 6.183162$$

This is still unacceptable, so we increase λ_1 to 10, and eventually to $\lambda_1 = 100$ with

$$\mathbf{v}_1 = \begin{bmatrix} 98.8778 \\ -303.391 \end{bmatrix}, \qquad \boldsymbol{\theta}_2 = \begin{bmatrix} 198.8778 \\ 1696.609 \end{bmatrix}, \qquad \Phi_2 = 4.979104$$

which is an acceptable starting point for the next iteration. The full process, which converges in 10 iterations, is also shown in Fig. 5-3.

If the initial guesses for the parameters are much too large, f or its derivatives vanish, and the process does not converge. We can use the linearization method to obtain good initial guesses. For this purpose, we observe that Eq. (5) is equivalent to

$$\log(-\log y) = \log \theta_1 - \theta_2/x_2 + \log x_1 \tag{5-21-12}$$

or

$$y^+ = \theta_1{}^+ + \theta_2{}^+ x^+ \tag{5-21-13}$$

where

$$y^+ \equiv \log(-\log y), \qquad x^+ \equiv -1/x_2, \qquad \theta_1{}^+ \equiv \log \theta_1, \qquad \theta_2{}^+ \equiv \theta_2 \tag{5-21-14}$$

We now have a model linear in the parameters $\theta_1{}^+$ and $\theta_2{}^+$, which may be estimated by linear least squares as

$$\theta_1{}^+ = 6.643963, \qquad \theta_2{}^+ = 928.6492$$

corresponding to

$$\theta_1 = \exp \theta_1{}^+ = 768.1331, \qquad \theta_2 = \theta_2{}^+ = 928.6492$$

These are obviously good initial guesses for estimating $\boldsymbol{\theta}$ by the nonlinear least squares procedure.

5-22. Adding Prior Information

Assume that prior to having obtained the data of Table 5-2 we had some knowledge concerning the values of the parameters. Let us suppose that this knowledge could be summarized in the following equations

$$\theta_1 = 1000 \pm 200, \qquad \theta_2 = 1000 \pm 200 \tag{5-22-1}$$

The quantity 200 is meant to represent the standard deviation of the distributions of θ_1 and θ_2. Let us elect to assign to $\boldsymbol{\theta}$ a normal prior distribution, so that apart from an additive constant

$$\log p_0(\boldsymbol{\theta}) = -\tfrac{1}{2}[(1/200^2)(\theta_1 - 1000)^2 + (1/200^2)(\theta_2 - 1000)^2] \tag{5-22-2}$$

Assuming that the observation errors in y are also normal with unknown variance v, then the log likelihood is

$$\log L(\boldsymbol{\theta}) = -(1/2v) \sum_{\mu=1}^{15} [y_\mu - f_\mu(\boldsymbol{\theta})]^2 \tag{5-22-3}$$

and the log posterior distribution is the sum of the two. As in Section 4-8, we can eliminate v and form the concentrated log posterior distribution, which with sign reversed, reduces to the following objective function to be minimized

$$\Phi(\boldsymbol{\theta}) = (15/2) \log S(\boldsymbol{\theta}) + (1/80{,}000)[(\theta_1 - 1000)^2 + (\theta_2 - 1000)^2] \tag{5-22-4}$$

where

$$S(\boldsymbol{\theta}) \equiv \sum_{\mu=1}^{15} [y_\mu - f_\mu(\boldsymbol{\theta})]^2 = \sum_{\mu=1}^{15} e_\mu^2(\boldsymbol{\theta}) \tag{5-22-5}$$

Hence

$$\mathbf{q} = \begin{bmatrix} -\dfrac{15}{S(\theta)} \displaystyle\sum_{\mu=1}^{15} e_\mu \frac{\partial f_\mu}{\partial \theta_1} + \frac{1}{40{,}000}(\theta_1 - 1000) \\ -\dfrac{15}{S(\theta)} \displaystyle\sum_{\mu=1}^{15} e_\mu \frac{\partial f_\mu}{\partial \theta_2} + \frac{1}{40{,}000}(\theta_2 - 1000) \end{bmatrix} \tag{5-22-6}$$

$$\mathbf{N} = \begin{bmatrix} \dfrac{15}{S(\boldsymbol{\theta})} \displaystyle\sum_{\mu=1}^{15} \left(\frac{\partial f_\mu}{\partial \theta_1}\right)^2 + \frac{1}{40{,}000} & \dfrac{15}{S(\boldsymbol{\theta})} \displaystyle\sum_{\mu=1}^{15} \frac{\partial f_\mu}{\partial \theta_1}\frac{\partial f_\mu}{\partial \theta_2} \\ \dfrac{15}{S(\boldsymbol{\theta})} \displaystyle\sum_{\mu=1}^{15} \frac{\partial f_\mu}{\partial \theta_1}\frac{\partial f_\mu}{\partial \theta_2} & \dfrac{15}{S(\boldsymbol{\theta})} \displaystyle\sum_{\mu=1}^{15} \left(\frac{\partial f_\mu}{\partial \theta_2}\right)^2 + \frac{1}{40{,}000} \end{bmatrix} \tag{5-22-7}$$

With this prior distribution it is natural to start with the initial guess $\theta_1 = [1000, 1000]^{\mathrm{T}}$.

From here, the Marquardt method converges in three iterations to

$$\boldsymbol{\theta}^* = \begin{bmatrix} 929.7134 \\ 990.8511 \end{bmatrix} \qquad (\sigma_0 = 200)$$

When the standard deviations of θ_1 and θ_2 are assumed to be 100 instead of 200, our estimates turn out to be

$$\boldsymbol{\theta}^* = \begin{bmatrix} 976.2349 \\ 1000.1695 \end{bmatrix} \qquad (\sigma_0 = 100)$$

The solution of Section 5-21 may be regarded as corresponding to a prior distribution with infinite standard deviations. We recall that the result was

$$\boldsymbol{\theta}^* = \begin{bmatrix} 813.4583 \\ 960.9063 \end{bmatrix} \qquad (\sigma_0 = \infty)$$

Observe how the solution progressively approaches the mode of the prior distribution $[1000,\ 1000]^T$ as the variance (i.e., uncertainty) of our prior information decreases.

5-23. A Two-Equation Maximum Likelihood Problem

We take a two equation econometric model which was used by Bodkin and Klein (1967) to fit U.S. production data for the years 1909–1949. The model is based on the constant elasticity of substitution (CES) theory of production, and it takes the form

$$\begin{aligned} g_1 &\equiv c_1 10^{c_2 z_4}[c_5 z_1^{-c_4} + (1 - c_5) z_2^{-c_4}]^{-c_3/c_4} - z_3 - 0 \\ g_2 &\equiv [c_5/(1 - c_5)](z_1/z_2)^{-1-c_4} - z_5 = 0 \end{aligned} \tag{5-23-1}$$

where z_1 is capital input, z_2 is labor input, z_3 is real output, z_4 is time (in years; 1929 taken as origin), z_5 is ratio of price of capital services to wage scale, and c_1, c_2, c_3, c_4, c_5 are unknown parameters.

The data, in the form of yearly values of $\mathbf{z}_\mu$ for $\mu = 1, 2, \ldots, 41$, are given in Table 5-6. Of the variables involved, z_1 and z_2 are considered dependent (endogenous) whereas z_3, z_4 and z_5 are independent (exogenous). The treatment given by Bodkin and Klein is the standard one in econometrics, i.e., the distribution of the measured values of z_1 and z_2 is such as to give rise to normally distributed errors in Eq. (1). The likelihood is formed as in Eq. (2-13-6). The details of those calculations are given in Eisenpress *et al.* (1966) and Eisenpress and Greenstadt (1966) .

For illustrative purposes, we shall adopt a different approach here. We note that the model equations can be solved explicitly for the dependent variables to give the reduced form equations

$$z_1 = A z_5^{-1/(1+c_4)}, \qquad z_2 = A[(1 - c_5)/c_5]^{1/(1+c_4)} \tag{5-23-2}$$

where

$$A \equiv (z_3/c_1 10^{c_2 z_4})^{1/c_3}\{c_5[((1 - c_5)/c_5)^{1/(1+c_4)} + z_5^{c_4/(1+c_4)}]\}^{1/c_4} \tag{5-23-3}$$

To cast these equations into more tractable form we introduce the following new variables

$$y_1 \equiv z_1, \qquad y_2 \equiv z_2, \qquad x_1 \equiv z_4, \qquad x_2 \equiv \log z_3, \qquad x_3 \equiv \log z_5 \tag{5-23-4}$$

and we reparametrize the problem by defining

$$\begin{aligned} &\theta_1 \equiv (1/c_4) \log c_5 - (1/c_3) \log c_1, \qquad \theta_2 \equiv -(c_2/c_3) \log 10 \\ &\theta_3 \equiv 1/c_3, \qquad \theta_4 \equiv 1/c_4, \qquad \theta_5 \equiv [(1 - c_5)/c_5]^{1/(1+c_4)} \end{aligned} \tag{5-23-5}$$

Table 5-6
U.S. Production Data[a]

μ	z_1	z_2	z_3	z_4	z_5
1	1.33135	0.64629	0.4026	−20	0.24447
2	1.39235	0.66302	0.4084	−19	0.23454
3	1.41640	0.65272	0.4223	−18	0.23206
4	1.48773	0.67318	0.4389	−17	0.22291
5	1.51015	0.67720	0.4605	−16	0.22487
6	1.43385	0.65175	0.4445	−15	0.21879
7	1.48188	0.65570	0.4387	−14	0.23203
8	1.67115	0.71417	0.4999	−13	0.23828
9	1.71327	0.77524	0.5264	−12	0.26571
10	1.76412	0.79465	0.5793	−11	0.23410
11	1.76869	0.71607	0.5492	−10	0.22181
12	1.80776	0.70068	0.5052	− 9	0.18157
13	1.54947	0.60764	0.4679	− 8	0.22931
14	1.66933	0.67041	0.5283	− 7	0.20595
15	1.93377	0.74091	0.5994	− 6	0.19472
16	1.95460	0.71336	0.5964	− 5	0.17981
17	2.11198	0.75159	0.6554	− 4	0.18010
18	2.26266	0.78838	0.6851	− 3	0.16933
19	2.33228	0.79600	0.6933	− 2	0.16279
20	2.43980	0.80788	0.7061	− 1	0.16906
21	2.58714	0.84547	0.7567	0	0.16239
22	2.54865	0.77232	0.6796	1	0.16103
23	2.26042	0.67880	0.6136	2	0.14456
24	1.91974	0.58529	0.5145	3	0.20079
25	1.80000	0.58065	0.5046	4	0.18307
26	1.86020	0.62007	0.5711	5	0.18352
27	1.88201	0.65575	0.6184	6	0.18847
28	1.97018	0.72433	0.7113	7	0.20415
29	2.08232	0.76838	0.7461	8	0.19006
30	1.94062	0.69806	0.6981	9	0.17800
31	1.98646	0.74679	0.7722	10	0.19979
32	2.07987	0.79083	0.8557	11	0.21115
33	2.28232	0.88462	0.9925	12	0.23453
34	2.52779	0.95750	1.0877	13	0.20937
35	2.62747	1.00285	1.1834	14	0.19843
36	2.61235	0.99329	1.2565	15	0.18898
37	2.52320	0.94857	1.2293	16	0.17203
38	2.44632	0.97853	1.1889	17	0.18140
39	2.56478	1.02591	1.2249	18	0.19431
40	2.64588	1.03760	1.2669	19	0.19492
41	2.69105	0.99669	1.2708	20	0.17912

[a] Data adapted from Solow (1957).

Our reduced equations now take the form

$$y_1 = f_1(\mathbf{x}, \boldsymbol{\theta}) = \exp[a - \theta_4 x_3/(1 + \theta_4)],$$
$$y_2 = f_2(\mathbf{x}, \boldsymbol{\theta}) = \exp(a + \log \theta_5) \tag{5-23-6}$$

where

$$a \equiv \theta_1 + \theta_2 x_1 + \theta_3 x_2 + \theta_4 \log\{\theta_5 + \exp[x_3/(1 + \theta_4)]\} \tag{5-23-7}$$

We shall solve the problem in terms of the $\boldsymbol{\theta}$, and then convert the answer into $\mathbf{c}$ using the inverse transformations

$$c_1 = [1 + \theta_5^{(1+\theta_4)/\theta_4}]^{-\theta_4/\theta_3} \exp(-\theta_1/\theta_3), \qquad c_2 = -\theta_2/(\theta_3 \log 10)$$
$$c_3 = 1/\theta_3, \qquad c_4 = 1/\theta_4, \qquad c_5 = 1/[1 + \theta_5^{(1+\theta_4)/\theta_4}] \tag{5-23-8}$$

Let us formulate some alternative likelihood functions to be maximized. Assuming the errors in the reduced equations to be normally distributed, independently for each year and with covariance matrix $\mathbf{V}$, we have (apart from irrelevant constants)

$$\log L(\boldsymbol{\theta}, \mathbf{V}) = -(n/2) \log \det \mathbf{V} - \tfrac{1}{2} \sum_{\mu=1}^{41} (\mathbf{y}_\mu - \mathbf{f}_\mu)^{\mathrm{T}} \mathbf{V}^{-1} (\mathbf{y}_\mu - \mathbf{f}_\mu) \tag{5-23-9}$$

We examine the following cases:

(a) Unknown V. The concentrated likelihood is equivalent to the objective function Eq. (4-9-9)

$$\Phi(\boldsymbol{\theta}) = (n/2) \log \det \mathbf{M}$$
$$= (41/2) \log \left\{ \sum_{\mu=1}^{41} (y_{\mu 1} - f_{\mu 1})^2 \sum_{\mu=1}^{41} (y_{\mu 2} - f_{\mu 2})^2 - \left[\sum_{\mu=1}^{41} (y_{\mu 1} - f_{\mu 1})(y_{\mu 2} - f_{\mu 2}) \right]^2 \right\} \tag{5-23-10}$$

(b) Unknown diagonal V. The objective function is given by Eq. (4-8-4)

$$\Phi(\boldsymbol{\theta}) = (n/2) \sum_{a=1}^{2} \log M_{aa}(\boldsymbol{\theta}) = (41/2) \log \left[\sum_{\mu=1}^{41} (y_{\mu 1} - f_{\mu 1})^2 \sum_{\mu=1}^{41} (y_{\mu 2} - f_{\mu 2})^2 \right] \tag{5-23-11}$$

(c) Covariance matrix proportional to $\mathbf{Q} \equiv \begin{bmatrix} 4 & 0 \\ 0 & 1 \end{bmatrix}$, i.e., the errors in y_1 are assumed twice as large, and independent of, the errors in y_2. The relevant objective function is given by Eq. (4-21-1)

$$\Phi(\boldsymbol{\theta}) = (nm/2) \log \mathrm{Tr}\, (\mathbf{Q}^{-1}\mathbf{M})$$
$$= (41 \times 2/2) \log \left[\tfrac{1}{4} \sum_{\mu=1}^{41} (y_{\mu 1} - f_{\mu 1})^2 + \sum_{\mu=1}^{41} (y_{\mu 2} - f_{\mu 2})^2 \right] \tag{5-23-12}$$

(d) As in (a) above, but $\log y_1$ and $\log y_2$ are the dependent variables. The objective function has the same form, with $\log y_{\mu a} - \log f_{\mu a}$ replacing $y_{\mu a} - f_{\mu a}$.

All these objective functions have the Gauss form, and the approximate Hessian has the form $\mathbf{N} = 2\sum_{\mu=1}^{41} \mathbf{B}_\mu^{\mathrm{T}} \mathbf{\Gamma} \mathbf{B}_\mu$. Let us examine case (d) first. Here

$$\mathbf{B}_\mu = \partial \log \mathbf{f}_\mu / \partial \boldsymbol{\theta}$$

Table 5-7

Elements of $\mathbf{B}_\mu^T$, Case (d)[a] $B_{\mu a \alpha} = \partial \log f_{\mu a}/\partial \theta_\alpha$

a / α	1	2
1	1	1
2	$x_{\mu 1}$	$x_{\mu 1}$
3	$x_{\mu 2}$	$x_{\mu 2}$
4	$\log\left(\theta_5 + \exp\dfrac{x_{\mu 3}}{(1+\theta_4)}\right) - \dfrac{x_{\mu 3}\theta_4 \exp\dfrac{x_{\mu 3}}{(1+\theta_4)}}{(1+\theta_4)^2\left(\theta_5 + \exp\dfrac{x_{\mu 3}}{(1+\theta_4)}\right)} - \dfrac{x_{\mu 3}}{(1+\theta_4)^2}$	$\log\left(\theta_5 + \exp\dfrac{x_{\mu 3}}{(1+\theta_4)}\right) - \dfrac{x_{\mu 3}\theta_4 \exp\dfrac{x_{\mu 3}}{(1+\theta_4)}}{(1+\theta_4)^2\left(\theta_5 + \exp\dfrac{x_{\mu 3}}{(1+\theta_4)}\right)}$
5	$\dfrac{\theta_4}{\theta_5 + \exp\dfrac{x_{\mu 3}}{(1+\theta_4)}}$	$\dfrac{\theta_4}{\theta_5 + \exp\dfrac{x_{\mu 3}}{(1+\theta_4)}} + \dfrac{1}{\theta_5}$

[a] In Cases (a), (b), and (c) Multiply Table Entries by $f_{\mu a}$.

The elements of $\mathbf{B}_\mu$ are given in Table 5-7. Referring to the third row of Table 5-1, we find that

$$\mathbf{\Gamma} = \frac{41}{2}\begin{bmatrix} \sum_{\mu=1}^{41} (\log y_{\mu 1} - \log f_{\mu 1})^2 & \sum_{\mu=1}^{41} (\log y_{\mu 1} - \log f_{\mu 1}) \times (\log y_{\mu 2} - \log f_{\mu 2}) \\ \sum_{\mu=1}^{41} (\log y_{\mu 1} - \log f_{\mu 1}) \times (\log y_{\mu 2} - \log f_{\mu 2}) & \sum_{\mu=1}^{41} (\log y_{\mu 2} - \log f_{\mu 2})^2 \end{bmatrix}^{-1} \tag{5-23-13}$$

In cases (a), (b), and (c) we obtain $B_{\mu a\alpha}$ by multiplying the corresponding entry in Table 5-7 by $f_{\mu a}$, since $\partial f_{\mu a}/\partial\theta_\alpha = f_{\mu a}\,\partial \log f_{\mu a}/\partial\theta_\alpha$.
The expression for $\mathbf{\Gamma}$ as given in Table 5-1 comes out to be:
case (a)

$$\mathbf{\Gamma} = \frac{41}{2}\begin{bmatrix} \sum_{\mu=1}^{41}(y_{\mu 1}-f_{\mu 1})^2 & \sum_{\mu=1}^{41}(y_{\mu 1}-f_{\mu 1})(y_{\mu 2}-f_{\mu 2}) \\ \sum_{\mu=1}^{41}(y_{\mu 1}-f_{\mu 1})(y_{\mu 2}-f_{\mu 2}) & \sum_{\mu=1}^{41}(y_{\mu 2}-f_{\mu 2})^2 \end{bmatrix}^{-1} \tag{5-23-14}$$

case (b)

$$\mathbf{\Gamma} = \frac{41}{2}\begin{bmatrix} \left[\sum_{\mu=1}^{41}(y_{\mu 1}-f_{\mu 1})^2\right]^{-1} & 0 \\ 0 & \left[\sum_{\mu=1}^{41}(y_{\mu 2}-f_{\mu 2})^2\right]^{-1} \end{bmatrix} \tag{5-23-15}$$

case (c)

$$\mathbf{\Gamma} = \frac{41}{\frac{1}{4}\sum_{\mu=1}^{41}(y_{\mu 1}-f_{\mu 1})^2 + \sum_{\mu=1}^{41}(y_{\mu 2}-f_{\mu 2})^2}\begin{bmatrix} \frac{1}{4} & 0 \\ 0 & 1 \end{bmatrix} \tag{5-23-16}$$

We shall omit the details of the calculation. The results for cases (a)–(d) are reported in Table 5-8 in the form of the final estimates $\boldsymbol{\theta}^*$ and the minimum

Table 5-8
Results of Parameter Estimation for Production Theory Problem (Estimates of $\boldsymbol{\theta}$ and Minimum of Objective Function)

Case	θ_1^*	θ_2^*	θ_3^*	θ_4^*	θ_5^*	Φ^*
(a)	−0.0758463	−0.0115747	0.790686	1.00224	0.859255	−82.71488
(a)[a]	−4.27288	−0.00882702	0.713410	4.85440	1.47252	−76.70853
(b)	−0.155586	−0.00978696	0.737126	1.05720	0.878024	−79.2353
(b)[a]	−4.27489	−0.00834795	0.696653	4.85221	1.47180	−75.8870
(c)	−4.76163	−0.00569320	0.600316	5.27596	1.49747	−66.5601
(c)[a]	−2.45064	−0.00556030	0.598743	3.13535	1.30821	−66.0913
(d)	−0.0409260	−0.0118384	0.802121	0.96870	0.850246	−99.3714
(d)[a]	−3.54201	−0.00882492	0.724913	4.18994	1.41740	−95.2669

[a] Local minimum of objective function.

objective function values Φ^*. In Table 5-9 the results are given in terms of the original variables $\mathbf{c}^*$. Also reported are the results of Bodkin and Klein (1967), who used an objective function of the form Eq. (2-13-6).

Table 5-9
Results of Parameter Estimation for Production Theory Problem.

Case	c_1^*	c_2^*	c_3^*	c_4^*	c_5^*
(a)	0.5460	0.00636	1.265	0.998	0.5752
(a)[a]	0.6074	0.00537	1.402	0.206	0.3854
(b)	0.5417	0.00577	1.357	0.946	0.5629
(b)[a]	0.6051	0.00520	1.435	0.206	0.3855
(c)	0.5935	0.00412	1.666	0.189	0.3822
(c)[a]	0.5791	0.00403	1.670	0.319	0.4123
(d)	0.5473	0.00640	1.247	1.027	0.5806
(d)[a]	0.6049	0.00529	1.379	0.239	0.3936
(e)	0.5839	0.00589	1.362	0.475	0.4471
(f)	0.5420	0.00643	1.238	1.130	0.6037

[a] Local minimum of objective function.

(e) The model equations take the form of Eq. (1).

(f) The model equations are written as

$$g_1 \equiv \log c_1 + c_2 z_4 \log 10 - (c_3/c_4) \log[c_5 z_1^{-c_4} + (1 - c_5) z_2^{-c_4}] - \log z_3 = 0$$
$$g_2 \equiv \log c_5 - \log(1 - c_5) - (1 + c_4)(\log z_1 - \log z_2) - \log z_5 = 0$$

It is revealing to find such discrepancies in the estimates (particularly of c_4), since in all cases the same model equations were fitted to the same data, the only difference being in the assumptions concerning the distribution of errors. We must defer further discussion of this problem to the end of Chapter VII.

Matters are further complicated by the question of convergence. In attempting to solve these problems using various algorithms and starting values, we found that in each case there exists at least one local minimum of the objective function other than the global minimum. These local minima are also recorded in Tables 5-8 and 5-9. The other minima are, we think, global, but we have no way of proving that this is so. The performance of various algorithms is summarized in Table 5-10.

Table 5-10
Convergence of Various Algorithms for Problem 2

Case	Starting point[a]	Algorithm[b]	Convergence[c]	Iterations	Objective function evaluation for interpolation–extrapolation.
(a)	1	1	1	15	28
		2	1	27	55
		3	1	28	56
		4	1	12	16
		5	1	35	73
		6	1	36	350
	2	3	2	28	56
	3	1	1	15	32
(b)	1	3	2	8	15
	4	3	1	8	17
(c)	4	3	2	15	27
(d)	1	1	1	38	78
		2	2	8	15
		3	2	8	15
		4	1	10	15

[a] Starting points:

$\boldsymbol{\theta}_1 =$
1. [0, 0, 0, 0, 1]
2. [4.27489, −0.00834795, 0.696653, 4.85221, 1.47180]
3. [−1.29970, −0.00995700, 0.734317, 2.10535, 1.15490]
4. [−0.0758463, −0.0115747, 0.790686, 1.00224, 0.859255]

[b] Algorithms:
1. Gauss, directional discrimination, using Eq. (5-7-9).
2. Gauss, directional discrimination, using Eq. (5-7-10), $\varepsilon = 10^{-5}$, $a = 1$.
3. Gauss, directional discrimination, using Eq. (5-7-11), $\varepsilon = 10^{-5}$, $\beta = 1$.
4. Gauss, Marquardt.
5. Variable metric, ROC.
6. Variable metric, DFP.

[c] Convergence:
1. To best known solution.
2. To a local minimum.

5-24. Problems

1. Verify that the Gauss method is invariant under reparametrization; i.e., the sequence of iterations is the same whether we use the original set of variables $\boldsymbol{\theta}$ or a transformed set $\mathbf{c} = \mathbf{c}(\boldsymbol{\theta})$, provided only that $\mathbf{c}_1 = \mathbf{c}(\boldsymbol{\theta}_1)$, and that $\mathbf{c}$ is a linear function of $\boldsymbol{\theta}$.

2. Determine what conditions must be fulfilled for the Marquardt and variable metric methods to be invariant under reparametrization.

3. Suppose $\Phi(\boldsymbol{\theta}) = \theta_1^2 + 10000\,\theta_2^2$. Let $\boldsymbol{\theta}_1^{\mathrm{T}} = [100, 1]$. Compare the progress towards the minimum of Φ that can be made in a single iteration of the steepest descent and Newton methods.

4. For the objective function of Problem 3, find initial guesses for which the upper and lower limits of Eq. (5-6-5) are satisfied. Show that these limits cannot be violated.

5. Devise an algorithm similar to the Gauss method for finding $\boldsymbol{\theta}$ and $\hat{\mathbf{X}}$ to minimize the objective function of Eq. (4-10-12).

Chapter

VI

Computation of the Estimates II: Problems with Constraints

A. Inequality Constraints

6-1. Penalty Functions

Inequality constraints of the form Eq. (5-1-1) limit the domain of parameter values within which the estimate is to be found. They often arise from prior information concerning the values of the parameters (see Section 2-16). The presence of inequality constraints, particularly in the form of upper and lower bounds on each parameter, often exerts a beneficial influence on the convergence of an optimization algorithm. In quite a few problems, convergence to a correct minimum results from imposition of somewhat arbitrary bounds, without which the algorithms bog down in irrelevant regions of parameter space. We would go so far as to recommend imposition of generous, though not unreasonably so, bounds in all nonlinear parameter estimation problems.

We possess several powerful algorithms for unconstrained optimization and would like to apply the same algorithms to the constrained problems. We need to modify the objective function in such a way that it remains almost unchanged well in the interior of the feasible region, but increases drastically as one approaches the constraints. To accomplish this, we assign a *penalty function* to each inequality constraint. This function is nearly zero when the constraint function is strongly positive, but increases sharply as the constraint function approaches zero from above. To the constraint

$$h_j(\boldsymbol{\theta}) \geqslant 0 \tag{6-1-1}$$

we assign, following Carroll (1961), the penalty function

$$\zeta_j(\boldsymbol{\theta}) \equiv \alpha_j / h_j(\boldsymbol{\theta}) \tag{6-1-2}$$

where α_j is a small positive constant. We now modify the objective function by adding to it the penalty functions for all the constraints‡

$$\Phi^{\dagger}(\boldsymbol{\theta}) \equiv \Phi(\boldsymbol{\theta}) + \sum_j \alpha_j / h_j(\boldsymbol{\theta}) \tag{6-1-3}$$

Let $\boldsymbol{\theta}^{\dagger}$ and $\boldsymbol{\theta}^*$ be the points at which $\Phi^{\dagger}$ and Φ attain their respective minima within the feasible region. Fiacco and McCormick (1964) have proven that under suitable conditions

$$\lim_{\alpha_j \to 0} \boldsymbol{\theta}^{\dagger} = \boldsymbol{\theta}^* \tag{6-1-4}$$

These concepts are employed in SUMT (sequential unconstrained maximization technique), originally presented in Carroll's paper but later amplified by Fiacco and McCormick:

1. Select the α_j and a feasible initial guess $\boldsymbol{\theta}_1$.
2. Find $\boldsymbol{\theta}^{\dagger}$ using one of the unconstrained optimization methods.
3. Reduce the values of the α_j, and return to step 2, using $\boldsymbol{\theta}^{\dagger}$ as the initial guess. The process is continued until $\boldsymbol{\theta}^{\dagger}$ does not change significantly upon reducing the α_j. Then we accept $\boldsymbol{\theta}^{\dagger}$ as our estimate of $\boldsymbol{\theta}^*$.

The search for the minimum of $\Phi^{\dagger}$ in step 2 must still be confined to the feasible region. It may appear, therefore, that nothing has been gained. Why not minimize Φ directly? The answer is that we have created a situation where the objective function always starts increasing before one has a chance to leave the feasible region. Therefore, the procedures for determining step length always succeed in producing an acceptable feasible step. If we happen to be near a constraint, it is quite possible for a minimization method when applied to Φ to direct one towards the infeasible region, even though there exist feasible directions in which the function decreases. By adding the penalty functions we deflect our step to a feasible direction. The point is illustrated in Fig. 6-1 where the contours of an objective function are drawn. The minimum occurs at point A. Starting out at point B, the steepest descent procedure carries us along the path $BCDA$§ to the minimum. If the feasible region is constrained to lie to the left of the line FG, the path is blocked at point C. Introducing a penalty function (Fig. 6-2) leaves the contours around A almost undisturbed, but distorts them near the constraint in the mannei shown. We now have a feasible path $BHIA$§ to the minimum. Although the example is given in terms of steepest descent, it applies equally well to other minimization methods.

This example illustrates the important point that the path from a feasible starting point to a feasible minimum may pass through infeasible territory.

‡ Alternatively, we may use $-\alpha \sum_j \log h_j(\boldsymbol{\theta})$. This has the benefit of being unaffected by scaling the functions $h_j(\boldsymbol{\theta})$.

§ The details of the hemstitching near the minimum were omitted from the Figures.

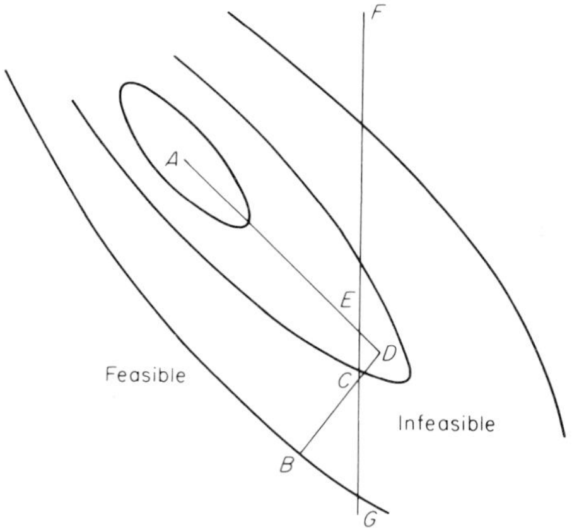

Fig. 6-1. Contours of Φ: Minimization without penalty functions.

If matters were always as simple as in our illustration, then we could simply do away with the constraint altogether. In practice it may happen, however, that there are local minima in the infeasible region, or that it is impossible even to compute the value of the function for infeasible parameter values (this occurs frequently with dynamic systems whose differential equations become unstable). Hence the importance of creating paths that lie entirely within the feasible region.

Suppose $\boldsymbol{\theta}^\dagger$ is well in the interior of the feasible region. This is recognized to be the case when $\sum_j \alpha_j / h_j(\boldsymbol{\theta}^\dagger)$ is very small. Then the minima with and without penalty functions nearly coincide. Having obtained $\boldsymbol{\theta}^\dagger$ we may take $\boldsymbol{\theta}^* = \boldsymbol{\theta}^\dagger$, or perhaps go through an additional iteration of the minimization

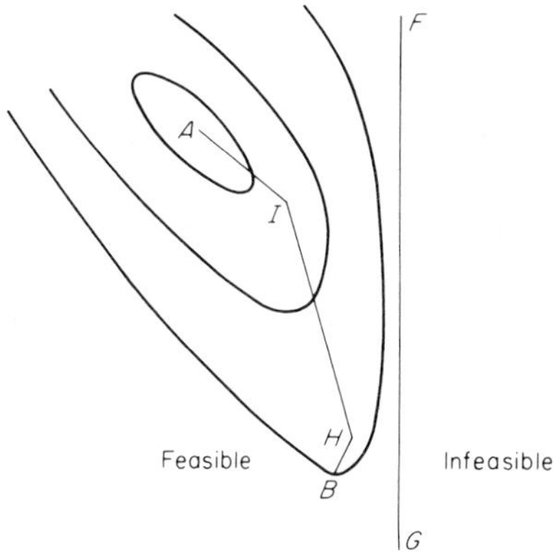

Fig. 6-2. Contours of $\Phi^\dagger$: Minimization with penalty functions.

procedure, starting at $\boldsymbol{\theta}^\dagger$ and omitting the penalty functions entirely. If $\boldsymbol{\theta}^\dagger$ turns out to be near some constraints [sizable value of $\sum_j \alpha_j/h_j(\boldsymbol{\theta}^\dagger)$], a gradual reduction in the α_j is called for. All iterates $\boldsymbol{\theta}_i$ must be restricted to the *interior* of the feasible region, except for the last iterate which should be allowed to fall on the boundary. This means that in Eq. (5-3-7) we must have

$$\rho_i < \rho_{i,\,\max} \tag{6-1-5}$$

in all iterations but the last, and

$$\rho_i \leqslant \rho_{i,\,\max} \tag{6-1-6}$$

in the last iteration, taken with no penalty functions. Here, $\rho_{i,\,\max}$ is the greatest lower bound on the positive values of ρ for which $\boldsymbol{\theta}_i + \rho\mathbf{v}_i$ is not feasible. The flowcharts of Section 5-14 assume that $\rho_{i,\,\max}$ can be calculated at each iteration.

If we use a gradient method to minimize $\Phi^\dagger(\boldsymbol{\theta})$ we need to compute first and sometimes second derivatives of the penalty functions. From Eq. (2):

$$\partial\zeta_j/\partial\theta_\alpha = -[\alpha_j/h_j^2(\boldsymbol{\theta})]\,\partial h_j/\partial\theta_\alpha \tag{6-1-7}$$

$$\partial^2\zeta_j/\partial\theta_\alpha\,\partial\theta_\beta = [\alpha_j/h_j^3(\boldsymbol{\theta})][2(\partial h_j/\partial\theta_\alpha)(\partial h_j/\partial\theta_\beta) - h_j(\boldsymbol{\theta})\,\partial^2 h_j/\partial\theta_\alpha\,\partial\theta_\beta] \tag{6-1-8}$$

When $\boldsymbol{\theta}$ is far from the jth constraint, the contribution of ζ_j to the objective function and its derivatives is very small. Near the jth constraint, h_j is nearly zero, and the second term of Eq. (8) may be neglected relative to the first term. In either case, it is safe to replace Eq. (8) with

$$\partial^2\zeta_j/\partial\theta_\alpha\,\partial\theta_\beta \approx [2\alpha_j/h_j^3(\boldsymbol{\theta})](\partial h_j/\partial\theta_\alpha)(\partial h_j/\partial\theta_\beta) \tag{6-1-9}$$

which does not require computation of the second derivatives of the constraint functions. This is analogous to the way in which the second derivatives of the model equations are suppressed in the Gauss method. Note also that Eq. (9) is at least positive semidefinite, and does not spoil the definiteness of $\mathbf{N}$ when added to the latter.

Frequently (e.g., with upper or lower bounds) the constraints are linear functions whose second derivatives vanish anyway. Eq. (9) is then exact.

The initial choice of α_j should be dictated by the range of values that $h_j(\boldsymbol{\theta})$ and $\Phi(\boldsymbol{\theta})$ can take in the feasible region. For instance, if we have two constraints reflecting the bounds $b_\alpha \leqslant \theta_\alpha \leqslant a_\alpha$

$$h_j(\boldsymbol{\theta}) \equiv a_\alpha - \theta_\alpha \geqslant 0 \tag{6-1-10}$$

$$h_{j+1}(\boldsymbol{\theta}) \equiv \theta_\alpha - b_\alpha \geqslant 0 \tag{6-1-11}$$

we might set $\alpha_j = \alpha_{j+1} = 0.001(a_\alpha - b_\alpha)\ \Phi(\boldsymbol{\theta}_1)$. If the initial guesses $\boldsymbol{\theta}_1$ are well in the interior of the feasible region, then a good choice for α_j may be

$$\alpha_j = 0.001 h_j(\boldsymbol{\theta}_1)\ \Phi(\boldsymbol{\theta}_1) \tag{6-1-12}$$

The penalty function method is easy to program, and has been found to work well when the solution is known (or expected) to be in the interior of the feasible region. A numerical illustration appears in Section 6-11. When the solution is likely to be on the boundary, then the projection method discussed below is preferable. Even an interior minimum may be reached faster by the projection method, but the complexity of the latter mitigates against its use.

Penalty function methods other than the one described here have been proposed, e.g., by Zangwill (1967a) and Fiacco and McCormick (1967). These methods possess the advantage (for general nonlinear programming problems) that the initial guess and intermediate iterates are not restricted to the interior of the feasible region; i.e., they are allowed to violate the constraints. In the case of parameter estimation problems, however, this is not at all an advantage. In the first place, it is usually easy to stay in the feasible region because of the simple nature of the constraints. Secondly, the objective function often behaves in an erratic manner, and may even be uncomputable outside the feasible region. The importance of staying within the feasible region was stressed above.

While penalty functions may appear to be merely a computational artifact, they do indeed possess a statistical interpretation. Suppose $\Phi(\boldsymbol{\theta})$ is a log likelihood, and let us assign to $\boldsymbol{\theta}$ a discontinuous prior distribution, which is zero outside, and uniform inside the feasible region. Then the posterior density is zero outside, and proportional to $\Phi(\boldsymbol{\theta})$ inside the region, and the probem can be formulated as being that of finding the maximum of the posterior density. Let us, however, try to smooth out the prior distribution so that its density approaches zero continuously (though rapidly) as one goes out to the boundary of the feasible region. This can be accomplished precisely by making

$$-\log p_0(\boldsymbol{\theta}) = \sum_j \zeta_j(\boldsymbol{\theta}) \tag{6-1-13}$$

As an example, suppose $0 \leqslant \theta_1 \leqslant 1$, and we use $\zeta_j = -\alpha \log h_j(\boldsymbol{\theta})$. In this case

$$-\log p_0(\boldsymbol{\theta}) = -\alpha \log \theta_1 - \alpha \log (1 - \theta_1) \tag{6-1-14}$$

so that

$$p_0(\boldsymbol{\theta}) = \theta_1{}^\alpha (1 - \theta_1)^\alpha \tag{6-1-15}$$

which is the $B_{1+\alpha,\,1+\alpha}$ (beta) distribution.

6-2. Projection Methods

Another class of methods for optimizing with inequality constraints is variously known as *gradient projection* and *reduced gradient* (Rosen, 1960, 1961; Wolfe, 1963; Faure and Huard, 1965; Abadie and Carpentier, 1966; Abadie, 1967b). These methods, which in Fig. 6-1 would take us along the path *BCEA*, may be summarized as follows:

At each iteration define the *normal step* as the one computed according to the gradient method of our choice, with the constraints ignored. We now face one of the following two situations:

1. If $\boldsymbol{\theta}_i$ is in the interior of the feasible region, apply the normal step (e.g., *BC* in Fig. 6-1). If this results in an infeasible point, the step is truncated so as to leave us on the boundary of the feasible region.
2. If $\boldsymbol{\theta}_i$ is on the boundary, take the normal step or a fraction thereof if this is feasible (e.g., *EA* in Fig. 6-1). Otherwise, treat some of the active constraints as equality constraints, and take a step along these constraints (e.g., *CE* in Figure 6-1).

The question of which ones of the active constraints should be retained in any given situation is a difficult one. Although Rosen (1960, 1961) gives a working solution to this problem, this solution is not necessarily optimal. The quadratic programming solution listed below is a good one, provided all the algebraic manipulations involved are less time-consuming than the function evaluations.

Efficient algorithms have been constructed by using these techniques in combination with variable metric methods for generating the step directions [Goldfarb and Lapidus (1968) and Murtagh and Sargent (1969) for linear constraints; Davies (1970) for nonlinear constraints]. These algorithms are usually superior to the penalty function method for finding minima that lie on a constraint. In many parameter estimation problems, where we hope to find an interior minimum, the penalty function method seems preferable because of its greater simplicity. Exceptions to this rule do occur, however, and therefore we indicate how the *i*th iteration of a gradient method may be modified in the presence of linear inequality constraints (e.g., upper and lower bounds). Note that when the penalty function method is used, all iterates $\boldsymbol{\theta}_i$ (except possibly the last) are in the interior of the feasible region, and hence the method is immediately applicable even when some of the constraints Eq. (6-1-1) are in the form of strict inequalities $h_j(\boldsymbol{\theta}) > 0$. Such situations arise, for instance, when terms like $1/\theta_\alpha$ or $\log \theta_\alpha$ appear in the model equations, requiring $\theta_\alpha > 0$. In the projection method some iterates may fall on the boundary. Hence the constraint $\theta_\alpha > 0$ must be replaced by $\theta_\alpha - \varepsilon \geqslant 0$, where ε is a small positive number.

In an unconstrained gradient method we take a step in the direction $\mathbf{v} = -\mathbf{R}_i \mathbf{q}_i$. The minimum would be attained in a single step if the objective function to be minimized were, in fact

$$Q_i(\mathbf{v}) \equiv \Phi_i + \mathbf{v}^T \mathbf{q}_i + \tfrac{1}{2} \mathbf{v}^T \mathbf{R}_i^{-1} \mathbf{v} \tag{6-2-1}$$

which is close to Φ provided $\mathbf{R}_i$ is a good approximation to $\mathbf{H}_i^{-1}$.

Suppose the constraints take the form

$$a_j^T \boldsymbol{\theta} \geqslant b_j \qquad (j = 1, 2, \ldots) \tag{6-2-2}$$

Let $\mathbf{A}_i$ be the matrix whose columns are precisely those vectors $\mathbf{a}_j$ such that $\mathbf{a}_j^T \boldsymbol{\theta}_i = b_j$, i.e., those constraints which are active at the point $\boldsymbol{\theta}_i$. Let p be the number of such constraints; $\mathbf{A}_i$ is $l \times p$. Any feasible step must satisfy

$$\mathbf{A}_i^T \mathbf{v} \geqslant \mathbf{0} \tag{6-2-3}$$

The following strategy will be adopted for finding the direction $\mathbf{v}_i$: Minimize the current approximation Eq. (1) to the objective function, subject to the currently active constraints (as will be seen later, currently inactive constraints will help determine the step length, but not its direction). The quadratic expression Eq. (1) therefore acts as a temporary objective function, and the problem of finding its minimum subject to the linear constraints Eq. (3) is called *quadratic programming* (QP). Since $\mathbf{R}_i$, $\mathbf{q}_i$, and $\mathbf{A}_i$ differ from iteration to iteration, it follows that each iteration of the original problem requires the solution of a different QP problem. The algorithm described below is very efficient, however, and in many estimation problems the computation of Φ and its derivatives is more time consuming than the solution of the QP problem.

Let $\mathbf{v}_i$ be the solution to the ith iteration QP problem. At $\boldsymbol{\theta} = \boldsymbol{\theta}_i + \mathbf{v}_i$ the gradient of $Q_i(\mathbf{v})$ is obtained from Eq. (1) as

$$\mathbf{q}(\mathbf{v}_i) = \mathbf{q}_i + \mathbf{R}_i^{-1} \mathbf{v}_i \tag{6-2-4}$$

According to the Kuhn–Tucker conditions, there must exist a vector of Lagrange multipliers $\boldsymbol{\lambda}_i$ satisfying Eq. (3-7-3), which becomes in our case

$$\mathbf{q}_i + \mathbf{R}_i^{-1} \mathbf{v}_i = \mathbf{A}_i \boldsymbol{\lambda}_i \tag{6-2-5}$$

Let $\mathbf{w}_i \equiv \mathbf{A}_i^T \mathbf{v}_i$ denote the vector of constraint functions evaluated at $\boldsymbol{\theta} = \boldsymbol{\theta}_i + \mathbf{v}_i$. Then Eqs. (3-7-4)–(3-7-6) take the form

$$\mathbf{w}_i \geqslant \mathbf{0} \qquad \boldsymbol{\lambda}_i \geqslant \mathbf{0} \qquad \boldsymbol{\lambda}_i^T \mathbf{w}_i = 0 \tag{6-2-6}$$

Let $\mathbf{z}_i \equiv -\mathbf{A}_i^T \mathbf{R}_i \mathbf{q}_i$ and $\mathbf{W}_i \equiv \mathbf{A}_i^T \mathbf{R}_i \mathbf{A}_i$. Premultiply Eq. (5) by $\mathbf{A}_i^T \mathbf{R}_i$. Conditions Eqs. (5) and (6) are now transformed into the following problem (we henceforth drop the subscript i):

Find $\boldsymbol{\lambda}$ and $\mathbf{w}$ satisfying

$$\mathbf{w} = \mathbf{z} + \mathbf{W}\boldsymbol{\lambda} \qquad \mathbf{w} \geqslant \mathbf{0} \qquad \boldsymbol{\lambda} \geqslant \mathbf{0} \qquad \boldsymbol{\lambda}^T \mathbf{w} = 0 \tag{6-2-7}$$

This is known as the *complementary pivot* problem (Cottle and Dantzig, 1968) and can be solved by an algorithm given by Dantzig and Cottle (1967). We present here a simpler and faster algorithm (Zoutendijk, 1960; Bard, 1971) whose convergence has not been proven, but which has not failed in hundreds of applications.

Observe that from Eq. (7) for each j either λ_j or w_j vanishes. Since w_j is the value of the jth constraint function at the solution to the QP problem, it follows that this constraint remains active if $w_j = 0$. In this case we refer to the jth constraint as *binding*. If $w_j > 0$ then the solution would not be affected by removal of this constraint, and it is called *nonbinding*. Let $\mathbf{w}_{\mathrm{B}} = \mathbf{0}$ and $\mathbf{w}_{\mathrm{N}} > \mathbf{0}$ be the vectors of binding and nonbinding constraints, respectively. Of course, we do not know as yet which constraints are going to be included in which set, but for the time being we ignore this difficulty. Let $\boldsymbol{\lambda}_{\mathrm{B}}$ and $\boldsymbol{\lambda}_{\mathrm{N}}$ be the corresponding partition of $\boldsymbol{\lambda}$. Then from Eq. (7) $\boldsymbol{\lambda}_{\mathrm{N}} = \mathbf{0}$.

Let $\mathbf{W}$ be partitioned along the same lines into

$$\begin{bmatrix} \mathbf{W}_{\mathrm{BB}} & \mathbf{W}_{\mathrm{BN}} \\ \mathbf{W}_{\mathrm{BN}}^{\mathrm{T}} & \mathbf{W}_{\mathrm{NN}} \end{bmatrix}$$

and $\mathbf{z}^{\mathrm{T}}$ into $\mathbf{z}_{\mathrm{B}}{}^{\mathrm{T}}$, $\mathbf{z}_{\mathrm{N}}{}^{\mathrm{T}}$. Then Eq. (7) becomes:

$$\mathbf{0} = \mathbf{z}_{\mathrm{B}} + \mathbf{W}_{\mathrm{BB}}\boldsymbol{\lambda}_{\mathrm{B}} \tag{6-2-8}$$

$$\mathbf{w}_{\mathrm{N}} = \mathbf{z}_{\mathrm{N}} + \mathbf{W}_{\mathrm{BN}}^{\mathrm{T}}\boldsymbol{\lambda}_{\mathrm{B}} \tag{6-2-9}$$

$$\mathbf{w}_{\mathrm{N}} > \mathbf{0}, \qquad \boldsymbol{\lambda}_{\mathrm{B}} \geqslant \mathbf{0} \tag{6-2-10}$$

From Eq. (8), $\boldsymbol{\lambda}_{\mathrm{B}} = -\mathbf{W}_{\mathrm{BB}}^{-1}\mathbf{z}_{\mathrm{B}}$, so that from Eq. (9) $\mathbf{w}_{\mathrm{N}} = \mathbf{z}_{\mathrm{N}} - \mathbf{W}_{\mathrm{BN}}^{\mathrm{T}}\mathbf{W}_{\mathrm{BB}}^{-1}\mathbf{z}_{\mathrm{B}}$. Let us form the tableau

$$\mathbf{E} = [\mathbf{W}, \mathbf{z}] = \begin{bmatrix} \mathbf{W}_{\mathrm{BB}} & \mathbf{W}_{\mathrm{BN}} & \mathbf{z}_{\mathrm{B}} \\ \mathbf{W}_{\mathrm{BN}}^{\mathrm{T}} & \mathbf{W}_{\mathrm{NN}} & \mathbf{z}_{\mathrm{N}} \end{bmatrix} \tag{6-2-11}$$

If Gauss-Jordan pivots (see Section A-3) were to be effected in turn on all the diagonal elements of $\mathbf{W}_{\mathrm{BB}}$, then the last column of $\mathbf{E}$ would be transformed into

$$\begin{bmatrix} \mathbf{W}_{\mathrm{BB}}^{-1}\mathbf{z}_{\mathrm{B}} \\ \mathbf{z}_{\mathrm{N}} - \mathbf{W}_{\mathrm{BN}}^{\mathrm{T}}\mathbf{W}_{\mathrm{BB}}^{-1}\mathbf{z}_{\mathrm{B}} \end{bmatrix} = \begin{bmatrix} -\boldsymbol{\lambda}_{\mathrm{B}} \\ \mathbf{w}_{\mathrm{N}} \end{bmatrix}$$

It follows that if the proper partitioning of the constraints into binding and nonbinding ones is at hand, and if we sweep those rows of $\mathbf{E}$ which correspond to the binding constraints, then the last elements in those rows must become nonpositive (since $\boldsymbol{\lambda}_{\mathrm{B}} \geqslant \mathbf{0}$). Conversely, the last element in each unswept row of $\mathbf{E}$ must become nonnegative (since $\mathbf{w}_{\mathrm{N}} > \mathbf{0}$, but zero elements may appear under certain conditions of degeneracy). In order to find the proper partition, the following algorithm is suggested:

1. Form the tableau **E** which has p rows and $p + 1$ columns (p is the number of constraints active at $\boldsymbol{\theta}_i$).

2. Assign to the jth row of **E** ($j = 1, 2, \ldots, p$) the indicator $k_j = 1$.

3. Let e_j denote the current value of the last element in the jth row. Find $a = \min_j k_j e_j$. If $a \geqslant -\varepsilon$ (where ε is a small positive constant, say $\varepsilon = 10^{-6}$ for single precision calculations), proceed to step 5. Otherwise:

4. Let r be an index for which $a = k_r e_r$. Sweep the rth row (i.e., execute a Gauss–Jordan pivot on E_{rr}) and change the sign of k_r. Return to step 3.

5. The solution is now at hand. Consider the jth row of **E**. If $k_j = 1$, the jth row is unswept (or swept an even number of times). Hence the jth constraint is nonbinding. Therefore, $e_j = w_j = (\mathbf{A}^{\mathrm{T}}\mathbf{v})_j \geqslant 0$, and $\lambda_j = 0$. If $k_j = -1$, the jth row is swept and the jth constraint is binding. Hence $e_j = -\lambda_j \leqslant 0$, and $w_j + (\mathbf{A}^{\mathrm{T}}\mathbf{v})_j = 0$. To compute $\mathbf{v}_i$ (we now restore the subscript i), solve Eq. (5)

$$\mathbf{v}_i = \mathbf{R}_i(\mathbf{A}_i\boldsymbol{\lambda}_i - \mathbf{q}_i) = \mathbf{R}_i(\mathbf{A}_{\mathrm{B}}\boldsymbol{\lambda}_{\mathrm{B}} - \mathbf{q}_i) \tag{6-2-12}$$

where $\mathbf{A}_{\mathrm{B}}$ consists of the columns of $\mathbf{A}_i$ corresponding to the swept rows of **E**, and $\boldsymbol{\lambda}_{\mathrm{B}}$ consists of the last elements of those rows, with signs changed.

6. The actual step $\boldsymbol{\sigma}_i = \rho_i \mathbf{v}_i$ is computed by interpolation–extrapolation along the ray $\boldsymbol{\theta}_i + \rho\mathbf{v}_i$ ($\rho > 0$), with the additional proviso that $\boldsymbol{\theta}_{i+1} = \boldsymbol{\theta}_i + \rho_i \mathbf{v}_i$ must also satisfy all the constraints inactive at $\boldsymbol{\theta}_i$, and therefore not included in $\mathbf{A}_i$. If we denote these constraints as $\mathbf{c}_j{}^{\mathrm{T}}\boldsymbol{\theta} - b_j \geqslant 0$ ($j = 1, 2, \ldots$), then we must have $\mathbf{c}_j{}^{\mathrm{T}}\boldsymbol{\theta}_i + \rho_i \mathbf{c}_j{}^{\mathrm{T}}\mathbf{v}_i - b_j \geqslant 0$. Since $\rho_i > 0$ and $\mathbf{c}_j{}^{\mathrm{T}}\boldsymbol{\theta}_i - b_j > 0$ (inactivity at $\boldsymbol{\theta}_i$), it follows that only constraints for which $\mathbf{c}_j{}^{\mathrm{T}}\mathbf{v}_i < 0$ threaten to become violated. Hence ρ_i must satisfy the inequality

$$\rho_i \leqslant \rho_{\max} \equiv \min_{j \in J} \left[(b_j - \mathbf{c}_j{}^{\mathrm{T}}\boldsymbol{\theta}_i)/\mathbf{c}_j{}^{\mathrm{T}}\mathbf{v}_i\right] \tag{6-2-13}$$

where J is the set of indices j for which $\mathbf{c}_j{}^{\mathrm{T}}\mathbf{v}_i < 0$. If J is vacuous, then $\rho_{\max} = \infty$.

The following simple example illustrates the computation of $\mathbf{v}_i$. Assume that the objective function is approximated locally by the quadratic function

$$Q(\mathbf{v}) = \tfrac{1}{2}(v_1 + 1)^2 + \tfrac{1}{2}v_2{}^2 \tag{6-2-14}$$

We start at $v_1 = v_2 = 0$, where the active constraints are

$$v_2 \geqslant 0, \qquad v_1 + v_2 \geqslant 0, \qquad -v_1 + v_2 \geqslant 0 \tag{6-2-15}$$

The constraints and the contours of the objective function are displayed in Fig. 6-3. From Eq. (14) and Eq. (15) we deduce $\mathbf{q}^T = [1, 0]$, $\mathbf{H} = \mathbf{I}_2$; $\mathbf{R} = \mathbf{H}^{-1} = \mathbf{I}_2$, and

$$\mathbf{A} = \begin{bmatrix} 0 & 1 & -1 \\ 1 & 1 & 1 \end{bmatrix} \tag{6-2-16}$$

We form $\mathbf{z} = -\mathbf{A}^T\mathbf{R}\mathbf{q} = [0, -1, 1]^T$ and

$$\mathbf{W} = \mathbf{A}^T\mathbf{R}\mathbf{A} = \begin{bmatrix} 1 & 1 & 1 \\ 1 & 2 & 0 \\ 1 & 0 & 2 \end{bmatrix}.$$

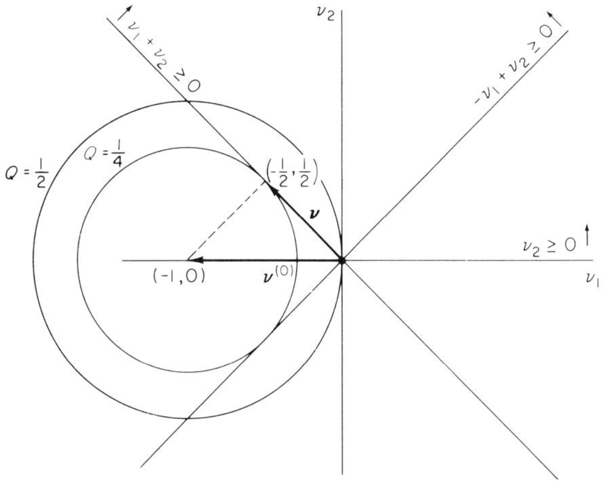

Fig. 6.3. Projection method with $Q(\nu) = \frac{1}{2}(\nu_1 + 1)^2 + \frac{1}{2}\nu_2{}^2$.

The algorithm proceeds as follows:

1. $$\mathbf{E} = \begin{bmatrix} 1 & 1 & 1 & 0 \\ 1 & 2 & 0 & -1 \\ 1 & 0 & 2 & 1 \end{bmatrix}.$$
2. $\mathbf{k}^T = [1, 1, 1]$.
3. $a = e_2 k_2 = -1 < -10^{-6}$.
4. $r = 2$, hence we pivot on $E_{2,2}$. The resulting tableau is

$$\mathbf{E} = \begin{bmatrix} \frac{1}{2} & -\frac{1}{2} & 1 & \frac{1}{2} \\ \frac{1}{2} & \frac{1}{2} & 0 & -\frac{1}{2} \\ 1 & 0 & 2 & 1 \end{bmatrix}, \qquad \mathbf{k} = \begin{bmatrix} 1 \\ -1 \\ 1 \end{bmatrix}$$

3. $a = e_1 k_1 = e_2 k_2 = \frac{1}{2} > -10^{-6}$.

5. Since $k_2 = -1$, the second constraint is binding, whereas the first and third are not. Hence $\mathbf{A}_B = \begin{bmatrix} 1 \\ 1 \end{bmatrix}$ (the second column of **A**) and $\boldsymbol{\lambda}_B = [\frac{1}{2}]$ (minus the second element in the last column of **E**). Hence, according to Eq. (12)

$$\mathbf{v} = \begin{bmatrix} 1 & 0 \\ 0 & 1 \end{bmatrix} \left(\begin{bmatrix} 1 \\ 1 \end{bmatrix} \times \tfrac{1}{2} - \begin{bmatrix} 1 \\ 0 \end{bmatrix} \right) = \begin{bmatrix} -\frac{1}{2} \\ \frac{1}{2} \end{bmatrix}$$

which, as is evident from Fig. 6-3, is the correct solution.

An extension of this algorithm to the case of nonlinear constraints is given by Bard (1971).

6-3. Projection with Bounded Parameters

The reader may wonder at the need for a complicated algorithm when bounds are the only constraints imposed on the parameters. Why not simply suppress those components of the step **v** which would violate the bounds? A simple example shows that such a procedure can produce erroneous results. Let, for instance

$$\boldsymbol{\theta}_i = \begin{bmatrix} 0 \\ 0 \end{bmatrix}, \qquad \mathbf{q}_i = \begin{bmatrix} -1 \\ 2 \end{bmatrix}, \qquad \mathbf{R}_i = \begin{bmatrix} 5 & 4 \\ 4 & 5 \end{bmatrix}$$

and suppose both components of $\boldsymbol{\theta}$ have zero lower bounds. Since $q_1 = -1$, it is clear that the objective function can be reduced by increasing θ_1 while keeping θ_2 constant. However, if we compute **v** we find $-\mathbf{R}_i\mathbf{q}_i = [-3, \ -6]$. According to the above suggestion, since both components of **v** are to be reduced below their lower bounds, we would conclude erroneously that we are at the minimum and take no step at all. We shall proceed therefore to apply the algorithm of the preceding section.

At the ith iteration, several of the components of $\boldsymbol{\theta}$ (we omit the subscript i for convenience) may be at their lower bounds, and others at their upper bounds, while the remainder are free to move in either direction. For simplicity in the algebraic exposition to follow, we shall treat lower bounds only, but later we give the arithmetic details for both cases. If θ_α is at its lower bound, the corresponding constraint takes the form

$$v_\alpha \geqslant 0 \tag{6-3-1}$$

Let $\mathbf{v}^{(0)} = -\mathbf{Rq}$ be the step that would be taken in the absence of constraints. It is easy to see that forming $\mathbf{A}^T\mathbf{R}$ merely picks out of **R** those rows which correspond to variables which are at a bound; postmultiplying this by **A** picks out the corresponding columns. Thus, $\mathbf{W} = \mathbf{A}^T\mathbf{R}\mathbf{A} = \tilde{\mathbf{R}}$ where $\tilde{\mathbf{R}}$ is the set of elements of $\mathbf{R}$ at the junction of active rows and columns. Similarly, $\mathbf{z} = -\mathbf{A}^T\mathbf{R}\mathbf{q}$ is obtained by taking the active elements of $\mathbf{v}^{(0)}$. Hence, $\mathbf{E}$ can be formed by inspection from $\mathbf{R}$ and $\mathbf{v}^{(0)}$

$$\mathbf{E} = [\tilde{\mathbf{R}}, \tilde{\mathbf{v}}^{(0)}] \tag{6-3-2}$$

It is easily verified that the following procedure is equivalent to the algorithm of Section 6-2 in the present context:

1. Set up the tableau $\mathbf{E}$, which has p rows and $p + 1$ columns.
2. For $j = 1, 2, \ldots, p$ let $m_j = \{\begin{smallmatrix}+\\-\end{smallmatrix}1\}$ if the jth row corresponds to a variable at its $\left\{\begin{smallmatrix}\text{lower}\\\text{upper}\end{smallmatrix}\right\}$ bound.
3. For $j = 1, 2, \ldots, p$ let $k_j = 1$.
4. Let e_j denote the current value of the last element in the jth row of $\mathbf{E}$. Find $a = \min_j m_j k_j e_j$. If $a \geqslant -\varepsilon$, proceed to step 6. Otherwise:
5. Let r be an index for which $a = m_r k_r e_r$. Sweep the rth row (Gauss-Jordan pivot on E_{rr}) and change the sign of k_r. Return to step 4.
6. The solution is now at hand. Bounds for which $k_j = -1$ are *binding*, others are not. We construct $\boldsymbol{\lambda}_B$ by taking the elements of $-\mathbf{e}$ in the binding rows, and we construct $\mathbf{R}\mathbf{A}_B$ by taking the columns of $\mathbf{R}$ corresponding to variables which are at a binding bound. It is thus easy to compute $\mathbf{v}$ using Eq. (6-2-12)

$$\mathbf{v} = (\mathbf{R}\mathbf{A}_B)\boldsymbol{\lambda}_B + \mathbf{v}^{(0)} \tag{6-3-3}$$

Since variables at binding bounds cannot change their values, the corresponding elements of $\mathbf{v}$ automatically come out 0.

We illustrate the procedure by means of a numerical example (a further illustration appears in Section 6-12):

Suppose the following values are current in the ith iteration

$$\boldsymbol{\theta}_i = \begin{bmatrix} 0 \\ 0.5 \\ 1 \end{bmatrix}, \qquad \mathbf{q}_i = -\begin{bmatrix} 3 \\ 1 \\ 2 \end{bmatrix}, \qquad \mathbf{R}_i = \begin{bmatrix} 5 & -10 & -4 \\ -10 & 30 & 8 \\ -4 & 8 & 5 \end{bmatrix}$$

Hence, $\mathbf{v}_i^{(0)} = -\mathbf{R}_i\mathbf{q}_i = [-3, 16, 6]^T$. Suppose all parameters are restricted to the range zero to one. Thus θ_1 and θ_3 are at their lower and upper bounds, respectively. The algorithm proceeds as follows:

1. $\mathbf{E} = \begin{bmatrix} 5 & -4 & -3 \\ -4 & 5 & 6 \end{bmatrix}.$
2. $\mathbf{m} = [1, -1]^T$.
3. $\mathbf{k} = [1, 1]^T$.
4. $a = m_2 k_2 e_2 = -6 < -10^{-6}$.

5. $r = 2$, hence we pivot on $E_{2,2}$ obtaining the new tableau

$$\mathbf{E} = \begin{bmatrix} 1.8 & 0.8 & 1.8 \\ -0.8 & 0.2 & 1.2 \end{bmatrix}, \qquad \mathbf{k} = \begin{bmatrix} 1 \\ -1 \end{bmatrix}$$

4. $a = m_2 k_2 e_2 = 1.2 > -10^{-6}$.

6. Since $k_2 = -1$, the second constraint (the bound on θ_3) is binding. We have $\boldsymbol{\lambda}_B = [-1.2]$ (second element of $-\mathbf{e}$) and $\mathbf{R}\mathbf{A}_B = [-4, 8, 5]^T$ (third column of $\mathbf{R}$). Hence

$$\mathbf{v} = \begin{bmatrix} -4 \\ 8 \\ 5^* \end{bmatrix} \times (-1.2) + \begin{bmatrix} -3 \\ 16 \\ 6^* \end{bmatrix} = \begin{bmatrix} 1.8 \\ 6.4 \\ 0 \end{bmatrix} \qquad \text{(6-3-4)}$$

As expected $v_3 = 0$, and we could have replaced the starred elements in Eq. (4) by zeros without altering the result.

The actual step $\boldsymbol{\sigma}$ will be some multiple of $\mathbf{v}$, i.e., $\boldsymbol{\sigma} = \rho\mathbf{v}$. The multiplier ρ is limited to the range $0 \leqslant \rho \leqslant \min(1, \rho_{max})$, with ρ_{max} to be determined by the requirement that $\boldsymbol{\theta}_i + \rho_{max}\mathbf{v}$ remains feasible. That is, we must have

$$0 + 1.8\rho_{max} \leqslant 1, \qquad 0.5 + 6.4\rho_{max} \leqslant 1$$

Hence

$$\rho_{max} = \min[(1 - 0)/1.8, (1 - 0.5)/6.4] = 0.3125$$

The actual value of ρ would be determined by interpolation–extrapolation (see Section 5-14) to guarantee a decrease in the objective function value.

6-4. Transformation of Variables

Sometimes a change of variables can transform a constrained problem into an unconstrained one. For instance, to minimize $\Phi(\theta)$ with θ required to be positive is equivalent to minimizing $\Phi(\rho^2)$ with ρ free to assume any value. Similarly, if θ must satisfy $\beta + \theta \geqslant \alpha$, then we minimize $\Phi((\alpha + \beta)/2 + [(\alpha - \beta)/2] \sin \rho)$ with ρ unconstrained, since as ρ varies from $-\infty$ to ∞, the quantity $(\alpha + \beta)/2 + [(\alpha - \beta)/2] \sin \rho$ remains within the bounds α and β. Box (1966) has demonstrated that with ingenuity even more complicated constraints can sometimes be eliminated by means of such transformations. We have some numerical evidence that the use of transformations is no more efficient than the use of penalty functions, and we prefer the latter because of their greater generality.

6-5. Minimax Problems

Some estimation problems (see Sections 4-13 and 4-17) led us to seek the value of $\boldsymbol{\theta}$ that minimizes

$$\Phi(\boldsymbol{\theta}) \equiv \max_{\mu} e_{\mu}(\boldsymbol{\theta}) \tag{6-5-1}$$

This was shown to be equivalent to finding z, $\boldsymbol{\theta}$ so as to minimize

$$\Psi(\boldsymbol{\theta}, z) \equiv z \tag{6-5-2}$$

subject to

$$z - |e_{\mu}(\boldsymbol{\theta})| \geqslant 0 \qquad (\mu = 1, 2, \ldots, n) \tag{6-5-3}$$

An iterative scheme which is analogous to the Gauss method for least squares problems has been suggested by Osborne and Watson (1969). Let $\boldsymbol{\theta}_i$ be the value of $\boldsymbol{\theta}$ at the ith iteration. We approximate $e_{\mu}(\boldsymbol{\theta}_i + \mathbf{v}_i)$ by a linear expression $e_{\mu}(\boldsymbol{\theta}_i) + \mathbf{b}_{\mu i}^T \mathbf{v}_i$. The nonlinear programming problem Eq. (2), Eq. (3) is replaced by the *linear programming problem*:

Find z, $\mathbf{v}_i$ so as to minimize Eq. (2) subject to

$$z - e_{\mu}(\boldsymbol{\theta}_i) - \mathbf{b}_{\mu i}^T \mathbf{v}_i \geq 0, \qquad z + e_{\mu}(\boldsymbol{\theta}_i) + \mathbf{b}_{\mu i}^T \mathbf{v}_i \geqslant 0 \qquad (\mu = 1, 2, \ldots, n) \tag{6-5-4}$$

This problem can be solved by means of standard linear programming(LP) methods. Once $\mathbf{v}_i$ has been computed, the length of the step taken in this direction is determined by interpolation–extrapolation.

B. Equality Constraints

6-6. Exact Structural Models

In Section 4-11 we have formulated some estimation problems as requiring the minimization of a function $\Phi(\hat{\mathbf{W}})$ subject to the constraints $\mathbf{G}(\hat{\mathbf{W}}, \boldsymbol{\theta}) = \mathbf{0}$. The "true" data $\hat{\mathbf{W}}$ and the parameters $\boldsymbol{\theta}$ are the unknowns. In the ensuing discussion we will treat $\mathbf{G}$ and $\hat{\mathbf{W}}$ as vectors $\mathbf{g}$ and $\hat{\mathbf{w}}$. We suppose that at the ith iteration we have current values of $\hat{\mathbf{w}}_i$ and $\boldsymbol{\theta}_i$. The method of solution given below follows suggestions by Deming (1943).

We denote the current values of Φ and $\mathbf{g}$ by Φ_i and $\mathbf{g}_i$, respectively, and adopt the following notation:

$$\mathbf{q} \equiv \partial\Phi/\partial\hat{\mathbf{w}}, \qquad \mathbf{H} \equiv \partial^2\Phi/\partial\hat{\mathbf{w}}\,\partial\hat{\mathbf{w}}$$
$$\mathbf{A} \equiv \partial\mathbf{g}/\partial\hat{\mathbf{w}}, \qquad \mathbf{B} \equiv \partial\mathbf{g}/\partial\boldsymbol{\theta} \tag{6-6-1}$$

with subscript i denoting the value at $\hat{\mathbf{W}} = \hat{\mathbf{W}}_i$, $\boldsymbol{\theta} = \boldsymbol{\theta}_i$. We define the following functions:

$$\tilde{\Phi}_i(\delta\hat{\mathbf{W}}) \equiv \Phi_i + \mathbf{q}_i^{\mathrm{T}}\,\delta\hat{\mathbf{W}} + \tfrac{1}{2}\,\delta\hat{\mathbf{W}}^{\mathrm{T}}\,\mathbf{H}_i\,\delta\hat{\mathbf{W}} \tag{6-6-2}$$

$$\tilde{\mathbf{G}}_i(\delta\hat{\mathbf{W}}, \delta\boldsymbol{\theta}) \equiv \mathbf{G}_i + \mathbf{A}_i\,\delta\hat{\mathbf{W}} + \mathbf{B}_i\,\delta\boldsymbol{\theta}_i \tag{6-6-3}$$

$\tilde{\Phi}_i$ is the second-order Taylor series approximation to Φ, and $\tilde{\mathbf{G}}_i$ is the first-order Taylor series approximation to $\mathbf{G}$. We now replace our original problem by the following:

Find $\delta\hat{\mathbf{W}}$ and $\delta\boldsymbol{\theta}$ so as to minimize $\tilde{\Phi}_i$, while satisfying the constraints $\tilde{\mathbf{G}}_i = \mathbf{0}$. We introduce a vector of Lagrange multipliers $\boldsymbol{\lambda}$ and seek the stationary point of the Lagrangian

$$\Lambda(\delta\hat{\mathbf{W}}, \delta\boldsymbol{\theta}, \boldsymbol{\lambda}) \equiv \tilde{\Phi}_i + \boldsymbol{\lambda}^{\mathrm{T}}\tilde{\mathbf{G}}_i \tag{6-6-4}$$

Accordingly, we form the normal equations:

$$\partial\Lambda/\partial(\delta\hat{\mathbf{W}}) = \mathbf{q}_i + \mathbf{H}_i\,\delta\hat{\mathbf{W}} + \mathbf{A}_i^{\mathrm{T}}\boldsymbol{\lambda} = \mathbf{0} \tag{6-6-5}$$

$$\partial\Lambda/\partial\boldsymbol{\lambda} = \mathbf{G}_i + \mathbf{A}_i\,\delta\hat{\mathbf{W}} + \mathbf{B}_i\,\delta\boldsymbol{\theta} = \mathbf{0} \tag{6-6-6}$$

$$\partial\Lambda/\partial(\delta\boldsymbol{\theta}) = \mathbf{B}_i^{\mathrm{T}}\boldsymbol{\lambda} = \mathbf{0} \tag{6-6-7}$$

From Eq. (5)

$$\delta\hat{\mathbf{W}} = -\mathbf{H}_i^{-1}(\mathbf{q}_i + \mathbf{A}_i^{\mathrm{T}}\boldsymbol{\lambda}) \tag{6-6-8}$$

So that, from Eq (6)

$$\mathbf{G}_i - \mathbf{A}_i\mathbf{H}_i^{-1}\mathbf{q}_i - \mathbf{A}_i\mathbf{H}_i^{-1}\mathbf{A}_i^{\mathrm{T}}\boldsymbol{\lambda} + \mathbf{B}_i\,\delta\boldsymbol{\theta} = 0 \tag{6-6-9}$$

Solving for $\boldsymbol{\lambda}$ we obtain

$$\boldsymbol{\lambda} = \mathbf{C}_i^{-1}(\mathbf{B}_i\,\delta\boldsymbol{\theta} - \mathbf{A}_i\mathbf{H}_i^{-1}\mathbf{q}_i + \mathbf{G}_i) \tag{6-6-10}$$

where

$$\mathbf{C}_i \equiv \mathbf{A}_i\mathbf{H}_i^{-1}\mathbf{A}_i^{\mathrm{T}} \tag{6-6-11}$$

Substituting Eq. (10) in Eq. (7) and solving for $\delta\boldsymbol{\theta}$ we obtain

$$\delta\boldsymbol{\theta} = \mathbf{D}_i^{-1}\mathbf{B}_i^{\mathrm{T}}\mathbf{C}_i^{-1}(\mathbf{A}_i\mathbf{H}_i^{-1}\mathbf{q}_i - \mathbf{G}_i) \tag{6-6-12}$$

where

$$\mathbf{D}_i = \mathbf{B}_i^{\mathrm{T}}\mathbf{C}_i^{-1}\,\mathbf{B}_i \tag{6-6-13}$$

The matrix $\mathbf{D}_i$ plays a role analogous to that of $\mathbf{N}_i$ in the Gauss method, and the same "almost inversion" methods (e.g., directional discrimination or Marquardt's method) should be used where $\mathbf{D}_i^{-1}$ is required.

Eq. (12) enables one to compute $\boldsymbol{\theta}_{i+1} = \boldsymbol{\theta}_i + \delta\boldsymbol{\theta}$. Then Eq. (10) can be used to compute $\boldsymbol{\lambda}$, which in turn can be substituted in Eq. (8) to compute $\delta\hat{\mathbf{W}}$ and the new approximation $\hat{\mathbf{W}}_{i+1} = \hat{\mathbf{W}}_i + \delta\hat{\mathbf{W}}$. Usually $\hat{\mathbf{W}}$ is close to the observed values $\mathbf{W}$, so that we naturally take $\hat{\mathbf{W}} = \mathbf{W}$ as the initial guess.

6-7. Convergence Monitoring

When we apply the Deming procedure there is no natural way of telling whether or not progress towards the solution has occurred in any given iteration. There is no way of telling in advance whether the final value of the objective function or the Lagrangian must be less or greater than the current value. At the solution, however, the equations $\mathbf{G} = \mathbf{0}$ must be satisfied, so that if $\mathbf{Q}_i$ is some positive definite matrix, it is natural to require that the ith iteration must cause the value of $\mathbf{G}^{\mathrm{T}}\mathbf{Q}_i\mathbf{G}$ to decrease. Now we have:

$$\partial(\mathbf{G}^{\mathrm{T}}\mathbf{Q}_i\mathbf{G})/\partial\boldsymbol{\theta} = 2\mathbf{B}_i^{\mathrm{T}}\mathbf{Q}_i\mathbf{G} \tag{6-7-1}$$

$$\partial(\mathbf{G}^{\mathrm{T}}\mathbf{Q}_i\mathbf{G})/\partial\hat{\mathbf{W}} = 2\mathbf{A}_i^{\mathrm{T}}\mathbf{Q}_i\mathbf{G} \tag{6-7-2}$$

Hence, if $\delta\boldsymbol{\theta}$ and $\delta\hat{\mathbf{W}}$ are given by Eq. (6-6-12) and Eq. (6-6-8), it turns out after much arithmetic that

$$\delta(\mathbf{G}^{\mathrm{T}}\mathbf{Q}_i\mathbf{G}) = -2\mathbf{G}^{\mathrm{T}}\mathbf{Q}_i\mathbf{G} < 0 \tag{6-7-3}$$

This means that the quantity $\mathbf{G}^{\mathrm{T}}\mathbf{Q}_i\mathbf{G}$ decreases initially as we take a small step in the prescribed direction. A natural choice of $\mathbf{Q}_i$ is the inverse of the covariance matrix of $\mathbf{G}$. If $\mathbf{V}_{\mathbf{W}}$ is the covariance matrix of the data $\mathbf{W}$, then the covariance of $\mathbf{G}$ is approximately $\mathbf{A}_i\mathbf{V}_{\mathbf{W}}\mathbf{A}_i^{\mathrm{T}}$. Usually $\mathbf{V}_{\mathbf{W}} = \mathbf{H}_i^{-1}$, so we choose

$$\mathbf{Q}_i = (\mathbf{A}_i\mathbf{H}_i^{-1}\mathbf{A}_i^{\mathrm{T}})^{-1} = \mathbf{C}_i^{-1} \tag{6-7-4}$$

Our strategy at the ith iteration is the following:

1. Compute $Z_0 \equiv \mathbf{G}_i^{\mathrm{T}}\mathbf{C}_i^{-1}\mathbf{G}_i$.
2. Compute $\delta\boldsymbol{\theta}$ and $\delta\hat{\mathbf{W}}$ using Eq. (6-6-12), Eq. (6-6-10), and Eq. (6-6-8).
3. Compute $Z_1 \equiv \mathbf{G}^{\mathrm{T}}(\boldsymbol{\theta}_i + \delta\boldsymbol{\theta}, \hat{\mathbf{W}}_i + \delta\hat{\mathbf{W}})\mathbf{C}_i^{-1}\mathbf{G}(\boldsymbol{\theta}_i + \delta\boldsymbol{\theta}, \hat{\mathbf{W}}_i + \delta\hat{\mathbf{W}})$.
4. If $Z_1 < Z_0$, set $\boldsymbol{\theta}_{i+1} = \boldsymbol{\theta}_i + \delta\boldsymbol{\theta}$, $\hat{\mathbf{W}}_{i+1} = \hat{\mathbf{W}}_i + \delta\hat{\mathbf{W}}$ and proceed to the next iteration. Otherwise, use interpolation (Section 5.14) to shorten $\delta\boldsymbol{\theta}$ and $\delta\hat{\mathbf{W}}$, and return to step 3.

Note that a different weighting matrix $\mathbf{C}_i^{-1}$ is used in each iteration. Therefore, we can compare Z_1 and Z_0 in any given iteration, but not from one iteration to the next.

At the end of the calculations, we can apply the necessary condition Eq. (3-6-6) as a test of convergence. In terms of our present variables, the conditions take the form

$$\mathbf{B}^{\mathrm{T}}\boldsymbol{\lambda} = \mathbf{0}, \qquad \mathbf{q} + \mathbf{A}^{\mathrm{T}}\boldsymbol{\lambda} = \mathbf{0} \tag{6-7-5}$$

These, along with the original equations

$$\mathbf{G} = \mathbf{0} \tag{6-7-6}$$

must be satisfied by the final values $\boldsymbol{\theta}^*$, $\hat{\mathbf{W}}^*$, $\boldsymbol{\lambda}^*$.

6-8. Some Special Cases

Consider an objective function of the weighted least squares form

$$\Phi(\hat{\mathbf{W}}) = \tfrac{1}{2} \sum_{\mu=1}^{n} (\hat{\mathbf{w}}_\mu - \mathbf{w}_\mu)^T \mathbf{V}_\mu^{-1} (\hat{\mathbf{w}}_\mu - \mathbf{w}_\mu) \tag{6-8-1}$$

Therefore:

$$\mathbf{q} = \begin{pmatrix} \mathbf{V}_1^{-1}(\hat{\mathbf{w}}_1 - \mathbf{w}_1) \\ \mathbf{V}_2^{-1}(\hat{\mathbf{w}}_2 - \mathbf{w}_2) \\ \vdots \\ \mathbf{V}_n^{-1}(\hat{\mathbf{w}}_n - \mathbf{w}_n) \end{pmatrix} \tag{6-8-2}$$

$$\mathbf{H} = \begin{bmatrix} \mathbf{V}_1^{-1} & \mathbf{0} & \cdots & \mathbf{0} \\ \mathbf{0} & \mathbf{V}_2^{-1} & & \\ \vdots & & & \\ \mathbf{0} & \cdots & & \mathbf{V}_n^{-1} \end{bmatrix} \tag{6-8-3}$$

The other interesting objective function is typified by Eq. (4-10-12), and occurs when a subset of variables $\mathbf{y}_\mu$ (not exceeding m in number) have unknown covariance, while the remaining (if any) variables $\mathbf{x}_\mu$ have known covariance $\mathbf{P}_\mu$. We have then (assuming zero correlation between $\mathbf{x}_\mu$ and $\mathbf{y}_\mu$)

$$\Phi(\hat{\mathbf{x}}, \hat{\mathbf{y}}) = (n/2) \log \det \sum_{\mu=1}^{n} (\hat{\mathbf{y}}_\mu - \mathbf{y}_\mu)(\hat{\mathbf{y}}_\mu - \mathbf{y}_\mu)^T + \tfrac{1}{2} \sum_{\mu=1}^{n} (\hat{\mathbf{x}}_\mu - \mathbf{x}_\mu)^T \mathbf{P}_\mu^{-1} (\hat{\mathbf{x}}_\mu - \mathbf{x}_\mu)$$

The vector $\mathbf{q}$ is made up of elements

$$\partial\Phi/\partial\hat{\mathbf{x}}_\mu = \mathbf{P}_\mu^{-1}(\hat{\mathbf{x}}_\mu - \mathbf{x}_\mu), \qquad \partial\Phi/\partial\hat{\mathbf{y}}_\mu = n\mathbf{M}_{yy}^{-1}(\hat{\mathbf{y}}_\mu - \mathbf{y}_\mu) \tag{6-8-4}$$

The Hessian $\mathbf{H}$ has a complicated form, but we may use the Gauss approximation in which $(1/n)\mathbf{M}_{yy}$ replaces the covariance of $\mathbf{y}_\mu$ where required. We take then

$$\partial^2\Phi/\partial\mathbf{x}_\mu\,\partial\mathbf{x}_\mu = \mathbf{P}_\mu^{-1}, \qquad \partial^2\Phi/\partial\mathbf{x}_\mu\,\partial\mathbf{y}_\mu = \mathbf{0}$$

$$\partial^2\Phi/\partial\mathbf{y}_\mu\,\partial\mathbf{y}_\mu \approx n\mathbf{M}_{yy}^{-1} = n\left[\sum_{\mu=1}^{n} (\hat{\mathbf{y}}_\mu - \mathbf{y}_\mu)(\hat{\mathbf{y}}_\mu - \mathbf{y}_\mu)^T\right]^{-1} \tag{6-8-5}$$

In the sequel, then, we shall assume that Eqs. (2) and (3) apply always, with (in the case of partly unknown covariance)

$$\mathbf{V}_\mu = \begin{bmatrix} \mathbf{P}_\mu & \mathbf{0} \\ \mathbf{0} & (1/n)\mathbf{M}_{yy} \end{bmatrix} \tag{6-8-6}$$

Note that $\mathbf{P}_\mu$ remains the same, but $(1/n)\mathbf{M}_{yy}$ varies from iteration to iteration. If the unknown portions of $\mathbf{V}$ are assumed diagonal, then we only use the corresponding diagonal elements of $(1/n)\mathbf{M}$, and substitute zeroes for the off-diagonal elements.

We now work out the details of the algorithm. Letting

$$\mathbf{e}_\mu \equiv \hat{\mathbf{w}}_\mu - \mathbf{w}_\mu \tag{6-8-7}$$

we note that:

$$\mathbf{q} = \begin{bmatrix} \mathbf{q}_1 \\ \mathbf{q}_2 \\ \vdots \\ \mathbf{q}_n \end{bmatrix} \quad \text{where} \quad \mathbf{q}_\mu \equiv \mathbf{V}_\mu^{-1}\mathbf{e}_\mu \tag{6-8-8}$$

$$\mathbf{H} = \begin{bmatrix} \mathbf{V}_1^{-1} & \mathbf{0} & \cdots & \mathbf{0} \\ \mathbf{0} & & \ddots & \\ \vdots & & & \\ \mathbf{0} & & & \mathbf{V}_n^{-1} \end{bmatrix} \tag{6-8-9}$$

$$\mathbf{G} = \begin{bmatrix} \mathbf{g}_1 \\ \mathbf{g}_2 \\ \vdots \\ \mathbf{g}_n \end{bmatrix} \tag{6-8-10}$$

$$\mathbf{A} = \begin{bmatrix} \mathbf{A}_1 & \mathbf{0} & \cdots & \mathbf{0} \\ \mathbf{0} & \mathbf{A}_2 & & \\ \vdots & & \ddots & \\ \mathbf{0} & & & \mathbf{A}_n \end{bmatrix} \quad \text{where} \quad \mathbf{A}_\mu \equiv \partial \mathbf{g}_\mu / \partial \hat{\mathbf{w}}_\mu \tag{6-8-11}$$

$$\mathbf{B} = \begin{bmatrix} \mathbf{B}_1 \\ \mathbf{B}_2 \\ \vdots \\ \mathbf{B}_n \end{bmatrix} \quad \text{where} \quad \mathbf{B}_\mu \equiv \partial \mathbf{g}_\mu / \partial \boldsymbol{\theta} \tag{6-8-12}$$

so that:

$$\mathbf{C} = \mathbf{A}\mathbf{H}^{-1}\mathbf{A}^{\mathrm{T}} = \begin{bmatrix} \mathbf{C}_1 & \mathbf{0} & \cdots & \mathbf{0} \\ \mathbf{0} & \mathbf{C}_2 & & \\ \vdots & & \ddots & \\ \mathbf{0} & & & \mathbf{C}_n \end{bmatrix} \quad \text{where} \quad \mathbf{C}_\mu \equiv \mathbf{A}_\mu \mathbf{V}_\mu {\mathbf{A}_\mu}^{\mathrm{T}} \tag{6-8-13}$$

$$\mathbf{D} = \mathbf{B}^{\mathrm{T}}\mathbf{C}^{-1}\mathbf{B} = \sum_{\mu=1}^{n} {\mathbf{B}_\mu}^{\mathrm{T}}\mathbf{C}_\mu^{-1}\mathbf{B}_\mu \tag{6-8-14}$$

Computations which we leave as an exercise result in:

$$\delta\boldsymbol{\theta} = \mathbf{D}^{-1} \sum_{\mu=1}^{n} {\mathbf{B}_\mu}^{\mathrm{T}}\mathbf{C}_\mu^{-1}(\mathbf{A}_\mu\mathbf{e}_\mu - \mathbf{g}_\mu) \tag{6-8-15}$$

$$\boldsymbol{\lambda}_\mu = \mathbf{C}_\mu^{-1}(\mathbf{B}_\mu\, \delta\boldsymbol{\theta} - \mathbf{A}_\mu\mathbf{e}_\mu + \mathbf{g}_\mu) \tag{6-8-16}$$

$$\delta\hat{\mathbf{w}}_\mu = -\mathbf{e}_\mu - \mathbf{V}_\mu{\mathbf{A}_\mu}^{\mathrm{T}}\boldsymbol{\lambda}_\mu \tag{6-8-17}$$

The pseudo-objective function of Section 6.7 takes the form

$$Z = \mathbf{G}^{\mathrm{T}}\mathbf{Q}\mathbf{G} = \sum_{\mu=1}^{n} \mathbf{g}_\mu{}^{\mathrm{T}}\mathbf{C}_\mu^{-1}\mathbf{g}_\mu \tag{6-8-18}$$

Further simplifications occur in the single equation case. Here $\mathbf{g}_\mu$ is a single number g_μ, while $\mathbf{A}_\mu$ and $\mathbf{B}_\mu$ are row vectors $\mathbf{a}_\mu{}^{\mathrm{T}}$ and $\mathbf{b}_\mu{}^{\mathrm{T}}$ respectively

$$\mathbf{a}_\mu \equiv \partial g_\mu / \partial \hat{\mathbf{w}}_\mu, \qquad \mathbf{b}_\mu \equiv \partial g_\mu / \partial \boldsymbol{\theta} \tag{6-8-19}$$

Hence:

$$\mathbf{C} = \mathrm{diag}(c_\mu), \qquad \text{where} \quad c_\mu \equiv \mathbf{a}_\mu{}^{\mathrm{T}}\mathbf{V}_\mu \mathbf{a}_\mu \tag{6-8-20}$$

$$\mathbf{D} = \sum_{\mu=1}^{n} (1/c_\mu)\mathbf{b}_\mu \mathbf{b}_\mu{}^{\mathrm{T}} \tag{6-8-21}$$

$$\delta\boldsymbol{\theta} = \mathbf{D}^{-1} \sum_{\mu=1}^{n} (1/c_\mu)(\mathbf{a}_\mu{}^{\mathrm{T}}\mathbf{e}_\mu - g_\mu)\mathbf{b}_\mu \tag{6-8-22}$$

$$\lambda_\mu = (1/c_\mu)(\mathbf{b}_\mu{}^{\mathrm{T}}\,\delta\boldsymbol{\theta} - \mathbf{a}_\mu{}^{\mathrm{T}}\mathbf{e}_\mu + g_\mu) \tag{6-8-23}$$

$$\delta\hat{\mathbf{w}}_\mu = -\,\mathbf{e}_\mu - \lambda_\mu \mathbf{V}_\mu \mathbf{a}_\mu \tag{6-8-24}$$

$$Z = \mathbf{g}^{\mathrm{T}}\mathbf{Q}\mathbf{g} = \sum_{\mu=1}^{n} (1/c_\mu) g_\mu{}^2 \tag{6-8-25}$$

The algorithm is always started with some initial guess, $\boldsymbol{\theta}_1$, and with $\hat{\mathbf{w}}_\mu = \mathbf{w}_\mu$. A difficulty arises when the covariance matrices $\mathbf{V}_\mu$ are at least partly unknown. For then, from Eq. (6), certain rows and columns of $\mathbf{V}_\mu$ are taken from $(1/n)\mathbf{M}$; but in the first iteration $\mathbf{M} = \mathbf{0}$ causing $\mathbf{V}_\mu$ to be singular. Furthermore, a glance at Eq. (17) reveals that all components of $\delta\hat{\mathbf{w}}_\mu$ corresponding to the y variables will be zero. The difficulty is easily overcome by arbitrarily assigning to the unknown elements of $\mathbf{V}$ some reasonable initial guesses for the first iteration only. The method is illustrated in Sections 6-13–6-14.

6-9. Penalty Functions

The idea of penalty functions can also be applied to equality constraints. Here we penalize values of the unknown $\boldsymbol{\phi}$ (including both $\boldsymbol{\theta}$ and $\hat{\mathbf{w}}$) according to their deviations from the constraints. To the objective function $\Phi(\boldsymbol{\phi})$ we add a term proportional to $g_j{}^2(\boldsymbol{\phi})$ for each equality constraint $g_j = 0$. SUMT (sequential unconstrained maximizing technique) as applied to problems including both equality and inequality constraints consists of defining the objective functions (Fiacco and McCormick 1965, 1967)

$$\Phi_k{}^{\dagger}(\boldsymbol{\phi}) \equiv \Phi(\boldsymbol{\phi}) + \alpha_k \sum_j 1/h_j(\boldsymbol{\phi}) + \alpha_k^{-1/2} \sum_j g_j{}^2(\boldsymbol{\phi}) \tag{6-9-1}$$

where $\alpha_1, \alpha_2, \ldots$ is a sequence of decreasing positive numbers converging to zero. Let $\boldsymbol{\phi}_k^*$ be the value of $\boldsymbol{\phi}$ which minimizes $\Phi_k^\dagger$; then, under suitable assumptions on the convexity of the feasible region and the concavity of the functions, Fiacco and McCormick prove the convergence of the sequence $\boldsymbol{\phi}_1^*, \boldsymbol{\phi}_2^*, \ldots$ to the minimum of $\Phi(\boldsymbol{\phi})$ satisfying the constraints

$$h_j(\boldsymbol{\phi}) \geqslant 0 \quad (j = 1, 2, \ldots), \qquad g_j(\boldsymbol{\phi}) = 0 \quad (j = 1, 2, \ldots) \tag{6-9-2}$$

The application of the Gauss method to the minimization of $\Phi_k^\dagger$ is obvious.

6-10. Linear Equality Constraints

If the unknown parameters are supposed to satisfy linear equality relationships, these can be handled by means of the projection method of Section 6-2. All we need to do is include permanently the equality constraints in the binding set by sweeping the corresponding rows of **E**. All tests on the sign of the last element in a row belonging to an equality constraint are omitted. The initial guess must be chosen so that it satisfies all the constraints.

6-11. Least Squares Problem with Penalty Functions

We return to the single equation least squares Problem of Section 5-21. We recall that we encountered some difficulties in converging to the solution when starting from the initial guess $\boldsymbol{\theta}_1 = [100, 2000]^{\mathrm{T}}$. We shall attempt to overcome these difficulties by imposing bounds on the parameters. Specifically let us require that

$$0 \leqslant \theta_1 \leqslant 100{,}000, \qquad 0 \leqslant \theta_2 \leqslant 2{,}000{,}000$$

corresponding to the constraints:

$$\begin{aligned} h_1(\boldsymbol{\theta}) &\equiv \theta_1 \geqslant 0, & h_2(\boldsymbol{\theta}) &\equiv 100{,}000 - \theta_1 \geqslant 0, \\ h_3(\boldsymbol{\theta}) &\equiv \theta_2 \geqslant 0, & h_4(\boldsymbol{\theta}) &\equiv 2{,}000{,}000 - \theta_2 \geqslant 0 \end{aligned} \tag{6-11-1}$$

According to Eq. (6-1-2) we form the penalty function

$$\zeta(\boldsymbol{\theta}) \equiv \sum_{j=1}^{4} \zeta_j(\boldsymbol{\theta}) = 0.01/\theta_1 + 0.01/(100{,}000 - \theta_1) + 0.2/\theta_2 + 0.2/(2{,}000{,}000 - \theta_2) \tag{6-11-2}$$

The coefficients α_j were determined as $10^{-5}\boldsymbol{\theta}_1$. Our new objective function is

$$\Phi^\dagger(\boldsymbol{\theta}) = \sum_{\mu=1}^{15} e_\mu^2(\boldsymbol{\theta}) + \zeta(\boldsymbol{\theta}) \tag{6-11-3}$$

Using Eqs. (6-1-7) and (6-1-9) we find:

$$\mathbf{q}^\dagger = \begin{bmatrix} -2\sum_{\mu=1}^{15} e_\mu \dfrac{\partial f_\mu}{\partial \theta_1} - \dfrac{0.01}{\theta_1{}^2} + \dfrac{0.01}{(100{,}000 - \theta_1)^2} \\ -2\sum_{\mu=1}^{15} e_\mu \dfrac{\partial f_\mu}{\partial \theta_2} - \dfrac{0.2}{\theta_2{}^2} + \dfrac{0.2}{(2{,}000{,}000 - \theta_2)^2} \end{bmatrix} \tag{6-11-4}$$

$$\mathbf{N}^\dagger = \begin{bmatrix} 2\sum_{\mu=1}^{15} \left(\dfrac{\partial f_\mu}{\partial \theta_1}\right)^2 + \dfrac{0.02}{\theta_1{}^3} + \dfrac{0.02}{(100{,}000 - \theta_1)^3} & 2\sum_{\mu=1}^{15} \dfrac{\partial f_\mu}{\partial \theta_1}\dfrac{\partial f_\mu}{\partial \theta_2} \\ 2\sum_{\mu=1}^{15} \dfrac{\partial f_\mu}{\partial \theta_1}\dfrac{\partial f_\mu}{\partial \theta_2} & 2\sum_{\mu=1}^{15} \left(\dfrac{\partial f_\mu}{\partial \theta_2}\right)^2 + \dfrac{0.4}{\theta_2{}^3} + \dfrac{0.4}{(2{,}000{,}000 - \theta_2)^3} \end{bmatrix} \tag{6-11-5}$$

For the first iteration, using the data of Eq. (5-21-10), we find:

$\Phi^\dagger = 5.299702$

$$\mathbf{q}_1{}^\dagger = \begin{bmatrix} -0.0007098080 - \dfrac{0.01}{100^2} \\ 0.0002442936 - \dfrac{0.2}{2000^2} \end{bmatrix} = \begin{bmatrix} -0.0007108080 \\ 0.0002442436 \end{bmatrix}$$

$$\mathbf{N}_1{}^\dagger = \begin{bmatrix} 0.7036033 \times 10^{-7} + \dfrac{0.02}{100^3} & -0.2354773 \times 10^{-7} \\ -0.2354773 \times 10^{-7} & 0.07896382 \times 10^{-7} + \dfrac{0.4}{2000^3} \end{bmatrix}$$

$$= \begin{bmatrix} 0.9036033 & -0.2354773 \\ -0.2354773 & 0.07946382 \end{bmatrix} \times 10^{-7}$$

$$\mathbf{v}_1{}^\dagger = -(\mathbf{N}_1{}^\dagger)^{-1}\mathbf{q}_1{}^\dagger = \begin{bmatrix} -629.9713 \\ -32604.29 \end{bmatrix}$$

Comparing $\mathbf{v}_1{}^\dagger$ to $\mathbf{v}_1$ given by Eq. (5-21-11) we see that although the penalty function has but a small effect on the value of Φ, it has the power to turn the step direction away from the troublesome $\theta_1 = 0$ axis.

We compute now the largest value of ρ for which $\rho\mathbf{v}_1{}^\dagger$ is feasible

$$\rho_{\max} = \min\{(100 - 0)/629.9713, \quad (2000 - 0)/32604.29\} = 0.06134162$$

Following the flowchart of Fig. 5-2a, we initially try

$$\rho^{(0)} = 0.5\rho_{\max} = 0.03067081$$

for which

$$\boldsymbol{\theta} = \begin{bmatrix} 100 - 0.03067081 \times 629.9713 \\ 2000 - 0.03067081 \times 32604.29 \end{bmatrix} = \begin{bmatrix} 80.67827 \\ 1000 \end{bmatrix}$$

where $\Phi^{\dagger} = 3.652150$, an acceptable value. We proceed, however, to extrapolate according to Fig. 5-2b, and obtain

$$\boldsymbol{\theta} = \begin{bmatrix} 80.67828 - 0.5 \times 0.03067081 \times 629.9713 \\ 1000 - 0.5 \times 0.03067081 \times 32604.29 \end{bmatrix} = \begin{bmatrix} 71.01740 \\ 500 \end{bmatrix}$$

with $\Phi^{\dagger} = 0.5349855$.

This much improved value is the basis for starting the second iteration. We converge after nine iterations (plotted in Fig. 5-3) of the Gauss method to

$$\boldsymbol{\theta} = \begin{bmatrix} 814.4814 \\ 961.1797 \end{bmatrix}, \qquad \Phi^{\dagger} = 0.04002851$$

At this point, $\Phi = 0.03980605$. One further iteration without penalty functions leads to

$$\boldsymbol{\theta} = \begin{bmatrix} 813.8381 \\ 960.9944 \end{bmatrix}, \qquad \Phi = 0.03980603$$

which is close to the solution obtained in 24 iterations without penalty functions.

6-12. Least Squares Problem—Projection Method

Suppose we try to solve the problem of the previous section, but using the projection method in place of penalty functions. A glance at Fig. 5-3 shows that even in the absence of any constraints, the Gauss method never carries one actually as far as the $\theta_1 = 0$ axis. Hence, no occasion to project a step into any of the constraints Eq. (6-11-1) arises, and the projection method does not affect the course of the iterations (other than avoiding some of the futile function evaluations in the unfeasible region). Let us, however, change the constraints to read

$$\theta_1 \geqslant 100, \qquad \theta_2 \geqslant 0 \qquad (6\text{-}12\text{-}1)$$

We now apply the algorithm of Section 6-3:

1. $\mathbf{R} = \mathbf{N}^{-1}$ is obtained by inversion from Eq. (5-21-10) and $\mathbf{v}^{(0)}$ from Eq. (5-21-11). Since only θ_1 is at a bound, only the first row and column are taken from $\mathbf{R}$, and only the first element from $\mathbf{v}^{(0)}$, to form $\mathbf{E} = [0.7199572 \times 10^{10}, -134608.0]$.

2. Since θ_1 is at its lower bound, $m_1 = 1$.
3. $k_1 = 1$.
4. $a = m_1 k_1 e_1 = -134608 < -10^{-6}$.
5. $r = 1$. Pivot on E_{11} to obtain $\mathbf{E} = [0.1388971 \times 10^{-9}, -0.1869666 \times 10^{-4}]$, $k_1 = -1$.
4. $a = m_1 k_1 e_1 = 0.1869666 \times 10^{-4} > -10^{-6}$.
7. Since $k_1 = -1$, the bound on θ_1 is binding. Hence $v_1 = 0$ (θ_1 cannot change) and $v_2 = R_{12}\lambda_1 - v_2^{(0)} = 2.146975 \times 10^{10} \times 0.1869666 \times 10^{-4} - 432361.0 = -30948.37$. We now find $\rho_{\max} = 2000/30948.37$, and try $\boldsymbol{\sigma} = \rho_{\max}\mathbf{v}_1 = [0, -2000]^{\mathrm{T}}$. This brings us to $\boldsymbol{\theta} = [100, 0]^{\mathrm{T}}$ and $\Phi = 6.028293$, which is unacceptable. Interpolation forces $\rho = 0.25\rho_{\max}$ with consequent $\boldsymbol{\theta} = [100, 1500]^{\mathrm{T}}$, $\Phi = 4.975522$, which is acceptable. At this point the tableau reads $\mathbf{E} = [0.5366083 \times 10^{8}, -7951.062]$. Row 1 must be swept to produce $\mathbf{E} = [0.1863556 \times 10^{-7}, -0.1481725 \times 10^{-3}]$ and $\mathbf{v}_2 = [0, -6144.531]^{\mathrm{T}}$.

Only in the fourth iteration with $\boldsymbol{\theta} = [100, 656.7756]^{\mathrm{T}}$ do we obtain the tableau $\mathbf{E} = [93635.87, 196.3826]$ which does not require sweeping, so that the lower bound on θ_1 ceases to be binding. In this iteration $\mathbf{v}_4 = [196.3826, 269.1018]^{\mathrm{T}}$. Convergence occurs in ten iterations, which are also plotted in Fig. 5-3.

6-13. Independent Variables Subject to Error

Let us take once more the model of Eq. (5-21-5), but assume now that in addition to y, also x_1 and x_2 are subject to measurement errors. We shall use the exact structural model approach, writing

$$g_\mu \equiv g(\hat{w}_{\mu 1}, \hat{w}_{\mu 2}, \hat{w}_{\mu 3}, \theta_1, \theta_2) = \exp[-\theta_1 \hat{w}_{\mu 1} \exp-(\theta_2/\hat{w}_{\mu 2})] - \hat{w}_{\mu 3} \quad (6\text{-}13\text{-}1)$$

where $\hat{w}_{\mu 1}, \hat{w}_{\mu 2}, \hat{w}_{\mu 3}$ represent the "true" but unknown values of the measured quantities $w_{\mu 1} \equiv x_{\mu 1}$, $w_{\mu 2} \equiv x_{\mu 2}$, $w_{\mu 3} \equiv y$. We assume all measurement errors to be independent, and form the objective function

$$\tfrac{1}{2} \sum_{a=1}^{3} (1/v_a) \sum_{\mu=1}^{15} (w_{\mu a} - \hat{w}_{\mu a})^2$$

Since we have only one equation, it follows from the discussion in Section 4-10 that a maximum likelihood estimate can be obtained for only one of the variances v_a. Let us then assign standard deviations of 0.01 and 0.5 to the time ($w_{\mu 1}$) and temperature ($w_{\mu 2}$) measurements, respectively, and let the variance of the $w_{\mu 3}$ measurements remain unknown. Following Eq. (4-10-12), the concentrated objective function, from which v_3 has been eliminated, takes the form

$$\Phi(\hat{\mathbf{W}}) = (15/2) \log S_3 + \tfrac{1}{2}((1/0.0001)S_1 + (1/0.25)S_2) \quad (6\text{-}13\text{-}2)$$

where

$$S_a \equiv \sum_{\mu=1}^{15} (w_{\mu a} - \hat{w}_{\mu a})^2 \qquad (a = 1, 2, 3) \tag{6-13-3}$$

Because we take as initial guesses $\hat{w}_{\mu a} = w_{\mu a}$, it follows that $S_3 = 0$ initially, and neither Φ nor its derivatives can be evaluated. The solution suggested in Section 6-8 is to take for the first iteration only the objective function

$$\Phi^{(0)}(\mathbf{W}^*) = \tfrac{1}{2}((1/0.0001)S_1 + (1/0.25)S_2 + (1/v_3^{(0)})S_3) \tag{6-13-4}$$

where $v_3^{(0)}$ is an initial guess for the variance of $w_{\mu 3}$. We take $v_3^{(0)} = 0.01^2 = 0.0001$. As our initial guess for $\boldsymbol{\theta}$ we take $\boldsymbol{\theta}_1 = [750, 1200]^{\mathrm{T}}$.

The solution is obtained iteratively, using Eqs. (6-8-19)–(6-8-25). We present details of the first iteration.

First we compute the vectors $\mathbf{a}_\mu = (\partial g_\mu/\partial \hat{\mathbf{w}}_\mu)$ and $\mathbf{b}_\mu = (\partial g_\mu/\partial \boldsymbol{\theta})$ for $\mu = 1, 2, \ldots, 15$. The values of $\mathbf{b}_\mu$ already appear in the last two columns of Table 5-3. In Table 6-1 we list the values of $\mathbf{a}_\mu$, as well as g_μ (from Eq. (1)) and of c_μ which according to Eq. (6-8-20) is given by

$$c_\mu = a_\mu^{\mathrm{T}} V_\mu a_\mu = 0.0001\, a_{\mu 1}^2 + 0.25\, a_{\mu 2}^2 + v_3^{(0)}\, a_{\mu 3}^2$$

Table 6-1
First Iteration Data

μ	$g_\mu = g(\boldsymbol{\theta}, \hat{\mathbf{w}}_\mu)$	$\mathbf{a}_\mu = \partial g_\mu/\partial \hat{\mathbf{w}}_\mu$			$10^3 c_\mu$
		$a_{\mu 1}$	$10^4 a_{\mu 2}$	$a_{\mu 3}$	
1	0.01953936	−0.004606035	−0.5527238	−1	0.1000029
2	0.01607883	−0.004603911	−1.104938	−1	0.1000052
3	0.04361850	−0.004601792	−1.656644	−1	0.1000090
4	0.01915848	−0.004599672	−2.207842	−1	0.1000143
5	0.004698634	−0.004597552	−2.758531	−1	0.1000211
6	0.2852362	−1.694046	−25.41069	−1	0.3885932
7	0.2863515	−1.543676	−46.31024	−1	0.3436550
8	0.3016463	−1.406653	−63.29931	−1	0.3078843
9	0.4644834	−1.281794	−76.90754	−1	0.2790862
10	0.4612821	−1.168016	−87.60117	−1	0.2556110
11	0.1937739	−10.43681	−27.83145	−1	10.9946
12	0.2602563	−7.929612	−42.29121	−1	6.392341
13	0.4045841	−6.024710	−48.19766	−1	3.735518
14	0.3172247	−4.577417	−48.82574	−1	2.201234
15	0.1871758	−3.477806	−46.37070	−1	1.314888

We now compute $\mathbf{D} = \sum_{\mu=1}^{15} (1/c_\mu)\mathbf{b}_\mu \mathbf{b}_\mu{}^{\mathrm{T}}$. The first term ($\mu = 1$) in the sum is

$$(1/0.1000029 \times 10^{-3})$$
$$\times \begin{bmatrix} (0.6141379)^2 \times 10^{-12} & 0.6141379 \times 0.4606032 \times 10^{-11} \\ 0.6141379 \times 0.4606032 \times 10^{-11} & (0.4606032)^2 \times 10^{-10} \end{bmatrix}$$

and the result is

$$\mathbf{D} = \begin{bmatrix} 0.001794073 & -0.006267238 \\ -0.006367238 & 0.02235472 \end{bmatrix}$$

Using Eq. (6-8-22) we compute $\delta\boldsymbol{\theta}$. In the first iteration $\mathbf{e}_\mu = 0$. The first term of the sum is thus $-(g_1/c_1)\mathbf{b}_1$, and

$$\delta\boldsymbol{\theta} = \begin{bmatrix} -846.9727 \\ -563.7969 \end{bmatrix}$$

It is now easy to compute λ_μ and $\delta\hat{\mathbf{w}}_\mu$ using Eqs (6-8-23) and (6-8-24) in turn. Table 6-2 contains the results.

Finally, we evaluate our test function Z (Eq. (6-8-25)), whose value turns out to be 2508.496. After modifying $\boldsymbol{\theta}$ and $\hat{\mathbf{w}}_\mu$ by adding $\delta\boldsymbol{\theta}$ and $\delta\hat{\mathbf{w}}_\mu$ we recompute the value of Z, which has now increased to 59729.57. The step is unacceptable, and we must interpolate. The method of Fig. 5-2 should be used,

Table 6-2
First Iteration Results

		$\delta\hat{\mathbf{w}}_\mu$		
μ	λ_μ	$\delta\hat{w}_{\mu 1}$	$\delta\hat{w}_{\mu 2}$	$\delta\hat{w}_{\mu 3}$
1	174.6214	0.00008043119	0.002412934	0.01746213
2	119.2671	0.00005490948	0.003294569	0.01192671
3	373.9075	0.0001720644	0.01548579	0.03739074
4	108.6156	0.00004995962	0.005995154	0.01086156
5	−56.64597	−0.00002604326	−0.003906488	−0.005664594
6	365.7187	0.06195441	0.2323291	0.03657187
7	74.25694	0.01146286	0.08597142	0.007425692
8	−178.2293	−0.02507068	−0.2820446	−0.01782293
9	112.2150	0.01438364	0.2157544	0.01122149
10	−125.6333	−0.01467417	−0.2751405	−0.01256333
11	3.384982	0.003532839	0.002355224	0.0003384980
12	3.497939	0.002773728	0.003698302	0.0003497938
13	35.72807	0.02152512	0.04305023	0.003572807
14	19.33928	0.008852392	0.02360636	0.001933928
15	−56.02644	−0.01948491	−0.06494963	−0.005602643

but for simplicity we just cut the step in half ($\rho = 0.5$) to obtain the new value $\boldsymbol{\theta}_2 = [533.1387, 894.4258]^{\mathrm{T}}$, and $\hat{\mathbf{w}}_\mu$ as given in Table 6-3 along with the new values of g_μ. The corresponding value of Z is 908.2344, which is smaller than the initial value and hence acceptable. We are ready now for the second iteration for which we first compute the new value of

$$v_3 = (1/15) \sum_{\mu=1}^{15} (\hat{w}_{\mu 3} - w_{\mu 3})^2 = 0.6729098.$$

Table 6-3
Start of Second Iteration

μ	g_μ	$\hat{\mathbf{w}}_\mu$		
		$\hat{w}_{\mu 1}$	$\hat{w}_{\mu 2}$	$\hat{w}_{\mu 3}$
1	0.007910669	0.1000401	100.0012	0.9897310
2	0.004332781	0.2000274	100.0016	0.9889632
3	0.01625860	0.3000859	100.0077	0.9736953
4	0.002205729	0.4000249	100.0030	0.9844307
5	−0.006835222	0.4999869	99.99805	0.9901676
6	0.1198401	0.08097714	200.1161	0.6442859
7	0.1565019	0.1057313	200.0430	0.5477127
8	0.1889961	0.1374646	199.8590	0.4460884
9	0.2718676	0.2071918	200.1079	0.2306107
10	0.2879401	0.2426628	199.8624	0.1607183
11	0.1505129	0.02176642	300.0010	0.5661692
12	0.2136052	0.04138686	300.0017	0.3171748
13	0.3027226	0.07076252	300.0212	0.03578640
14	0.2576900	0.08442611	300.0115	0.01696696
15	0.1881614	0.09025747	299.9625	0.06319863

Convergence is obtained in 21 iterations. The final results are given in Table 6-4. The value of the objective function Eq. (2) at the solution is −32.09045.

The final value of **D**, which we shall need in Section 7-23, is

$$\mathbf{D} = \begin{bmatrix} 0.0005472710 & -0.003062132 \\ -0.003062132 & 0.01747675 \end{bmatrix} \tag{6-13-5}$$

We shall also need later the moment matrix of the final residuals

$$\mathbf{M} \equiv \sum_{\mu=1}^{15} \mathbf{e}_\mu{}^* \mathbf{e}_\mu^{*\mathrm{T}} = \sum_{\mu=1}^{15} (\hat{\mathbf{w}}_\mu{}^* - \mathbf{w}_\mu)(\hat{\mathbf{w}}_\mu{}^* - \mathbf{w}_\mu)^{\mathrm{T}}$$

which, using the data of Table 6-4 we find to be

$$\mathbf{M} = \begin{bmatrix} 0.001781781 & 0.009602152 & 0.001960129 \\ 0.009602152 & 0.07176667 & 0.01202689 \\ 0.001960129 & 0.01202689 & 0.0041455559 \end{bmatrix} \tag{6-13-6}$$

Table 6-4

Final Result, with $\boldsymbol{\theta}^* = \begin{bmatrix} 1170.862 \\ 1027.773 \end{bmatrix}$

μ		$\hat{\mathbf{w}}_\mu{}^*$		$\lambda_\mu{}^*$
1	0.1002329	100.0060	0.9959696	57.79277
2	0.2001303	100.0067	0.9919683	32.46887
3	0.3004779	100.0369	0.9879283	119.1740
4	0.4000721	100.0074	0.9840074	18.15489
5	0.4998157	99.97632	0.9801232	−46.54858
6	0.06580788	200.0668	0.6359499	36.13177
7	0.09067100	199.9456	0.5370256	−25.30307
8	0.1235170	199.7895	0.4301884	−90.03149
9	0.2081143	200.1083	0.2386000	49.34750
10	0.2534879	200.0567	0.1749817	28.97121
11	0.01497819	299.9978	0.5653576	−23.33610
12	0.03035963	299.9915	0.3147850	−8.046125
13	0.07608849	300.0349	0.05511629	76.52017
14	0.08861703	300.0217	0.03421704	65.96593
15	0.08537388	299.9644	0.03879805	−98.72794

The slow convergence of the algorithm in this problem is not typical. In many different cases relating to the same problem convergence occurred in fewer than ten iterations.

6-14. An Implicit Equations Model

In Section 5-23 we estimated the parameters $\mathbf{c}$ appearing in Eq. (5-23-1) by solving these equations explicitly for the dependent variables z_1 and z_2. Suppose, however, that explicit solutions were impossible. We could then use the methods of this chapter, by introducing as unknowns the "true" values $\hat{z}_{\mu 1}$ and $\hat{z}_{\mu 2}$, and using the model equations as constraints. Since we do not know the covariance of the errors, we take as our objective function

$$\begin{aligned}\Phi(\hat{\mathbf{Z}}) &= (41/2) \log \det \mathbf{M} \\ &= (41/2) \log \left\{ \sum_{\mu=1}^{41} (\hat{z}_{\mu 1} - z_{\mu 1})^2 \sum_{\mu=1}^{41} (\hat{z}_{\mu 2} - z_{\mu 2})^2 \right. \\ &\quad \left. - \left[\sum_{\mu=1}^{41} (\hat{z}_{\mu 1} - z_{\mu 1})(\hat{z}_{\mu 2} - z_{\mu 2}) \right]^2 \right\}\end{aligned}$$

The $\hat{\mathbf{z}}_\mu$ and $\mathbf{c}$ must satisfy the model equations which become after taking logarithms:

$$\log c_1 + z_{\mu 4} c_2 \log 10 - (c_3/c_4) \log[c_5 \hat{z}_{\mu 1}^{-c_4} + (1 - c_5)\hat{z}_{\mu 2}^{-c_4}] - \log z_{\mu 3} = 0$$
$$\log c_5 - \log(1 - c_5) - (1 + c_4) \log \hat{z}_{\mu 1} + (1 + c_4) \log \hat{z}_{\mu 2} - \log z_{\mu 5} = 0$$
$$(\mu = 1, 2, 3, \ldots, 41)$$

Application of the method of Section 6-8 is now straightforward. We have

$$\mathbf{A}_\mu \equiv \frac{\partial \mathbf{g}}{\partial \hat{\mathbf{z}}_\mu} = \begin{bmatrix} \dfrac{c_3 c_5 \hat{z}_{\mu 1}^{-c_4 - 1}}{\zeta_\mu} & \dfrac{c_3(1 - c_5)\hat{z}_{\mu 2}^{-c_4 - 1}}{\zeta_\mu} \\ -\dfrac{1 + c_4}{\hat{z}_{\mu 1}} & \dfrac{1 + c_4}{\hat{z}_{\mu 2}} \end{bmatrix}$$

where $\zeta_\mu \equiv c_5 \hat{z}_{\mu 1}^{-c_4} + (1 - c_5)\hat{z}_{\mu 2}^{-c_4}$, and

$$\mathbf{B}_\mu^{\mathrm{T}} \equiv \left(\frac{\partial \mathbf{g}}{\partial \mathbf{c}}\right)^{\mathrm{T}}$$

$$= \begin{bmatrix} \dfrac{1}{c_1} & 0 \\ z_{\mu 4} \log 10 & 0 \\ -\dfrac{1}{c_4} \log \zeta_\mu & 0 \\ \dfrac{c_3}{c_4^2} \log \zeta_\mu + \dfrac{c_3[c_5 \hat{z}_{\mu 1}^{-c_4} \log \hat{z}_{\mu 1} + (1 - c_5)\hat{z}_{\mu 2}^{-c_4} \log \hat{z}_{\mu 2}]}{c_4 \zeta_\mu} & \log \dfrac{\hat{z}_{\mu 2}}{\hat{z}_{\mu 1}} \\ -\dfrac{c_3(\hat{z}_{\mu 1}^{-c_4} - \hat{z}_{\mu 2}^{-c_4})}{c_4 \zeta_\mu} & \dfrac{1}{c_5} + \dfrac{1}{1 - c_5} \end{bmatrix}$$

For $\mathbf{V}$ we take the value of $(1/41)\mathbf{M}$ from the previous iteration. In the first iteration, we use $\mathbf{V}_1 = \mathbf{I}$. Starting from the initial guess $\mathbf{c} = [1, 1, 1, 1, 0.5]^{\mathrm{T}}$, we converged to the same value of $\mathbf{c}^*$ as given under case (a), Table 5-8, in twelve iterations.

6-15. Problems

1. Verify Eqs. (6-8-15)–(6-8-17).
2. Prove that if the model equations take the form

$$\mathbf{g}(\hat{\mathbf{w}}_\mu, \boldsymbol{\theta}) = \hat{\mathbf{w}}_\mu - \mathbf{f}_\mu(\boldsymbol{\theta}) = \mathbf{0}$$

then the iterations produced by the method of Section 6-8 are identical to those produced by the Gauss method.

3. Using the data of Section 5-21, find the parameter estimates which minimize the maximum residual (MLE for uniform error distribution). Compare the results to the least squares estimates.

Chapter

VII

Interpretation of the Estimates

7-1. Introduction

It is not enough to compute a vector $\boldsymbol{\theta}^*$ and to state that this is the estimated value of the unknown parameters $\boldsymbol{\theta}$. We must also investigate the reliability and precision of our estimates. We wish to answer questions such as "what are the chances that the estimate is off by no more than 1%?" or "how much can we change the estimates and still fit the data well?" There are several ways in which one can go about answering these questions; some of these are of a heuristic nature, while others depend on statistical considerations. We shall present several alternative approaches in the succeeding sections.

Even more important than the question of the reliability of the estimate is that of the reliability of the model itself. This question is answered by goodness of fit criteria and statistical hypothesis testing. We cover these topics only very briefly here, since extensive treatments can be found in the statistical texts. In particular, the reader may consult Anderson (1958) on the topics that are of direct interest to us here, and Lehman (1959) for a more general treatment.

Some of the statistical tests and estimates of variability that we discuss here apply only approximately to nonlinear models. Refinement of these approximations is often possible [see, e.g., Beale (1960), Hartley (1964), and Guttman and Meeter (1965), but even with linear models, the tests are exact only if the measurement errors do indeed follow whatever distribution was assumed for them. Since this is rarely if ever so, even so called "exact" tests are only approximate in practice. Furthermore, we do not feel that the statement "the probability that model A is incorrect is exactly 5%" has greater practical utility than the statement "the probability that model A is incorrect is approximately 5%." For these reasons, we present only the simplest approximate tests, and leave the reader interested in more exact formulations to consult the cited references.

7-2. Response Surface Techniques

The estimate $\boldsymbol{\theta}^*$ is usually obtained by minimizing or maximizing some function $\Phi(\boldsymbol{\theta})$. Then $\Phi^* \equiv \Phi(\boldsymbol{\theta}^*)$ is the " best " attainable value of the objective function Φ. Suppose for a moment that $\Phi(\boldsymbol{\theta})$ is a risk function of decision theory (see Section 4-16), i.e., the value of $\Phi(\boldsymbol{\theta})$ represents the economic loss that we expect to sustain if we act on the assumption that the parameters have the value $\boldsymbol{\theta}$. In this case Φ^* is the minimum possible expected loss. However, should some parameter values $\tilde{\boldsymbol{\theta}} \neq \boldsymbol{\theta}^*$ give rise to a risk $\Phi(\tilde{\boldsymbol{\theta}})$ that is only insignificantly larger than Φ^*, then we have no compelling reason to prefer $\boldsymbol{\theta}^*$ over $\tilde{\boldsymbol{\theta}}$. In fact, let ε be the largest difference between risks that we are willing to consider insignificant. Then we have no reason to prefer $\boldsymbol{\theta}^*$ over any other value of $\boldsymbol{\theta}$ for which

$$|\Phi(\boldsymbol{\theta}) - \Phi^*| \leqslant \varepsilon \tag{7-2-1}$$

We refer to the set of values of $\boldsymbol{\theta}$ which satisfy Eq. (1) as the *ε-indifference region.*

The argument used here may be applied heuristically to any other objective function. The fact that we have elected to minimize a function $\Phi(\boldsymbol{\theta})$ means that we set some store by obtaining a low value of this function. It is not unreasonable to suppose that values of Φ almost as low as Φ^* would satisfy us almost as much as Φ^*. This gives rise to an indifference region in $\boldsymbol{\theta}$ space as described by Eq. (1). The choice of a suitable ε may be more arbitrary when Φ is a sum of squares or a likelihood than when it is an economic risk. Once ε is chosen, however, the analysis is the same in all cases.

When Φ is continuous and $\boldsymbol{\theta}^*$ its unique unconstrained minimum, the ε-indifference region for a sufficiently small positive ε is a simply-connected domain surrounding $\boldsymbol{\theta}^*$ in the l-dimensional $\boldsymbol{\theta}$ space. The region is bounded by the l-1 dimensional hypersurface whose equation is

$$\Phi(\boldsymbol{\theta}) = \Phi^* + \varepsilon \tag{7-2-2}$$

We shall restrict our attention to regions of this nature; i.e., we shall ignore the possibility that for a given ε there may be regions surrounding local minima other than $\boldsymbol{\theta}^*$ in which Eq. (1) holds.

In a sufficiently small neighborhood of $\boldsymbol{\theta}^*$ we may approximate Φ by means of the first few terms of its Taylor series expansion

$$\Phi(\boldsymbol{\theta}) \approx \Phi^* + \mathbf{q}^{*\mathrm{T}}\, \delta\boldsymbol{\theta} + \tfrac{1}{2}\, \delta\boldsymbol{\theta}^{\mathrm{T}}\, \mathbf{H}^*\, \delta\boldsymbol{\theta} \tag{7-2-3}$$

where $\delta\boldsymbol{\theta} \equiv \boldsymbol{\theta} - \boldsymbol{\theta}^*$, and $\mathbf{q}^*$ and $\mathbf{H}^*$ are, respectively, the gradient and Hessian of Φ at $\boldsymbol{\theta} = \boldsymbol{\theta}^*$. If $\boldsymbol{\theta}^*$ is an unconstrained optimum of Φ, then $\mathbf{q}^* = \mathbf{0}$ and Eq. (3) becomes

$$\Phi(\boldsymbol{\theta}) \approx \Phi^* + \tfrac{1}{2}\, \delta\boldsymbol{\theta}^{\mathrm{T}}\, \mathbf{H}^*\, \delta\boldsymbol{\theta} \tag{7-2-4}$$

so that the ε-indifference region is defined, approximately, by

$$|\delta\boldsymbol{\theta}^{\mathrm{T}}\, \mathbf{H}^*\, \delta\boldsymbol{\theta}| \leqslant 2\varepsilon \tag{7-2-5}$$

Let $\mathbf{A} = \mathbf{H}^*$ if $\boldsymbol{\theta}^*$ is a minimum ($\mathbf{H}^*$ positive definite), and $\mathbf{A} = -\mathbf{H}^*$ if $\boldsymbol{\theta}^*$ is a maximum ($\mathbf{H}^*$ negative definite). In either case, $\mathbf{A}$ is positive definite (semidefinite in exceptional cases), and Eq. (5) becomes

$$\delta\boldsymbol{\theta}^{\mathrm{T}}\, \mathbf{A}\, \delta\boldsymbol{\theta} \leqslant \mathbf{2}\varepsilon \tag{7-2-6}$$

which is the equation of an l-dimensional ellipsoid whose volume is $(2\varepsilon\pi)^{l/2} \det^{-1/2} \mathbf{A}/\Gamma(l/2+1)$. The ellipsoids corresponding to different values of ε are concentric and similar in shape and orientation, so that much information can be gained from the analysis of the matrix $\mathbf{A}$, without regard to the actual value of ε.

We can now answer the question of how much the individual parameter θ_α can be varied from its optimal value $\theta_\alpha{}^*$. If we let $\delta\theta_\beta = 0$ for all $\beta \neq \alpha$, Eq. (6) reduces to

$$A_{\alpha\alpha}\, \delta\theta_\alpha{}^2 \leqslant 2\varepsilon \tag{7-2-7}$$

so that

$$\theta_\alpha{}^* - (2\varepsilon/A_{\alpha\alpha})^{1/2} \leqslant \theta_\alpha \leqslant \theta_\alpha{}^* + (2\varepsilon/A_{\alpha\alpha})^{1/2} \tag{7-2-8}$$

This is often written in shorthand notation as $\theta_\alpha = \theta_\alpha{}^* \pm (2\varepsilon/A_{\alpha\alpha})^{1/2}$.

We say that θ_α is *well-determined* if the quantity $(2\varepsilon/A_{\alpha\alpha})^{1/2}$ is small on the scale by which θ_α is measured; θ_α is *ill-determined* if $(2\varepsilon/A_{\alpha\alpha})^{1/2}$ is large. Usually we wish the parameters to be well-determined. There are exceptions, though; a design parameter may advantageously be ill-determined for greater flexibility in the implementation of the design. It is important, however, that the ill-determination should be inherent in the design, and not merely the result of poor data.

It is not enough to determine how well the individual parameters are determined. Consider, for instance, the two-dimensional case depicted in Fig. 7-1, with

$$A = \begin{bmatrix} 0.505 & -0.495 \\ -0.495 & 0.505 \end{bmatrix}$$

and $\varepsilon = 0.5$. Here Eq. (6) reduces to

$$0.505\, \delta\theta_1{}^2 - 0.99\, \delta\theta_1\, \delta\theta_2 + 0.505\, \delta\theta_2{}^2 \leqslant 1 \tag{7-2-9}$$

so that with $\delta\theta_2 = 0$ we have $|\delta\theta_1| \leqslant 1.407$, and with $\delta\theta_1 = 0$ we have $|\delta\theta_2| \leqslant 1.407$. Thus θ_1 and θ_2 may be varied individually by ± 1.407 without leaving the indifference region. It is clear from Fig. 7-1, however, that if we increase (or decrease) θ_1 and θ_2 simultaneously, much larger changes can be tolerated. In fact, $\delta\theta_1 = \delta\theta_2 = 7.701$ satisfies Eq. (9). On the other hand, if the changes in θ_1 and θ_2 are taken in opposite directions, their bound is

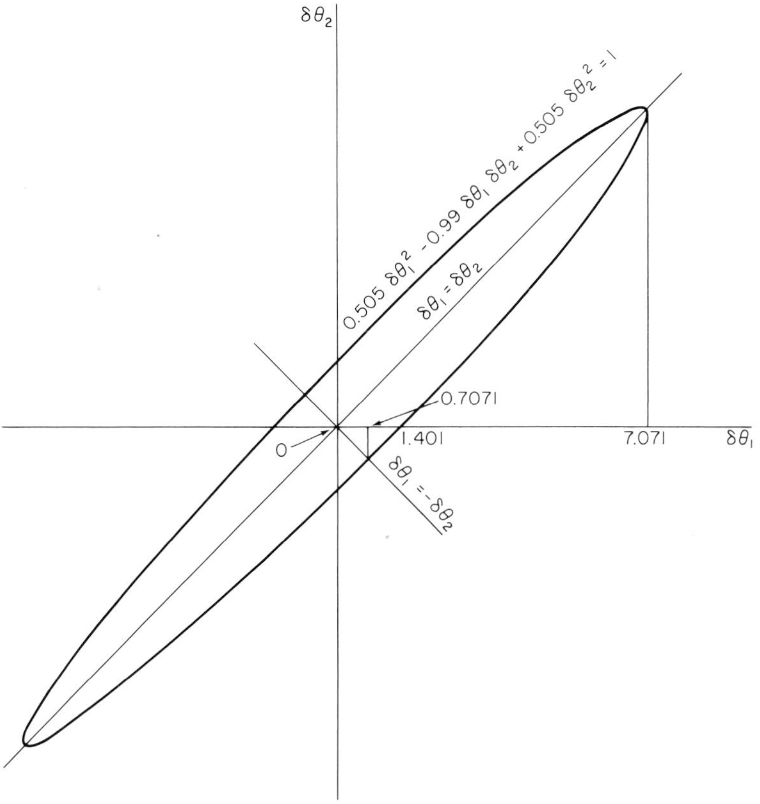

Fig. 7-1. $\varepsilon = 0.5$ uncertainty region for $A = \begin{bmatrix} 0.505 & -0.495 \\ -0.495 & 0.505 \end{bmatrix}$

lower $\delta\theta_1 = -\delta\theta_2 = 0.7071$. While θ_1 and θ_2 appear individually well-determined, the quantity $\theta_1 - \theta_2$ is even better determined, but $\theta_1 + \theta_2$ is relatively ill-determined. This implies that we have a wide latitude in choosing, say, a value of θ_1 as long as we adjust θ_2 so that $\theta_1 - \theta_2$ is nearly equal to $\theta_1{}^* - \theta_2{}^*$.

Another numerical example appears in Section 7-21.

7-3. Canonical Form

We wish to find the points on the ellipsoid $\delta\boldsymbol{\theta}^{\mathrm{T}}\,\mathbf{A}\,\delta\boldsymbol{\theta} = 2\varepsilon$ which are furthest away from the origin, and also those which are closest. These points determine, respectively, the least-determined and best determined linear combinations of the parameters.

To find the vector $\delta\boldsymbol{\theta}$ satisfying $\delta\boldsymbol{\theta}^{\mathrm{T}}\,\mathbf{A}\,\delta\boldsymbol{\theta} = 2\varepsilon$ which maximizes or minimizes $\delta\boldsymbol{\theta}^{\mathrm{T}}\delta\boldsymbol{\theta}$, i.e., the squared distance from the origin, we introduce the Lagrange multiplier μ, and look for the stationary points of

$$\Pi(\delta\boldsymbol{\theta}, \mu) \equiv \delta\boldsymbol{\theta}^{\mathrm{T}}\,\delta\boldsymbol{\theta} - \mu(\delta\boldsymbol{\theta}^{\mathrm{T}}\,\mathbf{A}\,\delta\boldsymbol{\theta} - 2\varepsilon) \qquad (7\text{-}3\text{-}1)$$

Therefore

$$\partial\Pi/\partial(\delta\boldsymbol{\theta}) = 2\,\delta\boldsymbol{\theta} - 2\mu\,\mathbf{A}\,\delta\boldsymbol{\theta} = \mathbf{0} \qquad (7\text{-}3\text{-}2)$$

Letting $\lambda = 1/\mu$ and rearranging, we have

$$\mathbf{A}\,\delta\boldsymbol{\theta} = \lambda\,\delta\boldsymbol{\theta} \qquad (7\text{-}3\text{-}3)$$

Premultiplying by $\delta\boldsymbol{\theta}^{\mathrm{T}}$

$$\delta\boldsymbol{\theta}^{\mathrm{T}}\,\mathbf{A}\,\delta\boldsymbol{\theta} = \lambda\,\delta\boldsymbol{\theta}^{\mathrm{T}}\,\delta\boldsymbol{\theta} \qquad (7\text{-}3\text{-}4)$$

Hence

$$\delta\boldsymbol{\theta}^{\mathrm{T}}\,\delta\boldsymbol{\theta} = 2\varepsilon/\lambda \qquad (7\text{-}3\text{-}5)$$

Eq. (3) states that the desired vector $\delta\boldsymbol{\theta}$ is an (unnormalized) eigenvector of $\mathbf{A}$ with eigenvalue λ. Eq. (5) states that the length of the vector is $(2\varepsilon/\lambda)^{1/2}$. The l eigenvectors form the l principal axes of the ellipsoid. The longest axis, corresponding to the smallest eigenvalue, defines the worst-determined direction in $\boldsymbol{\theta}$ space, and the shortest axis (largest eigenvalue) defines the best-determined direction.

Let the eigenvalue decomposition (see Section A-5) of $\mathbf{A}$ be given by

$$\mathbf{A} = \mathbf{U}\boldsymbol{\Lambda}\mathbf{U}^{\mathrm{T}} \qquad (7\text{-}3\text{-}6)$$

where $\mathbf{U}$ is the unitary matrix whose columns are the normalized eigenvectors of $\mathbf{A}$, and $\boldsymbol{\Lambda}$ is the diagonal matrix of eigenvalues. Then

$$\delta\boldsymbol{\theta}^{\mathrm{T}}\,\mathbf{A}\,\delta\boldsymbol{\theta} = \delta\boldsymbol{\theta}^{\mathrm{T}}\,\mathbf{U}\boldsymbol{\Lambda}\mathbf{U}^{\mathrm{T}}\,\delta\boldsymbol{\theta} \qquad (7\text{-}3\text{-}7)$$

Letting $\boldsymbol{\psi} = \mathbf{U}^{\mathrm{T}}\,\delta\boldsymbol{\theta}$ we obtain

$$\delta\boldsymbol{\theta}^{\mathrm{T}}\,\mathbf{A}\,\delta\boldsymbol{\theta} = \boldsymbol{\psi}^{\mathrm{T}}\,\boldsymbol{\Lambda}\,\boldsymbol{\psi} = \sum_{i=1}^{l} \lambda_i \psi_i^2 \qquad (7\text{-}3\text{-}8)$$

Since $\mathbf{U}^{\mathrm{T}}$ is unitary, the transformation of coordinates given by $\boldsymbol{\psi} = \mathbf{U}^{\mathrm{T}}\,\delta\boldsymbol{\theta}$ is a rigid rotation (with possibly some reflections) which leaves distances and angles unaffected. The number ψ_i $(i = 1, 2, \ldots, l)$ is the ith component of the vector $\delta\boldsymbol{\theta}$ expressed in the system of coordinates whose axes are the eigenvectors of $\mathbf{A}$. Eq. (8) indicates that the principal axes of the ellipsoid coincide with the coordinate axes in the $\boldsymbol{\psi}$ space, and displays clearly the inverse relationship between the lengths of the axes and the square roots of the eigenvalues.

The ψ_i are referred to as the *canonical variables*, and the expression $\sum_{i=1}^{l} \lambda_i \psi_i^2$ as the *canonical form* of the quadratic expression $\delta\boldsymbol{\theta}^{\mathrm{T}}\,\mathbf{A}\,\delta\boldsymbol{\theta}$.

In the two-dimensional example of the preceding section, we had

$$\mathbf{A} = \begin{bmatrix} 0.505 & -0.495 \\ -0.495 & 0.505 \end{bmatrix}$$

whose normalized eigenvectors are [0.7071, −0.7071] and [0.7071, 0.7071] with eigenvalues 1 and 0.01. Therefore

$$\mathbf{U}^{\mathrm{T}} = \begin{bmatrix} 0.7071 & -0.7071 \\ 0.7071 & 0.7071 \end{bmatrix}$$

and

$$\boldsymbol{\psi} = \begin{bmatrix} 0.7071 & -0.7071 \\ 0.7071 & 0.7071 \end{bmatrix} \begin{bmatrix} \delta\theta_1 \\ \delta\theta_2 \end{bmatrix} = \begin{bmatrix} 0.7071(\delta\theta_1 - \delta\theta_2) \\ 0.7071(\delta\theta_1 + \delta\theta_2) \end{bmatrix}$$

The canonical form is $\psi_1^2 + 0.01\psi_2^2$. Thus, the principal axes have lengths 1 and 10, making the quantities $0.7071\ (\delta\theta_1 - \delta\theta_2)$ and $0.7071(\delta\theta_1 + \delta\theta_2)$ relatively well-and ill-determined, respectively.

Care must be taken that all the parameters be measured on compatible scales. It is evident from Fig. 7-1 that by drastically reducing or expanding the scale of one of the variables one can distort the ellipsoid to the point where it contains no useful information.

7-4. The Sampling Distribution

The sampling distribution of the estimates was defined in Section 3-1. It represents the manner in which the estimates would vary in response to the random variations we expect to occur from one data sample to another. The sampling distribution can shed light on the reliability of the estimates. A parameter is ill-determined if its estimated value can be affected strongly by seemingly insignificant variations in the data. Such a situation is characterized by the estimate having a large variance. The feature of the sampling distribution that is of most interest to us is, therefore, its covariance matrix.

We have remarked before (Chapter III) that we cannot generally hope to determine the true sampling distribution. The best that we can hope to do on the basis of a single data sample, and in the absence of a Monte Carlo study, is to arrive at a rough approximation to the covariance matrix. We would also like to know the mean of the sampling distribution, and how it is related to the true values of the parameters, and the actual estimate $\boldsymbol{\theta}^*$. Generally we accept $\boldsymbol{\theta}^*$ as an estimate of the mean of the sampling distribution, and we neglect the bias of this mean relative to the true value. There exist some methods for reducing this bias (see Section 7-9). For the moment, however, we restrict our attention to approximating the covariance matrix. In essence, then, we attempt to answer the question "If we were to repeat our series of experiments many times, how would the estimates differ from one replication to the next?"

7-5. The Covariance Matrix of the Estimates

Suppose our estimate is the unconstrained minimum of some objective function $\Phi(\boldsymbol{\theta})$. This objective function also depends on the data; in particular, it depends on the measured values $\mathbf{W}$ of the random variables $\boldsymbol{\Omega}$. We indicate this dependence by writing $\Phi(\boldsymbol{\theta}, \mathbf{W})$ in place of $\Phi(\boldsymbol{\theta})$. At the minimum we have

$$\partial\Phi(\boldsymbol{\theta}^*, \mathbf{W})/\partial\boldsymbol{\theta} = \mathbf{0} \tag{7-5-1}$$

Suppose we varied the data slightly, replacing $\mathbf{w}$ by $\mathbf{w} + \delta\mathbf{w}$. This would cause our minimum to shift from $\boldsymbol{\theta}^*$ to $\boldsymbol{\theta}^* + \delta\boldsymbol{\theta}^*$, where we must have

$$\partial\Phi(\boldsymbol{\theta}^* + \delta\boldsymbol{\theta}^*, \mathbf{W} + \delta\mathbf{W})/\delta\boldsymbol{\theta} = \mathbf{0} \tag{7-5-2}$$

Expanding Eq. (2) in Taylor series and retaining only terms up to first order, we find after subtracting Eq. (1)

$$(\delta^2\Phi/\delta\boldsymbol{\theta}^2)\,\delta\boldsymbol{\theta}^* + (\partial^2\Phi/\partial\boldsymbol{\theta}\,\partial\mathbf{w})\,\delta\mathbf{w} \approx \mathbf{0} \tag{7-5-3}$$

so that approximately

$$\delta\boldsymbol{\theta}^* = -\,\mathbf{H}^{*-1}(\partial^2\Phi/\partial\boldsymbol{\theta}\,\partial\mathbf{w})\,\delta\mathbf{w} \tag{7-5-4}$$

where as usual $\mathbf{H}^* = \partial^2\Phi/\partial\boldsymbol{\theta}^2)_{\boldsymbol{\theta}=\boldsymbol{\theta}^*}$.

The desired covariance matrix $\mathbf{V}_{\boldsymbol{\theta}}$ is defined by

$$\mathbf{V}_{\boldsymbol{\theta}} \equiv E(\delta\boldsymbol{\theta}^*\,\delta\boldsymbol{\theta}^{*\mathrm{T}}) \tag{7-5-5}$$

so that

$$\mathbf{V}_{\boldsymbol{\theta}} \approx E(\mathbf{H}^{*-1}\,(\partial^2\Phi/\partial\boldsymbol{\theta}\,\partial\mathbf{w})\,\delta\mathbf{w}\,\delta\mathbf{w}^{\mathrm{T}}\,(\partial^2\Phi/\partial\boldsymbol{\theta}\,\partial\mathbf{w})^{\mathrm{T}}\,\mathbf{H}^{*-1}) \tag{7-5-6}$$

The quantities $\mathbf{H}^*$ and $\partial^2\Phi/\partial\boldsymbol{\theta}\,\partial\mathbf{w}$ are evaluated at $\boldsymbol{\theta} = \boldsymbol{\theta}^*$ and at the actual sample $\mathbf{W}$. Hence they are constants, and can be taken outside the expectation sign in Eq. (6)

$$\mathbf{V}_{\boldsymbol{\theta}} \approx \mathbf{H}^{*-1}\,(\partial^2\Phi/\partial\boldsymbol{\theta}\,\partial\mathbf{w})\,\mathbf{V}_{\mathbf{W}}\,(\partial^2\Phi/\partial\boldsymbol{\theta}\,\partial\mathbf{w})^{\mathrm{T}}\,\mathbf{H}^{*-1} \tag{7-5-7}$$

where $\mathbf{V}_{\mathbf{W}}$ is the covariance matrix of the data, i.e.,

$$\mathbf{V}_{\mathbf{W}} \equiv E(\delta\mathbf{w}\,\delta\mathbf{w}^{\mathrm{T}}) \tag{7-5-8}$$

If we assume that $\mathbf{w}_\mu$ (i.e., the results of the μth experiment) has covariance matrix $\mathbf{V}_\mu$ and is independent of $\mathbf{w}_\eta$ $(\eta \neq \mu)$ then Eq. (7) reduces to

$$\mathbf{V}_{\boldsymbol{\theta}} \approx \mathbf{H}^{*-1}\left[\sum_{\mu=1}^{n} (\partial^2\Phi/\partial\boldsymbol{\theta}\,\partial\mathbf{w}_\mu)\,\mathbf{V}_\mu\,(\partial^2\Phi/\partial\boldsymbol{\theta}\,\partial\mathbf{w}_\mu)^{\mathrm{T}}\right]\mathbf{H}^{*-1} \tag{7-5-9}$$

This formula applies to any objective function, whether or not it has a basis in statistics. More specific results can be obtained when the objective function depends only on the moment matrix $\mathbf{M}$ of the residuals. This class of functions, which includes sums of squares and log-likelihood for normal distributions, was shown to admit the Gauss approximation Eq. (5-9-10) for $\mathbf{H}$. We derive a similar approximation to $\mathbf{V}_{\boldsymbol{\theta}}$. Eq. (5-9-10) can be rewritten as

$$\mathbf{H} \approx 2 \sum_{\mu=1}^{n} \mathbf{B}_\mu{}^{\mathrm{T}}\boldsymbol{\Gamma}\mathbf{B}_\mu \tag{7-5-10}$$

where

$$\mathbf{B}_\mu \equiv -\,\partial\mathbf{e}_\mu/\partial\boldsymbol{\theta} = \partial\mathbf{f}_\mu/\partial\boldsymbol{\theta}, \qquad \Psi(\mathbf{M}(\boldsymbol{\theta})) \equiv \Phi(\boldsymbol{\theta}), \qquad \boldsymbol{\Gamma} \equiv \partial\Psi/\partial\mathbf{M} \tag{7-5-11}$$

Under assumptions similar to those made in deriving Eq. (10) it can be shown that for standard reduced models with $\mathbf{w}_\mu = \mathbf{y}_\mu$

$$\partial^2\Phi/\partial\boldsymbol{\theta}\,\partial\mathbf{y}_\eta \approx -\,2\mathbf{B}_\eta{}^{\mathrm{T}}\boldsymbol{\Gamma} \tag{7-5-12}$$

Substituting Eq. (10) and Eq. (12) in Eq. (9), we obtain

$$\mathbf{V}_{\boldsymbol{\theta}} \approx \left(\sum_{\mu=1}^{n} \mathbf{B}_\mu{}^{\mathrm{T}}\boldsymbol{\Gamma}\mathbf{B}_\mu\right)^{-1}\left(\sum_{\mu=1}^{n} \mathbf{B}_\mu{}^{\mathrm{T}}\boldsymbol{\Gamma}\mathbf{V}_\mu\boldsymbol{\Gamma}\mathbf{B}_\mu\right)\left(\sum_{\mu=1}^{n} \mathbf{B}_\mu{}^{\mathrm{T}}\boldsymbol{\Gamma}\mathbf{B}_\mu\right)^{-1} \tag{7-5-13}$$

A derivation similar to the one given in Appendix E for the Gauss–Markov theorem, shows that if $\mathbf{V}_\mu = \mathbf{V}$ $(\mu = 1, 2, \ldots, n)$, then choosing $\boldsymbol{\Gamma}$ proportional to $\mathbf{V}^{-1}$ leads to the least possible value of det $\mathbf{V}_{\boldsymbol{\theta}}$. This, in fact, occurs in the cases listed below, and shows that these maximum likelihood and pseudomaximum likelihood estimates are at least approximately optimal.

In the case of single equation least squares, assuming observations with standard deviation σ, we have

$$\mathbf{B}_\mu{}^{\mathrm{T}} = \mathbf{b}_\mu \equiv \partial f_\mu/\partial\boldsymbol{\theta}, \qquad \boldsymbol{\Gamma} = 1, \qquad \mathbf{V}_\mu = \sigma^2$$

so that Eq. (13) reduces to

$$\mathbf{V}_\theta \approx \sigma^2\left(\sum_{\mu=1}^{n} \mathbf{b}_\mu \mathbf{b}_\mu^{\mathrm{T}}\right)^{-1} = \sigma^2\left(\sum_{\mu=1}^{n} (\partial f_\mu/\partial\boldsymbol{\theta})(\partial f_\mu/\partial\boldsymbol{\theta})^{\mathrm{T}}\right)^{-1} \tag{7-5-14}$$

Comparing this to Eq. (5-9-4) shows that here

$$\mathbf{V}_\theta \approx 2\sigma^2\mathbf{N}^{-1} \tag{7-5-15}$$

When σ is not known, we replace it with its estimate $[1/(n-l)]\sum_{\mu=1}^{n} e_\mu{}^2 = [1/(n-l)]\Phi^*$. A numerical illustration appears in Section 7-21.

We now treat some of the likelihood functions that were considered in Section 5-9.

1. Normal distribution with known $\mathbf{V}_\mu = \mathbf{V}$. According to row 1 of Table 5-1 $\boldsymbol{\Gamma} = \frac{1}{2}\mathbf{V}^{-1}$, so that Eq. (13) becomes in view of Eq. (10)

$$\mathbf{V}_\theta \approx \tfrac{1}{2}\left(\sum_{\mu=1}^{n} \mathbf{B}_\mu{}^{\mathrm{T}}\boldsymbol{\Gamma}\mathbf{B}_\mu\right)^{-1} = \mathbf{N}^{-1} \approx \mathbf{H}^{*-1} \tag{7-5-16}$$

2. As above, but with unknown $\mathbf{V}_\mu = \mathbf{V}$. From row 3 of Table 5-1 $\boldsymbol{\Gamma} = (n/2)\mathbf{M}^{-1}$. But according to Eq. (4-9-6), the maximum likelihood estimate for $\mathbf{V}$ is given by $(1/n)\mathbf{M}$, so that approximately $\boldsymbol{\Gamma} = \frac{1}{2}\mathbf{V}^{-1}$ and Eq. (16) is still valid.

For a wide class of maximum likelihood estimates with normal distributions we have then

$$\mathbf{V}_\theta \approx \mathbf{H}^{*-1} = -(\partial^2 \log L/\partial\boldsymbol{\theta}\,\partial\boldsymbol{\theta})^{-1}_{\boldsymbol{\theta}=\boldsymbol{\theta}^*} \tag{7-5-17}$$

The quality of this approximation improves as the variance of the measurements decreases and the fit of the model to the data gets better.

For most unconstrained maximum likelihood estimates it can be shown (Cramer, 1946, p. 500 *et seq.*) that asymptotically (as the series of experiments is repeated ad infinitum) the sampling distribution approaches (with probability 1) the normal form, with means equal to the true values of the parameters, and with covariance matrix given by

$$\mathbf{V}_\theta = -[E(\partial^2 \log L/\partial\boldsymbol{\theta}\,\partial\boldsymbol{\theta})]^{-1} \qquad (n \to \infty) \tag{7-5-18}$$

The computation of the required expectation is very tedious, if not altogether impossible; therefore, we generally replace the expected value by the most likely value, i.e., the value at $\boldsymbol{\theta} = \boldsymbol{\theta}^*$. This brings us back to Eq. (17). Again, the acceptability of this approximation depends on the goodness of fit. If the fit is very good, the likelihood function has a sharp peak, and the expected and most likely values nearly coincide.

The estimates given here and in the sequel for $\mathbf{V}_\theta$ and other statistical parameters are computed from the data. Hence they are in themselves random variables subject to sampling variations. As illustrated by the examples of Section 7-22 these variations can be quite large even when a good fit to the data can be obtained. We shall not concern ourselves here with the computation of the sampling variances of $\mathbf{V}_\theta$. Nevertheless, we point out that these variances can always be estimated by the Monte Carlo method (Section 3-3). Generally the $\mathbf{V}_\theta$ computed from any given data sample can be regarded as no more than a rough estimate, correct to within an order of magnitude.

The fact that the approximations may break down when the fit is poor need not worry us too much, since in this case we would not place much reliance on the model anyway, and would attempt to improve either the model or the data. Even a very rough approximation to $\mathbf{V}_\theta$ can be of considerable use, as will be seen in Chapter X.

7-6. Exact Structural Model

We can also derive an approximation to $\mathbf{V}_\theta$ for the case of structural equations acting as equality constraints. Suppose we have obtained estimates $\boldsymbol{\theta}^*$ and $\hat{\mathbf{W}}^*$ given the data $\mathbf{W}$ and using the method of Section 6-6. If the data are replaced by $\mathbf{W} + \delta\mathbf{W}$ there will be a correction $\delta\boldsymbol{\theta}^*$ in $\boldsymbol{\theta}^*$ given by Eq. (6-6-12). Now at the solution $\mathbf{G} = \mathbf{0}$, and (as the reader may verify for himself) $\mathbf{B}^T\mathbf{C}^{-1}\mathbf{A}\mathbf{H}^{-1}\mathbf{q} = \mathbf{0}$ so that Eq. (6-6-12) becomes approximately

$$\delta\boldsymbol{\theta}^* = \mathbf{D}^{-1}\mathbf{B}^T\mathbf{C}^{-1}\mathbf{A}\mathbf{H}^{-1}\,\delta\mathbf{q} \tag{7-6-1}$$

where $\delta\mathbf{q} = (\partial\mathbf{q}/\partial\mathbf{w})\,\delta\mathbf{w} = (\partial^2\Phi/\partial\hat{\mathbf{w}}\,\partial\mathbf{w})\,\delta\mathbf{w}$ is the change in $\mathbf{q}$ due to the change in $\mathbf{W}$. Setting

$$\mathbf{V}_\mathrm{w} \equiv E(\delta\mathbf{w}\,\delta\mathbf{w}^T) \tag{7-6-2}$$

we are led to

$$\begin{aligned}\mathbf{V}_\theta &\equiv E(\delta\boldsymbol{\theta}^*\,\delta\boldsymbol{\theta}^{*T})\\ &\approx \mathbf{D}^{-1}\mathbf{B}^T\mathbf{C}^{-1}\mathbf{A}\mathbf{H}^{-1}(\partial^2\Phi/\partial\hat{\mathbf{w}}\,\delta\mathbf{w})\mathbf{V}_\mathrm{w}\,(\partial^2\Phi/\partial\hat{\mathbf{w}}\,\partial\mathbf{w})^T\mathbf{H}^{-1}\mathbf{A}^T\mathbf{C}^{-1}\mathbf{B}\mathbf{D}^{-1}\end{aligned} \tag{7-6-3}$$

To achieve further progress we assume that $\Phi = \frac{1}{2}(\mathbf{w} - \hat{\mathbf{w}})^T\mathbf{V}_\mathrm{w}^{-1}(\mathbf{w} - \hat{\mathbf{w}})$. Hence $\mathbf{H}^{-1} = \mathbf{V}_\mathrm{w}$, and $\partial^2\Phi/\partial\hat{\mathbf{w}}\,\partial\mathbf{w} = -\mathbf{V}_\mathrm{w}^{-1}$. Since $\mathbf{C} = \mathbf{A}\mathbf{H}^{-1}\mathbf{A}^T = \mathbf{A}\mathbf{V}_\mathrm{w}\mathbf{A}^T$ and $\mathbf{D} = \mathbf{B}^T\mathbf{C}^{-1}\mathbf{B}$, it follows that

$$\begin{aligned}\mathbf{V}_\theta &\approx \mathbf{D}^{-1}\mathbf{B}^T\mathbf{C}^{-1}\mathbf{A}\mathbf{V}_\mathrm{w}\mathbf{V}_\mathrm{w}^{-1}\mathbf{V}_\mathrm{w}\mathbf{V}_\mathrm{w}^{-1}\mathbf{V}_\mathrm{w}\mathbf{A}^T\mathbf{C}^{-1}\mathbf{B}\mathbf{D}^{-1}\\ &= \mathbf{D}^{-1}\mathbf{B}^T\mathbf{C}^{-1}\mathbf{A}\mathbf{V}_\mathrm{w}\mathbf{A}^T\mathbf{C}^{-1}\mathbf{B}\mathbf{D}^{-1} = \mathbf{D}^{-1}\mathbf{B}^T\mathbf{C}^{-1}\mathbf{B}\mathbf{D}^{-1}\end{aligned}$$

so that finally

$$\mathbf{V}_\theta \approx \mathbf{D}^{-1} = [\mathbf{B}^T(\mathbf{A}\mathbf{V}_\mathrm{w}\mathbf{A}^T)^{-1}\mathbf{B}]^{-1} \tag{7-6-4}$$

In particular, for the model treated in Section 6-8 we obtain $\mathbf{V}_\theta$ by inverting Eq. (6-8-14). A numerical illustration appears in Section 7-23.

7-7. Constraints

Let us examine now how the covariance matrix is affected by inequality and equality constraints that do not depend on the data. Let $\mathbf{h}(\boldsymbol{\theta}) \geqslant \mathbf{0}$ and $\mathbf{g}(\boldsymbol{\theta}) = \mathbf{0}$ denote these constraints, which, of course, are satisfied by $\boldsymbol{\theta}^*$.

Let $\boldsymbol{\theta}^{(i)}$ be the value of $\boldsymbol{\theta}$ that optimizes $\Phi(\boldsymbol{\theta})$ with the constraint $h_i(\boldsymbol{\theta}) \geqslant 0$ removed, but all other constraints retained. Clearly, if it happens that $\boldsymbol{\theta}^{(i)}$ is feasible, i.e., if $h_i(\boldsymbol{\theta}^{(i)}) \geqslant 0$, then $\boldsymbol{\theta}^* = \boldsymbol{\theta}^{(i)}$. Actually, we distinguish the following four cases (see Fig. 7-2):

(a) $\boldsymbol{\theta}^* = \boldsymbol{\theta}^{(i)}$ is well in the interior of the feasible region relative to the constraint $h_i(\boldsymbol{\theta}) \geqslant 0$. Therefore, different values of $\boldsymbol{\theta}^{(i)}$ arising from different

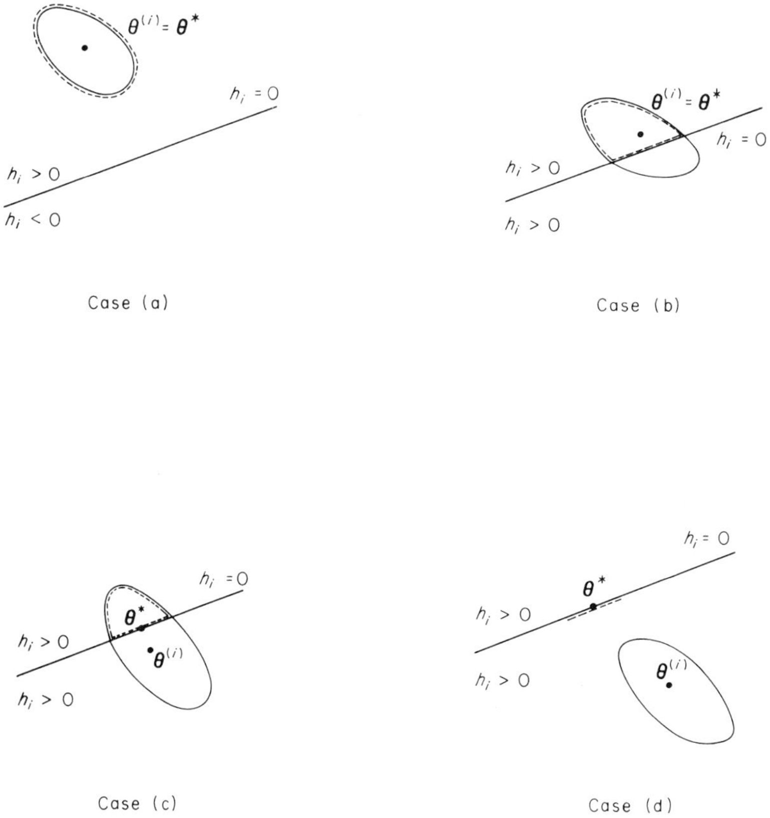

Fig. 7-2. Inequality constraint $h_i(\boldsymbol{\theta}) \geqslant 0$; $\boldsymbol{\theta}^{(i)}$, optimum without constraint; $\boldsymbol{\theta}^*$, optimum with constraint; —, boundary of region containing 90% of all realizations of $\boldsymbol{\theta}^{(i)}$; ---, boundary of region containing 90% of all realizations of $\boldsymbol{\theta}^*$.

data samples are likely to remain feasible, and the constraint $h_i(\boldsymbol{\theta}) \geqslant 0$ exerts no influence on $\mathbf{V}_{\boldsymbol{\theta}}$.

(b) $\boldsymbol{\theta}^* = \boldsymbol{\theta}^{(i)}$ is feasible, but it lies very close to the surface $h_i(\boldsymbol{\theta}) = 0$. In this case, a significant number of data samples may give rise to infeasible values of $\boldsymbol{\theta}^{(i)}$, causing $\boldsymbol{\theta}^*$ to lie on the constraint. The density of the sampling distribution of $\boldsymbol{\theta}^*$ is truncated: positive on one side of the constraint, infinite on the constraint, and zero on the other side. The computation of its covariance matrix is difficult, and will not be undertaken here.

(c) $\boldsymbol{\theta}^{(i)}$ is infeasible, but only slightly so. $\boldsymbol{\theta}^*$ is on the constraint. Some data samples make $\boldsymbol{\theta}^{(i)}$ feasible, and therefore some realizations of $\boldsymbol{\theta}^*$ fall in the interior of the feasible region. The distribution is similar to the one of case (b), and we will not treat it further.

(d) $\boldsymbol{\theta}^{(i)}$ is extremely infeasible, so that $\boldsymbol{\theta}^*$ remains on the constraint for all but a negligible proportion of all possible data samples. In this case, we may treat h_i as an equality constraint $h_i(\boldsymbol{\theta}) = 0$.

Let the vector $\mathbf{g}(\boldsymbol{\theta})$ now represent all the equality constraints, including the inequality constraints of type (d). As we know from Section 3-6, the Lagrangian conditions

$$\partial\Phi/\partial\boldsymbol{\theta} = \sum_i \lambda_i \, \partial g_i/\partial\boldsymbol{\theta} \tag{7-7-1}$$

must be satisfied at $\boldsymbol{\theta} = \boldsymbol{\theta}^*$. If we change the data by an amount $\delta\mathbf{w}$, we find $\boldsymbol{\theta}^*$ changed by $\delta\boldsymbol{\theta}^*$, and λ_i changed by $\delta\lambda_i$. At the new optimum, Eq. (1) takes the form (approximately to first-order terms)

$$\frac{\partial\Phi}{\partial\boldsymbol{\theta}} + \frac{\partial^2\Phi}{\partial\boldsymbol{\theta}\,\partial\boldsymbol{\theta}}\,\delta\boldsymbol{\theta}^* + \frac{\partial^2\Phi}{\partial\boldsymbol{\theta}\,\partial\mathbf{w}}\,\delta\mathbf{w} = \sum_i \left(\lambda_i \frac{\partial g_i}{\partial\boldsymbol{\theta}} + \delta\lambda_i \frac{\partial g_i}{\partial\boldsymbol{\theta}} + \lambda_i \frac{\partial^2 g_i}{\partial\boldsymbol{\theta}\,\partial\boldsymbol{\theta}}\,\delta\boldsymbol{\theta}^*\right) \tag{7-7-2}$$

with all derivatives evaluated at $\boldsymbol{\theta} = \boldsymbol{\theta}^*$. Subtracting Eq. (1) from Eq. (2) leaves

$$\mathbf{A}\,\delta\boldsymbol{\theta}^* = (\partial\mathbf{g}/\partial\boldsymbol{\theta})^{\mathrm{T}}\,\delta\boldsymbol{\lambda} - (\partial^2\Phi/\partial\boldsymbol{\theta}\,\partial\mathbf{w})\,\delta\mathbf{w} \tag{7-7-3}$$

where

$$\mathbf{A} \equiv \partial^2\Phi/\partial\boldsymbol{\theta}\,\partial\boldsymbol{\theta} - \sum_i \lambda_i\,\partial^2 g_i/\partial\boldsymbol{\theta}\,\partial\boldsymbol{\theta} = \mathbf{H}^* - \sum_i \lambda_i\,\partial^2 g_i/\partial\boldsymbol{\theta}\,\partial\boldsymbol{\theta} \tag{7-7-4}$$

so that

$$\delta\boldsymbol{\theta}^* = \mathbf{A}^{-1}\big((\partial\mathbf{g}/\partial\boldsymbol{\theta})^{\mathrm{T}}\,\delta\boldsymbol{\lambda} - (\partial^2\Phi/\partial\boldsymbol{\theta}\,\partial\mathbf{w})\,\delta\mathbf{w}\big) \tag{7-7-5}$$

The variation $\delta\boldsymbol{\theta}^*$ must leave the equations $\mathbf{g}(\boldsymbol{\theta}^* + \delta\boldsymbol{\theta}^*) = \mathbf{0}$ satisfied. Hence

$$\delta\mathbf{g} \equiv (\partial\mathbf{g}/\partial\boldsymbol{\theta})\,\delta\boldsymbol{\theta}^* = \mathbf{0} \tag{7-7-6}$$

Substituting Eq. (5) in Eq. (6) we find after solving for $\delta\boldsymbol{\lambda}$

$$\delta\boldsymbol{\lambda} = \left[\frac{\partial \mathbf{g}}{\partial \boldsymbol{\theta}} \mathbf{A}^{-1} \frac{\partial \mathbf{g}}{\partial \boldsymbol{\theta}}^{\mathrm{T}}\right]^{-1} \frac{\partial \mathbf{g}}{\partial \boldsymbol{\theta}} \mathbf{A}^{-1} \frac{\partial^2 \Phi}{\partial \boldsymbol{\theta}\, \partial \mathbf{w}} \delta\mathbf{w} \tag{7-7-7}$$

Finally, after inserting Eq. (7) into Eq. (5) we find

$$\delta\boldsymbol{\theta}^* = -\left\{\mathbf{I} - \mathbf{A}^{-1} \frac{\partial \mathbf{g}}{\partial \boldsymbol{\theta}}^{\mathrm{T}} \left[\frac{\partial \mathbf{g}}{\partial \boldsymbol{\theta}} \mathbf{A}^{-1} \frac{\partial \mathbf{g}}{\partial \boldsymbol{\theta}}^{\mathrm{T}}\right]^{-1} \frac{\partial \mathbf{g}}{\partial \boldsymbol{\theta}}\right\} \mathbf{A}^{-1} \frac{\partial^2 \Phi}{\partial \boldsymbol{\theta}\, \partial \mathbf{w}} \delta\mathbf{w} \tag{7-7-8}$$

Comparing Eq. (8) with Eq. (7-5-4) we find two changes

1. $\mathbf{H}^*$ is replaced by $\mathbf{A}$. From Eq. (4) we see that if all the constraints g_i are linear, then actually $\mathbf{A} = \mathbf{H}^*$. This occurs, e.g., when all the constraints are upper and lower bounds.

2. The expression $\mathbf{A}^{-1}(\partial^2\Phi/\partial\boldsymbol{\theta}\,\partial\mathbf{w})\,\delta\mathbf{w}$ is premultiplied by the projection matrix

$$\mathbf{P} = \mathbf{I} - \mathbf{A}^{-1} \frac{\partial \mathbf{g}}{\partial \boldsymbol{\theta}}^{\mathrm{T}} \left[\frac{\partial \mathbf{g}}{\partial \boldsymbol{\theta}} \mathbf{A}^{-1} \frac{\partial \mathbf{g}}{\partial \boldsymbol{\theta}}^{\mathrm{T}}\right]^{-1} \frac{\partial \mathbf{g}}{\partial \boldsymbol{\theta}} \tag{7-7-9}$$

which has the property that if $\mathbf{x}$ is any l-dimensional vector, then $\mathbf{y} = \mathbf{P}\mathbf{x}$ satisfies $(\partial\mathbf{g}/\partial\boldsymbol{\theta})\mathbf{y} = \mathbf{0}$, i.e., $\mathbf{y}$ lies in the tangent plane to the constraints. The matrix $\mathbf{P}$ thus projects any vector into this tangent plane.

The expression analogous to Eq. (7-5-7) is now

$$\mathbf{V}_{\boldsymbol{\theta}} = \mathbf{P}\mathbf{A}^{-1}(\partial^2\Phi/\partial\boldsymbol{\theta}\,\partial\mathbf{w})\mathbf{V}_{\mathbf{w}}\,(\partial^2\Phi/\partial\boldsymbol{\theta}\,\partial\mathbf{w})^{\mathrm{T}}\mathbf{A}^{-1}\mathbf{P}^{\mathrm{T}} \tag{7-7-10}$$

If all g_i are linear, i.e., $\mathbf{A} = \mathbf{H}^*$, then under the appropriate conditions Eq. (7-5-16) remains valid with the following modification

$$\mathbf{V}_{\boldsymbol{\theta}} \approx \mathbf{P}\mathbf{H}^{*-1}\mathbf{P}^{\mathrm{T}} \tag{7-7-11}$$

Let us compute $\mathbf{P}$ and $\mathbf{V}_{\boldsymbol{\theta}}$ for the simple though common case in which all active constraints are derived from upper or lower bounds on the parameters. For simplicity of representation, we assume that the parameters $\theta_1, \theta_2, \ldots, \theta_{l_1}$ are actively constrained to equal $a_1, a_2, \ldots, a_{l_1}$, respectively. The remaining parameters $\theta_{l_1+1}, \theta_{l_1+2}, \ldots, \theta_l$ are not actively constrained. The active constraints can be written as

$$[\mathbf{I} \quad \mathbf{0}]\boldsymbol{\theta} = \mathbf{a} \tag{7-7-12}$$

with $\mathbf{I}$ being the $l_1 \times l_1$ identity, and $\mathbf{0}$ the $l_1 \times (l - l_1)$ null matrix. We have, then

$$\partial\mathbf{g}/\partial\boldsymbol{\theta} = [\mathbf{I} \quad \mathbf{0}]$$

Since our constraints are linear, we have $\mathbf{A} = \mathbf{H}^*$. Let $\mathbf{H}^{*-1}$ be partitioned as follows

$$\mathbf{A}^{-1} = \mathbf{H}^{*-1} = \begin{bmatrix} \mathbf{B} & \mathbf{C}^{\mathrm{T}} \\ \mathbf{C} & \mathbf{D} \end{bmatrix} \tag{7-7-13}$$

where $\mathbf{B}$ is $l_1 \times l_1$, $\mathbf{C}$ is $(l - l_1) \times l_1$, and $\mathbf{D}$ is $(l - l_1) \times (l - l_1)$. Therefore:

$$\left[\frac{\partial \mathbf{g}}{\partial \boldsymbol{\theta}} \mathbf{A}^{-1} \frac{\partial \mathbf{g}}{\partial \boldsymbol{\theta}}^{\mathrm{T}}\right]^{-1} = \left([\mathbf{I} \quad \mathbf{0}] \begin{bmatrix} \mathbf{B} & \mathbf{C}^{\mathrm{T}} \\ \mathbf{C} & \mathbf{D} \end{bmatrix} \begin{bmatrix} \mathbf{I} \\ \mathbf{0} \end{bmatrix}\right)^{-1} = \mathbf{B}^{-1} \tag{7-7-14}$$

$$\begin{aligned} \mathbf{P} &= \begin{bmatrix} \mathbf{I} & \mathbf{0} \\ \mathbf{0} & \mathbf{I} \end{bmatrix} - \begin{bmatrix} \mathbf{B} & \mathbf{C}^{\mathrm{T}} \\ \mathbf{C} & \mathbf{D} \end{bmatrix} \begin{bmatrix} \mathbf{I} \\ \mathbf{0} \end{bmatrix} \mathbf{B}^{-1} [\mathbf{I} \quad \mathbf{0}] \\ &= \begin{bmatrix} \mathbf{I} & \mathbf{0} \\ \mathbf{0} & \mathbf{I} \end{bmatrix} - \begin{bmatrix} \mathbf{I} & \mathbf{0} \\ \mathbf{C}\mathbf{B}^{-1} & \mathbf{0} \end{bmatrix} = \begin{bmatrix} \mathbf{0} & \mathbf{0} \\ -\mathbf{C}\mathbf{B}^{-1} & \mathbf{I} \end{bmatrix} \end{aligned} \tag{7-7-15}$$

and finally

$$\mathbf{V}_{\boldsymbol{\theta}} = \begin{bmatrix} \mathbf{0} & \mathbf{0} \\ -\mathbf{C}\mathbf{B}^{-1} & \mathbf{I} \end{bmatrix} \begin{bmatrix} \mathbf{B} & \mathbf{C}^{\mathrm{T}} \\ \mathbf{C} & \mathbf{D} \end{bmatrix} \begin{bmatrix} \mathbf{0} & -\mathbf{B}^{-1}\mathbf{C}^{\mathrm{T}} \\ \mathbf{0} & \mathbf{I} \end{bmatrix} = \begin{bmatrix} \mathbf{0} & \mathbf{0} \\ \mathbf{0} & \mathbf{D} - \mathbf{C}\mathbf{B}^{-1}\mathbf{C}^{\mathrm{T}} \end{bmatrix} \tag{7-7-16}$$

As expected, the variances and covariances of $\theta_1, \theta_2, \ldots, \theta_{l_1}$ are zero, since these parameters are constrained to equal fixed values. The covariance matrix of the unconstrained θ's is reduced from $\mathbf{D}$ to $\mathbf{D} - \mathbf{C}\mathbf{B}^{-1}\mathbf{C}^{\mathrm{T}}$. The matrix $\mathbf{D} - \mathbf{C}\mathbf{B}^{-1}\mathbf{C}^{\mathrm{T}}$ is simply the inverse of the lower right-hand partition of $\mathbf{A}$, so that the covariance of the unconstrained parameters is obtained by inverting the corresponding part of $\mathbf{A}$.

7-8. Principal Components

Given the matrix $\mathbf{V}_{\boldsymbol{\theta}}$ or an approximation to it, we can determine which parameters, or linear combinations of parameters, are well determined (small variance), and which are poorly determined (large variance). The variance of the estimate for the αth parameter is given by $V_{\boldsymbol{\theta}\alpha\alpha}$, and its standard deviation by $\sigma_i \equiv V_{\boldsymbol{\theta}\alpha\alpha}^{1/2}$. As in Section 7-3, the full picture is obtained by finding the eigenvalue decomposition of $\mathbf{V}_{\boldsymbol{\theta}}$, say

$$\mathbf{V}_{\boldsymbol{\theta}} = \mathbf{U}\boldsymbol{\Pi}\mathbf{U}^{\mathrm{T}}, \quad \boldsymbol{\Pi} = \mathbf{U}^{\mathrm{T}}\mathbf{V}_{\boldsymbol{\theta}}\mathbf{U} \tag{7-8-1}$$

where $\boldsymbol{\Pi}$ is the diagonal matrix of the eigenvalues π_i of $\mathbf{V}_{\boldsymbol{\theta}}$.

Suppose we define a vector of new variables $\boldsymbol{\rho} = \mathbf{U}^{\mathrm{T}}\boldsymbol{\theta}$, $\delta\boldsymbol{\rho} = \mathbf{U}^{\mathrm{T}}\,\delta\boldsymbol{\theta}$. The covariance matrix of the ρ's is given by

$$\mathbf{V}_0 \equiv E(\delta\boldsymbol{\rho}\,\delta\boldsymbol{\rho}^{\mathrm{T}}) = E(\mathbf{U}^{\mathrm{T}}\,\delta\boldsymbol{\theta}\,\delta\boldsymbol{\theta}^{\mathrm{T}}\,\mathbf{U}) = \mathbf{U}^{\mathrm{T}}\mathbf{V}_{\boldsymbol{\theta}}\mathbf{U} = \boldsymbol{\Pi} \tag{7-8-2}$$

We have then a set of new variables $\rho_1, \rho_2, \ldots, \rho_l$ which replace the original parameters $\theta_1, \theta_2, \ldots, \theta_l$. Our estimated value of ρ_i is given by $\rho_i^* = \sum_{\alpha=1}^{\mu} U_{\alpha i}\theta_\alpha^*$. Since $\mathbf{V}_\rho = \mathbf{\Pi}$ is a diagonal matrix, the sampling variations in ρ_i^* and ρ_j^* are uncorrelated when $i \neq j$. The standard deviation of the estimate ρ_i^* is $\pi_i^{1/2}$. The ρ_i are called the *principal components*.

The advantage of dealing with the uncorrelated principal components rather than with the correlated original parameters is particularly great when (as in the normal distribution) lack of correlation implies statistical independence. For then we can establish confidence intervals and statistical tests for each component individually.

When Eq. (7-5-17) holds, we have $\mathbf{V}_\theta \approx \mathbf{H}^{*-1} = \mathbf{A}^{-1}$. The eigenvectors of $\mathbf{V}_\theta$ and $\mathbf{A}$ coincide, hence the $\delta\rho_i$ coincide with the canonical variables. The eigenvalues of $\mathbf{V}_\theta$ are the reciprocals of those of $\mathbf{A}$, i.e., $\pi_i = 1/\lambda_i$. Hence the ε-indifference interval of $\delta\rho_i = \psi_i$ is given by

$$\pm(2\varepsilon/\lambda_i)^{1/2} = \pm(2\varepsilon\pi_i)^{1/2}$$

The length of the interval is proportional to the standard deviation. For single-equation least squares, Eq. (7-5-15) implies that $\pi_i = 2\sigma^2/\lambda_i$.

Like the canonical variables, the principal components depend on the scaling of the variables. A natural scaling is one which adjusts each variable so as to have unit standard deviation. This can be achieved by defining $v_\alpha = \theta_\alpha V_{\theta\alpha\alpha}^{-1/2}$, so that

$$E(\delta v_\alpha\,\delta v_\beta) = V_{\theta\alpha\beta} V_{\theta\alpha\alpha}^{-1/2} V_{\theta\beta\beta}^{-1/2}$$

or

$$\mathbf{V}_v = \mathbf{D}\mathbf{V}_\theta\mathbf{D} \tag{7-8-3}$$

where $D_{\alpha\beta} = \delta_{\alpha\beta} V_{\theta\alpha\alpha}^{-1/2}$. The matrix $\mathbf{V}_v$, whose main diagonal consists entirely of ones, is the *correlation matrix* of $\boldsymbol{\theta}^*$. If $\mathbf{P}$ is the matrix of eigenvectors of $\mathbf{V}_v$, then the elements of the vector $\mathbf{P}^T\mathbf{v} = \mathbf{P}^T\mathbf{D}\boldsymbol{\theta}$ are the principal components of $\mathbf{V}_v$. They are uncorrelated, and their variances are the eigenvalues of $\mathbf{V}_v$. We call them the *scaled principal components* of $\mathbf{V}_\theta$.

In practice one is interested mostly in determining the principal components having unusually large variances, since these hold clues to inadequacies in model or data. This point is discussed further in Section 7-18. If $\mathbf{V}_\theta$ is known, but not its eigenvalues and vectors, then we can most easily determine its largest eigenvalue and corresponding eigenvector by means of the power method.

7-9. Confidence Intervals

Knowledge of the covariance matrix of an estimator gives an intuitive feeling of the degree to which the various parameters are well determined. We may wish however, to make more explicit statements such as "the true value of θ lies between the numbers a and b with 90% probability." From the point of view of classical statistics this statement is meaningless; θ is a constant, albeit unknown. The probability of its lying between a and b is unity if $a < \theta < b$, and zero otherwise; it can never be 90%. This difficulty is overcome by Neyman's (Neyman, 1937) theory of confidence intervals.

Suppose we had complete knowledge of the sampling distribution of the estimate θ^* (we discuss first the case of a single parameter θ). This distribution depends on the true value $\hat{\theta}$, and we denote its density function by $p(\theta^* \mid \hat{\theta})$. For any given value of $\hat{\theta}$ we can easily determine two numbers $a(\hat{\theta})$ and $b(\hat{\theta})$ such that, for a given $0 < \gamma < 1$ we have

$$\Pr[a(\hat{\theta}) \leqslant \theta^* \leqslant b(\hat{\theta})] = \gamma \tag{7-9-1}$$

This is equivalent to demanding that for each $\hat{\theta}$

$$\int_{a(\hat{\theta})}^{b(\hat{\theta})} p(\theta^* \mid \hat{\theta})\, d\theta^* = \gamma \tag{7-9-2}$$

It is clear that the choice of $a(\theta)$ is quite arbitrary; given any $\varepsilon < 1 - \gamma$, we can choose $a(\theta)$ so that $\Pr[\theta^* < a(\hat{\theta})] = \varepsilon$; $b(\theta)$ is then determined by the requirement that $\Pr[\theta^* > b(\hat{\theta})] = 1 - \gamma - \varepsilon$. Suppose we are able to choose $a(\hat{\theta})$ so that both it and $b(\hat{\theta})$ are monotonically increasing functions of $\hat{\theta}$. Then there exist inverse functions $\alpha(\theta^*)$ and $\beta(\theta^*)$ such that

$$\hat{\theta} = \alpha[a(\hat{\theta})] = \beta[b(\hat{\theta})]$$

Then the statement $\theta^* \geqslant a(\hat{\theta})$ is equivalent to $\hat{\theta} \leqslant \alpha(\theta^*)$ and the statement $\theta^* \leqslant b(\hat{\theta})$ is equivalent to $\hat{\theta} \geqslant \beta(\theta^*)$. It follows that

$$\Pr[\beta(\theta^*) \leqslant \hat{\theta} \leqslant \alpha(\theta^*)] = \Pr[a(\hat{\theta}) \leqslant \theta^* \leqslant b(\hat{\theta})] = \gamma \tag{7-9-3}$$

Note that θ^* as a function of the sample is a random variable, and therefore so are $\alpha(\theta^*)$ and $\beta(\theta^*)$. Eq. (3) asserts that the probability that the value of the random variable $\beta(\theta^*)$ does not exceed the true value $\hat{\theta}$, and that the random variable $\alpha(\theta^*)$ is no less than $\hat{\theta}$, is equal to γ. This is a perfectly meaningful statement within the confines of classical statistics. The interval $[\beta(\theta^*), \alpha(\theta^*)]$ is called a *γ-confidence interval* for θ.

The simple example of Section 3-1 will help clarify the situation. From observations w_μ on a normally distributed random variable ω with mean $\hat{\theta}$ and known variance σ^2 we obtain the maximum likelihood estimate Eq. (3-1-3)

$$\theta^* = (1/n) \sum_{\mu=1}^{n} w_\mu$$

It is well known that the sampling distribution of θ^* is normal with mean $\hat{\theta}$ and variance σ^2/n. From tables of the normal distribution [see, e.g., Cramer (1946, Table 2)], we note that

$$\Pr(|\theta^* - \hat{\theta}| \leqslant 1.6449\sigma/\sqrt{n}) = 0.9 \tag{7-9-4}$$

so that

$$\Pr(\hat{\theta} - 1.6449\sigma/\sqrt{n} \leqslant \theta^* \leqslant \hat{\theta} + 1.6449\sigma/\sqrt{n}) = 0.9 \qquad (7\text{-}9\text{-}5)$$

that is, $a(\hat{\theta}) = \hat{\theta} - 1.6449\sigma/\sqrt{n}$ and $b(\hat{\theta}) = \hat{\theta} + 1.6449\sigma/\sqrt{n}$. Both $a(\hat{\theta})$ and $b(\hat{\theta})$ are monotonic in $\hat{\theta}$ with inverses $\alpha(\theta^*) = \theta^* + 1.6449\sigma/\sqrt{n}$ and $\beta(\theta^*) = \theta^* - 1.6449\sigma/\sqrt{n}$. Clearly, Eq. (5) is equivalent to

$$\Pr(\theta^* - 1.6449\sigma/\sqrt{n} \leqslant \hat{\theta} \leqslant \theta^* + 1.6449\sigma/\sqrt{n}) = 0.9 \qquad (7\text{-}9\text{-}6)$$

so that $(\theta^* - 1.6449\sigma/\sqrt{n},\ \theta^* + 1.6449\sigma/\sqrt{n})$ is a 90% confidence interval for θ. This statement should be interpreted as follows: if the series of experiments were repeated one hundred times, each replication would yield a separate estimate θ^*. For each such estimate we could form the interval given by Eq. (6). Then the true value $\hat{\theta}$ should be contained in about ninety of these intervals.

If σ is unknown, the interval given by Eq. (6) is undefined. However, it is well known that in this case the quantity $\sqrt{n}(\theta^* - \hat{\theta})/s$, where

$$s \equiv \left\{[1/(n-1)] \sum_{\mu=1}^{n} (w_\mu - \theta^*)^2\right\}^{1/2} \qquad (7\text{-}9\text{-}7)$$

follows the t-distribution with $n - 1$ degrees of freedom. Hence, if $n = 10$, we have from tables [e.g., Cramer (1946, Table 3)]

$$\Pr[|10^{1/2}(\theta^* - \hat{\theta})/s| \leqslant 1.833] = 0.9 \qquad (7\text{-}9\text{-}8)$$

so that

$$\Pr(\theta^* - 1.833s/10^{1/2} \leqslant \hat{\theta} \leqslant \theta^* + 1.833\, s/10^{1/2}) = 0.9 \qquad (7\text{-}9\text{-}9)$$

This is a proper confidence interval, since s is computable from the sample. Since s is a consistent estimate for σ, Eq. (6) can be used with s replacing σ when n is large. In fact, the t-distribution approaches the normal as n increases.

If the exact form of the sampling distribution is unknown except for its variance V, we may derive conservative confidence intervals for the mean $\bar{\theta}$ of the sampling distribution. According to the Bienaymé–Chebyshev inequality (see Section 7-10 for the derivation of a more general result)

$$\Pr(|\theta^* - \bar{\theta}| \geqslant kV^{1/2}) \leqslant k^{-2} \qquad (7\text{-}9\text{-}10)$$

which is equivalent to (set $\gamma = 1 - k^{-2}$)

$$\Pr\{\bar{\theta} - [V/(1-\gamma)]^{1/2} \leqslant \theta^* \leqslant \bar{\theta} + [V/(1-\gamma)]^{1/2}\} \geqslant \gamma \qquad (7\text{-}9\text{-}11)$$

Hence

$$\Pr\{\theta^* - [V/(1-\gamma)]^{1/2} \leqslant \bar{\theta} \leqslant \theta^* + [V/(1-\gamma)]^{1/2}\} \geqslant \gamma \qquad (7\text{-}9\text{-}12)$$

We know then that a γ confidence interval for $\bar{\theta}$ is contained in the interval

$$\{\theta^* - [V/(1-\gamma)]^{1/2}, \theta^* + [V/(1-\gamma)]^{1/2}\}$$

If θ^* is an unbiased estimate, then $\hat{\theta}$ can be substituted for $\bar{\theta}$ in Eq. (12). Usually the bias is unknown, but when n is fairly large, maximum likelihood estimates can be treated as having a normal sampling distribution with variance given by Eq. (7-5-17), and bias proportional to $1/n$. In this case one may employ the method of Quenouille (1956) to reduce the bias. Let $\boldsymbol{\theta}^*$ be, as usual, the estimate of $\boldsymbol{\theta}$ based on n experiments. Let $\boldsymbol{\theta}_\mu^*$ $(\mu = 1, 2, \ldots, n)$ be the estimate based on the $n-1$ data points obtained by dropping the μth. Let $\tilde{\boldsymbol{\theta}} \equiv (\sum_{\mu=1}^{n} \boldsymbol{\theta}_\mu^*)/n$ be the mean of the $\boldsymbol{\theta}_\mu^*$. The bias in $\boldsymbol{\theta}^*$ is proportional to $1/n$, whereas the bias in each $\boldsymbol{\theta}_\mu^*$ is proportional to $1/(n-1)$, and so is the bias in $\tilde{\boldsymbol{\theta}}$. Hence the following relations hold approximately

$$\boldsymbol{\theta}^* \approx \hat{\boldsymbol{\theta}} + \alpha/n, \qquad \tilde{\boldsymbol{\theta}} \approx \hat{\boldsymbol{\theta}} + \alpha/(n-1) \tag{7-9-13}$$

where α is an unknown constant. Multiplying the first equation by n and the second by $n-1$, we obtain after subtraction

$$\hat{\boldsymbol{\theta}} \approx n\boldsymbol{\theta}^* - (n-1)\tilde{\boldsymbol{\theta}} \tag{7-9-14}$$

The bias of this estimate is of the order of $1/n^2$.

We have noted above that the choice of an interval for a given confidence level γ is somewhat arbitrary. Commonly employed criteria for choosing among all possible functions $a(\theta)$ and $b(\theta)$ that satisfy Eq. (1) are the following:

1. Minimum length. Choose $a(\theta)$ and $b(\theta)$ so that $b(\theta) - a(\theta)$ is minimum.
2. Symmetry around the true value

$$b(\theta) - \hat{\theta} = \hat{\theta} - a(\theta)$$

3. Symmetry in probability

$$\Pr[\theta^* < a(\hat{\theta})] = \Pr[\theta^* > b(\hat{\theta})] = (1-\gamma)/2.$$

4. Equal probability at ends. The density function of the sampling distribution assumes equal values at $\theta^* = a(\hat{\theta})$ and $\theta^* = b(\hat{\theta})$.

When $p(\theta^* \mid \hat{\theta})$ is unimodal and symmetric, criteria 1, 2, 3, and 4 coincide.

7-10. Confidence Regions

The idea of a confidence interval may be generalized to the case of several unknown parameters. Suppose that for any data sample $\mathbf{W}$ we are able to define a bounded closed subset $S(\mathbf{W})$ of the l-dimensional $\boldsymbol{\theta}$ space in such a way that $S(\mathbf{W})$ contains $\hat{\boldsymbol{\theta}}$ in a fraction γ of all possible data samples, i.e.,

$$\Pr[\hat{\boldsymbol{\theta}} \in S(\mathbf{W})] = \gamma \tag{7-10-1}$$

Then $S(\mathbf{W})$ is called a *γ-joint confidence region* for the parameters $\boldsymbol{\theta}$. In choosing a confidence region we have even more freedom than we had in the case of confidence intervals, since we may exercise control not only over the location, but also over the shape of the region. The most commonly used regions are l-dimensional rectangles or ellipsoids whose centers are at the estimate $\boldsymbol{\theta}^*$.

In Section 7-8 we have introduced the principal components $\boldsymbol{\rho} = \mathbf{U}^T\boldsymbol{\theta}$, which are uncorrelated linear combinations of the parameters. If (as would be implied when the sampling distribution were normal) the principal components are statistically independent, then confidence intervals for the individual components can be combined into a rectangular joint confidence region. Let (α_i, β_i) be a γ-confidence interval for ρ_i. Then

$$\Pr(\alpha_i \leqslant \rho_i \leqslant \beta_i,\ i = 1, 2, \ldots, l) = \gamma^l \tag{7-10-2}$$

so that we have a γ^l-joint confidence region for $\boldsymbol{\rho}$. Unfortunately, it is difficult to transform Eq. (2) into a simple statement in terms of the θ_α.

It seems reasonable to choose confidence regions which coincide with the indifference regions of the objective function that was used to obtain the estimate $\boldsymbol{\theta}^*$. We have seen in Section 7-2 that these take the approximate form

$$|(\boldsymbol{\theta} - \boldsymbol{\theta}^*)^{\mathrm{T}}\mathbf{H}^*(\boldsymbol{\theta} - \boldsymbol{\theta}^*)| \leqslant c \tag{7-10-3}$$

which in view of Eq. (7-5-16) may in most cases be approximated by the ellipsoid

$$(\boldsymbol{\theta} - \boldsymbol{\theta}^*)^{\mathrm{T}}\mathbf{V}_{\boldsymbol{\theta}}^{-1}(\boldsymbol{\theta} - \boldsymbol{\theta}^*) \leqslant c \tag{7-10-4}$$

For a given confidence level γ, we need only determine c in such a way that

$$\Pr[(\hat{\boldsymbol{\theta}} - \boldsymbol{\theta}^*)^{\mathrm{T}}\mathbf{V}_{\boldsymbol{\theta}}^{-1}(\hat{\boldsymbol{\theta}} - \boldsymbol{\theta}^*) \leqslant c] = \gamma \tag{7-10-5}$$

Such a value of c always exists, regardless of the accuracy of the assumptions made in estimating $\mathbf{V}_{\boldsymbol{\theta}}^{-1}$. The actual determination of the value of c does, however, depend on these assumptions, and on the form of the sampling distribution in general.

When the sampling distribution is normal, unbiased, and with known covariance matrix $\mathbf{V}_{\boldsymbol{\theta}}$, then as shown in Section 2-8 the quantity $(\boldsymbol{\theta}^* - \hat{\boldsymbol{\theta}})\mathbf{V}_{\boldsymbol{\theta}}^{-1}(\boldsymbol{\theta}^* - \hat{\boldsymbol{\theta}})$ is distributed as χ^2 with l degrees of freedom. Therefore, the constant c may be determined as the upper γ point of that distribution. For instance, let $l = 10$ and $\gamma = 0.9$. We find then (Cramer, 1946, Table 3) that

$$\Pr[(\hat{\boldsymbol{\theta}} - \boldsymbol{\theta}^*)^{\mathrm{T}}\mathbf{V}_{\boldsymbol{\theta}}^{-1}(\hat{\boldsymbol{\theta}} - \boldsymbol{\theta}^*) \leqslant 15.987] = 0.9 \tag{7-10-6}$$

When the sampling distribution cannot be assumed normal, we may generalize the Bienaymé–Chebyshev inequality Eq. (7-9-12) to obtain (provided $\bar{\boldsymbol{\theta}} = \hat{\boldsymbol{\theta}}$)

$$\Pr[(\hat{\boldsymbol{\theta}} - \boldsymbol{\theta}^*)^{\mathrm{T}}\mathbf{V}_{\boldsymbol{\theta}}^{-1}(\hat{\boldsymbol{\theta}} - \boldsymbol{\theta}^*) \leqslant l/(1 - \gamma)] \geqslant \gamma \tag{7-10-7}$$

We derive this relation as follows: Let

$$Q(\boldsymbol{\theta}^*) \equiv (\hat{\boldsymbol{\theta}} - \boldsymbol{\theta}^*)^{\mathrm{T}}\mathbf{V}_{\boldsymbol{\theta}}^{-1}(\hat{\boldsymbol{\theta}} - \boldsymbol{\theta}^*)$$

and let G_a be the region in which $Q(\boldsymbol{\theta}^*) \leqslant a$. Let

$$P_a(\boldsymbol{\theta}^*) \equiv 1 - (1/a)Q(\boldsymbol{\theta}^*)$$

Then $P_a(\boldsymbol{\theta}^*) \leqslant 0$ for $\boldsymbol{\theta}^*$ outside G_a, and $P_a(\boldsymbol{\theta}^*) \leqslant 1$ for $\boldsymbol{\theta}^*$ inside G_a. Hence

$$\int_{G_\infty} P_a(\boldsymbol{\theta}^*)p(\boldsymbol{\theta}^*)\,d\boldsymbol{\theta}^* \leqslant \int_{G_a} P_a(\boldsymbol{\theta}^*)p(\boldsymbol{\theta}^*)\,d\boldsymbol{\theta}^*$$
$$\leqslant \int_{G_a} p(\boldsymbol{\theta}^*)\,d\boldsymbol{\theta}^* = \Pr(\boldsymbol{\theta}^* \in G_a)$$

where G_∞ is the entire space and $p(\boldsymbol{\theta}^*)$ is the pdf of the sampling distribution. Now

$$\int_{G_\infty} P_a(\boldsymbol{\theta}^*)p(\boldsymbol{\theta}^*)\,d\boldsymbol{\theta}^* = \int_{G_\infty} [p(\boldsymbol{\theta}^*) - (1/a)Q(\boldsymbol{\theta}^*)p(\boldsymbol{\theta}^*)]\,d\boldsymbol{\theta}^*$$
$$= 1 - (1/a)EQ(\boldsymbol{\theta}^*)$$
$$= 1 - (1/a)E\,\mathrm{Tr}\,\mathbf{V}_{\boldsymbol{\theta}}^{-1}(\hat{\boldsymbol{\theta}} - \boldsymbol{\theta}^*)(\hat{\boldsymbol{\theta}} - \boldsymbol{\theta}^*)^{\mathrm{T}}$$
$$= 1 - (1/a)\,\mathrm{Tr}\,\mathbf{V}_{\boldsymbol{\theta}}^{-1}\mathbf{V}_{\boldsymbol{\theta}} = 1 - l/a$$

Hence

$$\Pr(\boldsymbol{\theta}^* \in G_a) \geqslant 1 - l/a$$

Taking $\gamma = 1 - l/a$ one obtains Eq. (7).

The method of Quenouille for reducing bias (see Section 7-9) is as applicable in the vector case as in the scalar case.

The volume of the region in $\boldsymbol{\theta}$ space defined by inequality Eq. (4) is

$$\mathscr{V}(c) = (c\pi)^{l/2} \det^{1/2} \mathbf{V}_{\boldsymbol{\theta}}/\Gamma\big((l/2) + 1\big) \tag{7-10-8}$$

7-11. Linearization

Somewhat more incisive results than those of the last section are often obtainable when the model equations can be approximated by linear ones in the vicinity of the estimate $\boldsymbol{\theta}^*$. The single equation model $f_\mu = f(\mathbf{x}_\mu, \boldsymbol{\theta})$ may be approximated by

$$f_\mu = f(\mathbf{x}_\mu, \boldsymbol{\theta}) \approx f(\mathbf{x}_\mu, \boldsymbol{\theta}^*) + \partial f_\mu/\partial\boldsymbol{\theta})_{\boldsymbol{\theta}=\boldsymbol{\theta}^*}(\boldsymbol{\theta} - \boldsymbol{\theta}^*) \tag{7-11-1}$$

This model resembles the multiple linear regression situation (Section 4-4), with **B** designating the $n \times l$ matrix whose μth row is $\partial f_\mu/\partial\boldsymbol{\theta}^{\mathrm{T}}$, F designating the n-vector whose μth element is $f_\mu - f(\mathbf{x}_\mu, \boldsymbol{\theta}^*)$, and $\boldsymbol{\theta} - \boldsymbol{\theta}^*$ replacing $\boldsymbol{\theta}$ in Eq. (4-4-2). According to Eq. (4-4-9), the vector $\boldsymbol{\theta} - \boldsymbol{\theta}^*$ has covariance $(\mathbf{B}^T\mathbf{V}^{-1}\mathbf{B})^{-1}$, where $\mathbf{V}$ is the covariance matrix of the observations y_μ [this corresponds to Eq. (7-5-16)]. If the errors in the observations are normally distributed, so are the estimates $\boldsymbol{\theta}^*$; and therefore, as stated above, the quantity

$$\mathcal{T} \equiv (\boldsymbol{\theta} - \boldsymbol{\theta}^*)^T\mathbf{B}^T\mathbf{V}^{-1}\mathbf{B}(\boldsymbol{\theta} - \boldsymbol{\theta}^*)$$

is distributed as χ^2 with l degrees of freedom. Hence, we can determine confidence regions for $\boldsymbol{\theta}^*$ provided $\mathbf{V}$ is known.

It is well known from the theory of multiple linear regression that the residual weighted sum of squares

$$\mathcal{S} \equiv [\mathbf{Y} - \mathbf{F}(\mathbf{X}, \boldsymbol{\theta}^*)]^T\mathbf{V}^{-1}[\mathbf{Y} - \mathbf{F}(\mathbf{X}, \boldsymbol{\theta}^*)] \tag{7-11-2}$$

is distributed independently of $\mathcal{T}$ as χ^2 with $n - l$ degrees of freedom. Hence $(n - l)\mathcal{T}/l\mathcal{S}$ has the $F_{l,n-l}$ distribution. Suppose it is known that $\mathbf{V} = \tau\mathbf{Q}$, where $\mathbf{Q}$ is a known matrix, and τ an unknown constant. Then

$$\frac{(n - l)\mathcal{T}}{l\mathcal{S}} = \frac{(n - l)(\boldsymbol{\theta} - \boldsymbol{\theta}^*)^T\mathbf{B}^T\mathbf{Q}^{-1}\mathbf{B}(\boldsymbol{\theta} - \boldsymbol{\theta}^*)}{l[\mathbf{Y} - \mathbf{F}(\mathbf{X}, \boldsymbol{\theta}^*)]^T\mathbf{Q}^{-1}[\mathbf{Y} - \mathbf{F}(\mathbf{X}, \boldsymbol{\theta}^*)]} \tag{7-11-3}$$

Thus $(n - l)\mathcal{T}/l\mathcal{S}$ may be computed without knowledge of τ, and tables of the F distribution [e.g., Scheffé (1959)] may be used to obtain confidence regions for $\boldsymbol{\theta}$.

An important special case occurs when all observations are independent, so that $\mathbf{V} = \sigma^2\mathbf{I}$ and

$$\frac{(n - l)\mathcal{T}}{l\mathcal{S}} = \frac{(n - l)(\boldsymbol{\theta} - \boldsymbol{\theta}^*)^T\mathbf{B}^T\mathbf{B}(\boldsymbol{\theta} - \boldsymbol{\theta}^*)}{l[\mathbf{Y} - \mathbf{F}(\mathbf{X}, \boldsymbol{\theta}^*)]^T[\mathbf{Y} - \mathbf{F}(\mathbf{X}, \boldsymbol{\theta}^*)]} \tag{7-11-4}$$

which corresponds to the case of unweighted least squares. For single equation least squares (illustrated in Section 7-21), this simplifies to

$$(n - l)\mathcal{T}/l\mathcal{S} = (\boldsymbol{\theta} - \boldsymbol{\theta}^*)^T\mathbf{N}^*(\boldsymbol{\theta} - \boldsymbol{\theta}^*)/2l\tilde{\sigma}^2 \tag{7-11-5}$$

where

$$\tilde{\sigma}^2 = \sum_{\mu=1}^{n} e_\mu{}^2/(n - l)$$

is the estimated variance of the residuals.

The foregoing discussion was based on the assumption that the model equations were nearly linear in the parameters around the estimate $\boldsymbol{\theta}^*$, at least for variations in $\boldsymbol{\theta}$ of the order of magnitude of the standard deviation of the estimates. It may happen that a change of variables will improve the validity of the linearity assumption. A systematic procedure for effecting such a change of variables is described by Hartley (1964). Procedures for determining the validity of the linearity assumption are described by Beale (1960) and illustrated by Guttman and Meeter(1965). A very simple method for testing the linearity assumption is the following:

Determine a confidence region based on that assumption. Calculate the actual values of the objective function at selected points on the boundary of the region, e.g., at the endpoints of the principal axes. If our assumption is valid, these values should differ only slightly from the approximation value

$$\Phi^* + \tfrac{1}{2}(\boldsymbol{\theta} - \boldsymbol{\theta}^*)^T\mathbf{B}^T\mathbf{V}^{-1}\mathbf{B}(\boldsymbol{\theta} - \boldsymbol{\theta}^*)$$

In fact, it may be worthwhile to determine the extent of the region in which the linearity assumption is valid by computing the values of the objective function on the boundaries of a series of successively larger confidence regions, until a serious mismatch between the true and approximate values occurs. Knowledge of this region may be useful in certain applications, such as sequential estimation (see Section 9-3).

7-12. The Posterior Distribution

If we accept the posterior distribution as the probability distribution of our parameters, we are able to define confidence regions in a straightforward way. Let $p^*(\boldsymbol{\theta})$ be the posterior density. Then for any measureable region $\mathscr{R}$ in parameter space, the quantity $\int_{\mathscr{R}} p^*(\boldsymbol{\theta})\, d\boldsymbol{\theta}$ is the probability that the true value of $\boldsymbol{\theta}$ lies in $\mathscr{R}$. Hence, any region $\mathscr{R}$ such that $\int_{\mathscr{R}} p^*(\boldsymbol{\theta})\, d\boldsymbol{\theta} = \gamma$ is a γ-confidence region (in the Bayesian sense) for $\boldsymbol{\theta}$. If $p^*(\boldsymbol{\theta})$ is approximately normal with covariance matrix $\mathbf{V}_{\boldsymbol{\theta}}$, then the methods of Section 7-10 can be used to construct such confidence regions. In most other cases, the task is a difficult one.

Suppose our estimate $\boldsymbol{\theta}^*$ has been arrived at by maximizing the logarithm of the posterior distribution. Around $\boldsymbol{\theta}^*$ we then have, approximately

$$\log p^*(\boldsymbol{\theta}) = \log p^*(\boldsymbol{\theta}^*) - \tfrac{1}{2}(\boldsymbol{\theta} - \boldsymbol{\theta}^*)^T\mathbf{H}^*(\boldsymbol{\theta} - \boldsymbol{\theta}^*) \qquad (7\text{-}12\text{-}1)$$

where $\mathbf{H}^*$ is the Hessian of $-\log p^*(\boldsymbol{\theta}^*)$. It follows that

$$p^*(\boldsymbol{\theta}) \approx c \exp[-\tfrac{1}{2}(\boldsymbol{\theta} - \boldsymbol{\theta}^*)^T\mathbf{H}^*(\boldsymbol{\theta} - \boldsymbol{\theta}^*)] \qquad (7\text{-}12\text{-}2)$$

which indicates that around $\boldsymbol{\theta}^*$ the posterior distribution looks like a normal distribution with mean $\boldsymbol{\theta}^*$ and covariance $\mathbf{V}^* = -(H^*)^{-1}$. If we had chosen a constant prior density, then the posterior density is proportional to the likelihood function. If then the model equations are linear in $\boldsymbol{\theta}$, Eq. (2) holds exactly. In this case the posterior distribution of $\boldsymbol{\theta}$ and the sampling distribution of the maximum likelihood estimate coincide, and inferences based on them are identical.

If the model equations are linear and the prior distribution is normal with covariance matrix $\mathbf{V}_0$, then Eq. (2) still holds exactly, but the covariance of the posterior distribution no longer coincides with that of the sampling distribution. Apart from constant terms, the posterior distribution has the form

$$\log p^*(\boldsymbol{\theta}) = -\tfrac{1}{2} \sum_{\mu=1}^{n} (\mathbf{y}_\mu - \mathbf{B}_\mu \boldsymbol{\theta})^{\mathrm{T}} \mathbf{V}^{-1} (\mathbf{y}_\mu - \mathbf{B}_\mu \boldsymbol{\theta}) - \tfrac{1}{2}(\boldsymbol{\theta} - \boldsymbol{\theta}_0)^{\mathrm{T}} \mathbf{V}_0^{-1} (\boldsymbol{\theta} - \boldsymbol{\theta}_0) \tag{7-12-3}$$

This achieves its maximum at

$$\boldsymbol{\theta}^* = \left[\sum_{\mu=1}^{n} (\mathbf{B}_\mu{}^{\mathrm{T}} \mathbf{V}^{-1} \mathbf{B}_\mu) + \mathbf{V}_0^{-1}\right]^{-1} \left[\sum_{\mu=1}^{n} (\mathbf{B}_\mu{}^{\mathrm{T}} \mathbf{V}^{-1} \mathbf{y}_\mu) + \mathbf{V}_0^{-1} \boldsymbol{\theta}_0\right] \tag{7-12-4}$$

and has the negative inverse Hessian

$$\mathbf{V}^* = \left[\sum_{\mu=1}^{n} (\mathbf{B}_\mu{}^{\mathrm{T}} \mathbf{V}^{-1} \mathbf{B}_\mu) + \mathbf{V}_0^{-1}\right]^{-1} \tag{7-12-5}$$

It can be shown that the sampling distribution of $\boldsymbol{\theta}^*$ [Eq. (4)] has mean

$$\bar{\boldsymbol{\theta}}^* = \mathbf{V}^* \left[\sum_{\mu=1}^{n} (\mathbf{B}_\mu{}^{\mathrm{T}} \mathbf{V}^{-1} \mathbf{B}_\mu) \hat{\boldsymbol{\theta}} + \mathbf{V}_0^{-1} \boldsymbol{\theta}_0\right] \tag{7-12-6}$$

where $\hat{\boldsymbol{\theta}}$ is the true value. This is unbiased only when $\boldsymbol{\theta}_0 = \hat{\boldsymbol{\theta}}$. Furthermore, the covariance of the sampling distribution is given by

$$\mathbf{V}_{\boldsymbol{\theta}} = \mathbf{V}^* \sum_{\mu=1}^{n} (\mathbf{B}_\mu{}^{\mathrm{T}} \mathbf{V}^{-1} \mathbf{B}_\mu) \mathbf{V}^* \tag{7-12-7}$$

which differs from $\mathbf{V}^*$.

7-13. The Residuals

After the estimates $\boldsymbol{\theta}^*$ for the parameters have been obtained, we can compute the final residuals

$$\mathbf{e}_\mu^* \equiv \mathbf{e}_\mu(\boldsymbol{\theta}^*) = \begin{cases} \hat{\mathbf{w}}_\mu^* - \mathbf{w}_\mu & \text{for exact structural model} \\ \mathbf{g}_\mu(\mathbf{z}_\mu, \boldsymbol{\theta}^*) & \text{for inexact structural model} \\ \mathbf{y}_\mu - \mathbf{f}(\mathbf{x}_\mu, \boldsymbol{\theta}^*) & \text{for reduced model} \end{cases}$$
$$(\mu = 1, 2, \ldots, n) \tag{7-13-1}$$

These residuals measure the departure of the data from the best curve or surface that could be fitted to them. If the model is exactly valid, these residuals must be related to the errors in the data. If such errors did not exist, there should be no residuals either. The residuals must on the average be smaller than the errors, because we have chosen the $\boldsymbol{\theta}^*$ so as to make the residuals as small as possible. The errors, which should be the residuals obtained with the true values $\hat{\boldsymbol{\theta}}$, must be larger (unless $\hat{\boldsymbol{\theta}} = \boldsymbol{\theta}^*$). The residuals, then, are biased estimates of the errors, the bias being toward smaller absolute values.

We now develop approximate expressions for this bias in some typical estimation situations. Suppose the errors in different experiments are uncorrelated, and have the same covariance matrix $\mathbf{V}$ in all experiments. Further, suppose $\boldsymbol{\theta}^*$ minimizes an objective function of the form

$$\Phi(\boldsymbol{\theta}) = \Psi(\mathbf{M}(\boldsymbol{\theta})) \tag{7-13-2}$$

as discussed in Section 5-9. At this minimum the gradient of Φ must vanish. Hence from Eq. (5-9-12) we have

$$\mathbf{q} = -2 \sum_{\mu=1}^{n} \mathbf{B}_\mu{}^{\mathrm{T}} \boldsymbol{\Gamma} \mathbf{e}_\mu{}^* = \mathbf{0} \tag{7-13-3}$$

where

$$\mathbf{B}_\mu \equiv -\,\partial \mathbf{e}_\mu / \partial \boldsymbol{\theta})_{\boldsymbol{\theta}=\boldsymbol{\theta}^*}, \qquad \boldsymbol{\Gamma} \equiv \partial \Psi / \partial \mathbf{M})_{\boldsymbol{\theta}=\boldsymbol{\theta}^*} \tag{7-13-4}$$

Assuming the model is correct, the error $\boldsymbol{\varepsilon}_\mu$ is the residual computed for the true $\boldsymbol{\theta}$, i.e.,

$$\boldsymbol{\varepsilon}_\mu = \mathbf{e}_\mu(\hat{\boldsymbol{\theta}}) \tag{7-13-5}$$

Assuming that $\boldsymbol{\theta}^*$ does not differ much from $\hat{\boldsymbol{\theta}}$, we have from Eq. (1), Eq. (4), and Eq. (5) approximately

$$\mathbf{e}_\mu{}^* \approx \boldsymbol{\varepsilon}_\mu - \mathbf{B}_\mu(\boldsymbol{\theta}^* - \hat{\boldsymbol{\theta}}) \tag{7-13-6}$$

Substituting Eq. (6) in Eq. (3) yields

$$\sum_{\mu=1}^{n} \mathbf{B}_\mu{}^{\mathrm{T}} \boldsymbol{\Gamma} [\boldsymbol{\varepsilon}_\mu - \mathbf{B}_\mu(\boldsymbol{\theta}^* - \hat{\boldsymbol{\theta}})] = \mathbf{0} \tag{7-13-7}$$

whence

$$\boldsymbol{\theta}^* - \hat{\boldsymbol{\theta}} = \mathbf{C}^{-1} \sum_{\eta=1}^{n} \mathbf{B}_\eta{}^{\mathrm{T}} \boldsymbol{\Gamma} \boldsymbol{\varepsilon}_\eta \tag{7-13-8}$$

where

$$\mathbf{C} \equiv \sum_{\mu=1}^{n} \mathbf{B}_\mu{}^{\mathrm{T}} \boldsymbol{\Gamma} \mathbf{B}_\mu \tag{7-13-9}$$

Substituting Eq. (8) in Eq. (6)

$$\mathbf{e}_\mu{}^* = \boldsymbol{\varepsilon}_\mu - \mathbf{B}_\mu \mathbf{C}^{-1} \sum_{\eta=1}^{n} \mathbf{B}_\eta{}^{\mathrm{T}} \boldsymbol{\Gamma} \boldsymbol{\varepsilon}_\eta \tag{7-13-10}$$

Since by assumption

$$E(\boldsymbol{\varepsilon}_\mu \boldsymbol{\varepsilon}_\eta{}^{\mathrm{T}}) = \delta_{\mu\eta} \mathbf{V}$$

we obtain from Eq. (10)

$$E(\mathbf{e}_\mu{}^*\mathbf{e}_\mu{}^{*\mathrm{T}}) = \mathbf{V} - \mathbf{B}_\mu\mathbf{C}^{-1}\mathbf{B}_\mu{}^{\mathrm{T}}\mathbf{\Gamma}\mathbf{V} - \mathbf{V}\mathbf{\Gamma}\mathbf{B}_\mu\mathbf{C}^{-1}\mathbf{B}_\mu{}^{\mathrm{T}} + \mathbf{B}_\mu\mathbf{C}^{-1}\left(\sum_{\eta=1}^{n}\mathbf{B}_\eta{}^{\mathrm{T}}\mathbf{\Gamma}\mathbf{V}\mathbf{\Gamma}\mathbf{B}_\eta\right)\mathbf{C}^{-1}\mathbf{B}_\mu{}^{\mathrm{T}} \tag{7-13-11}$$

An appropriate measure of the residuals is given by the (unadjusted) sample covariance matrix $\mathbf{V}^*$, defined by

$$\mathbf{V}^* \equiv (1/n)\mathbf{M}(\boldsymbol{\theta}^*) = (1/n)\sum_{\mu}\mathbf{e}_\mu{}^*\mathbf{e}_\mu{}^{*\mathrm{T}} \tag{7-13-12}$$

So that from Eq. (11)

$$E(\mathbf{V}^*) = \mathbf{V} - (1/n)\left(\sum_{\mu=1}^{n}\mathbf{B}_\mu\mathbf{C}^{-1}\mathbf{B}_\mu{}^{\mathrm{T}}\right)\mathbf{\Gamma}\mathbf{V} - (1/n)\mathbf{V}\mathbf{\Gamma}\left(\sum_{\mu=1}^{n}\mathbf{B}_\mu\mathbf{C}^{-1}\mathbf{B}_\mu{}^{\mathrm{T}}\right) + (1/n)\sum_{\mu=1}^{n}\mathbf{B}_\mu\mathbf{C}^{-1}\left(\sum_{\eta=1}^{n}\mathbf{B}_\eta{}^{\mathrm{T}}\mathbf{\Gamma}\mathbf{V}\mathbf{\Gamma}\mathbf{B}_\eta\right)\mathbf{C}^{-1}\mathbf{B}_\mu{}^{\mathrm{T}} \tag{7-13-13}$$

Of particular interest are the cases where $\mathbf{\Gamma}$ is proportional to $\mathbf{V}^{-1}$, as in rows 1 and 2 of Table 5-1 ($\mathbf{V}$ known or proportional to a known matrix). It is easy to verify that Eq. (13) remains unchanged if $\mathbf{\Gamma}$ is multiplied by a constant, since $\mathbf{C}$ would be multiplied by the same constant. Hence, we may simply substitute $\mathbf{\Gamma} = \mathbf{V}^{-1}$, and obtain

$$E(\mathbf{V}^*) = \mathbf{V} - (2/n)\sum_{\mu=1}^{n}\mathbf{B}_\mu\mathbf{C}^{-1}\mathbf{B}_\mu{}^{\mathrm{T}} + (1/n)\sum_{\mu=1}^{n}\mathbf{B}_\mu\mathbf{C}^{-1}\mathbf{C}\mathbf{C}^{-1}\mathbf{B}_\mu{}^{\mathrm{T}} = \mathbf{V} - (1/n)\sum_{\mu=1}^{n}\mathbf{B}_\mu\mathbf{C}^{-1}\mathbf{B}_\mu{}^{\mathrm{T}} \tag{7-13-14}$$

As expected, $E(\mathbf{V}^*)$ is "smaller" than $\mathbf{V}$, since $\mathbf{V} - E(\mathbf{V}^*)$ is positive definite.

Consider the case of a single equation model, with $\mathbf{V} = \sigma^2$. In this case $\mathbf{B}_\mu{}^{\mathrm{T}}$ is a vector $\mathbf{b}_\mu$. Hence from Eq. (9)

$$\mathbf{C} = (1/\sigma^2)\sum_{\mu=1}^{n}\mathbf{b}_\mu\mathbf{b}_\mu{}^{\mathrm{T}} \tag{7-13-15}$$

and, since $\mathbf{C}$ is an $l \times l$ matrix (l being the number of parameters)

$$\sum_{\mu=1}^{n}\mathbf{B}_\mu\mathbf{C}^{-1}\mathbf{B}_\mu{}^{\mathrm{T}} = \sum_{\mu=1}^{n}\mathbf{b}_\mu{}^{\mathrm{T}}\mathbf{C}^{-1}\mathbf{b}_\mu = \mathrm{Tr}\left(\mathbf{C}^{-1}\sum_{\mu=1}^{n}\mathbf{b}_\mu\mathbf{b}_\mu{}^{\mathrm{T}}\right) = \mathrm{Tr}(\mathbf{C}^{-1}\sigma^2\mathbf{C}) = \sigma^2\,\mathrm{Tr}\,\mathbf{I}_l = l\sigma^2 \tag{7-13-16}$$

Therefore, Eq. (14) becomes

$$E(\mathbf{V}^*) = (1 - l/n)\sigma^2 \tag{7-13-17}$$

If we want to estimate σ^2 (σ being the standard deviation of the errors) from the residuals, we use

$$\tilde{\sigma}^2 = \frac{1}{1 - l/n} \mathrm{V}^* = \frac{1}{1 - l/n} \frac{1}{n} M^* = \frac{1}{n - l} \sum_{\mu=1}^{n} \mathrm{e}_{\mu}^{*2} \tag{7-13-18}$$

This is a well known formula, which states that the variance of the errors is estimated by dividing the sum of squares of the residuals by the number of *degrees of freedom*, which is the number of observations n less the number of unknown parameters l.

If the model contains $m > 1$ equations, the situation is more complicated. Suppose, however, that we wish to estimate $\mathbf{V}$ as some multiple of $\mathbf{V}^*$; say we assume

$$E(\mathbf{V}^*) = \rho \mathbf{V} \tag{7-13-19}$$

Then, substituting Eq. (19) in Eq. (14) and multiplying by $\mathbf{V}^{-1}$, we obtain

$$\rho \mathbf{I}_m = \mathbf{I}_m - (1/n) \sum_{\mu=1}^{m} \mathbf{B}_{\mu} \mathbf{C}^{-1} \mathbf{B}_{\mu}^{\mathrm{T}} \mathbf{V}^{-1} \tag{7-13-20}$$

Taking traces on both sides, and remembering that if $\mathbf{AB}$ is a square matrix then

$$\mathrm{Tr}(\mathbf{AB}) = \mathrm{Tr}(\mathbf{BA})$$

we obtain

$$\rho m = m - (1/n) \,\mathrm{Tr}\, \mathbf{C}^{-1} \left(\sum_{\mu=1}^{n} \mathbf{B}_{\mu}^{\mathrm{T}} \mathbf{V}^{-1} \mathbf{B}_{\mu} \right) = m - (1/n) \,\mathrm{Tr}\, \mathbf{C}^{-1} \mathbf{C} = m - l/n$$

so that

$$\rho = 1 - l/mn \tag{7-13-21}$$

Hence, to estimate $\mathbf{V}$ we use the (adjusted) sample covariance matrix

$$\tilde{\mathbf{V}} = \frac{1}{\rho} \mathbf{V}^* = \frac{1}{1 - (l/mn)} \frac{1}{n} \mathbf{M}(\boldsymbol{\theta}^*) = \frac{1}{n - l/m} \sum_{\mu=1}^{n} \mathbf{e}_{\mu}^* \mathbf{e}_{\mu}^{*\mathrm{T}} \tag{7-13-22}$$

Once again, to estimate the covariance matrix of the errors, we take the moment matrix of the residuals and divide by the number of degrees of freedom per equation, that is, the number of observations n per variable less the "average" number of parameters per equation l/m. Clearly, Eq. (22) reduces to Eq. (18) in the single equation case $m = 1$.

In the case where $\mathbf{V}$ is completely unknown, we have from row 3 of Table 5-1 $\boldsymbol{\Gamma}$ proportional to $\mathbf{M}^{-1}$. If we now make the further assumption that $\mathbf{M}$ is proportional to $\mathbf{V}$ we have once more $\boldsymbol{\Gamma}$ proportional to $\mathbf{V}$, and Eq. (22) is still valid. The maximum likelihood estimate $\tilde{\mathbf{V}} = (1/n)\mathbf{M}(\boldsymbol{\theta}^*)$ is biased by the factor $1 - l/mn$. Since this factor approaches unity as $n \to \infty$, the maximum likelihood estimate is consistent.

The formulas derived here should be viewed with caution, since the residuals for one equation may turn out much smaller than predicted, while another equation has much larger ones. It is only on the average, in a certain sense, that our bias factors apply.

7-14. The Independent Variables Subject to Error

Even consistency does not hold when the independent variables are subject to error, so the number of unknown parameters is essentially proportional to n. A very rough indication of the bias to be expected even when the number of experiments is very large can be obtained as follows:

Suppose we have an m-equation model with r variables subject to error. Let the observed, estimated, and true values of these variables for the μth experiment be designated $\mathbf{w}_\mu$, $\mathbf{w}_\mu^*$, and $\hat{\mathbf{w}}_\mu$, respectively. We assume these three values are not far apart. By definition, the residuals are

$$\mathbf{e}_\mu^* \equiv \mathbf{w}_\mu^* - \mathbf{w}_\mu \tag{7-14-1}$$

and the errors are

$$\boldsymbol{\varepsilon}_\mu \equiv \hat{\mathbf{w}}_\mu - \mathbf{w}_\mu \tag{7-14-2}$$

Since we are dealing with the asymptotic case, we assume that our estimates $\boldsymbol{\theta}^*$ are errorless, i.e., $\boldsymbol{\theta}^* = \hat{\boldsymbol{\theta}}$. The estimates $\mathbf{w}_\mu^*$ must then minimize

$$\tfrac{1}{2}(\mathbf{w}_\mu^* - \mathbf{w}_\mu)^{\mathrm{T}}\mathbf{V}_\mu^{-1}(\mathbf{w}_\mu^* - \mathbf{w}_\mu)$$

subject to $\mathbf{g}(\mathbf{w}_\mu^*, \boldsymbol{\theta}^*) = \mathbf{0}$.

Letting

$$\mathbf{A}_\mu \equiv \partial \mathbf{g}/\partial \mathbf{w}_\mu)_{\boldsymbol{\theta}=\boldsymbol{\theta}^*}$$

we have approximately

$$\mathbf{g}(\mathbf{w}_\mu^*, \boldsymbol{\theta}^*) = \mathbf{g}(\mathbf{w}_\mu, \boldsymbol{\theta}^*) + \mathbf{A}_\mu(\mathbf{w}_\mu^* - \mathbf{w}_\mu) = \mathbf{0} \tag{7-14-3}$$

Since $\boldsymbol{\theta}^* = \hat{\boldsymbol{\theta}}$, the true values $\hat{\mathbf{w}}_\mu$ also satisfy $\mathbf{g}(\hat{\mathbf{w}}_\mu, \boldsymbol{\theta}^*) = \mathbf{0}$, i.e., approximately

$$\mathbf{g}(\hat{\mathbf{w}}_\mu, \boldsymbol{\theta}^*) = \mathbf{g}(\mathbf{w}_\mu, \boldsymbol{\theta}^*) + \mathbf{A}_\mu(\hat{\mathbf{w}}_\mu - \mathbf{w}_\mu) = \mathbf{g}(\mathbf{w}_\mu, \boldsymbol{\theta}^*) + \mathbf{A}_\mu\boldsymbol{\varepsilon}_\mu = \mathbf{0} \tag{7-14-4}$$

Solving Eq. (4) for $\mathbf{g}(\mathbf{w}_\mu, \boldsymbol{\theta}^*)$ and substituting in Eq. (3)

$$-\mathbf{A}_\mu \boldsymbol{\varepsilon}_\mu + \mathbf{A}_\mu(\mathbf{w}_\mu^* - \mathbf{w}_\mu) = \mathbf{0} \tag{7-14-5}$$

Because of its nature as the solution to an equality-constrained minimization problem, $\mathbf{w}_\mu^*$ must be a stationary point of the Lagrangian

$$\tfrac{1}{2}(\mathbf{w}_\mu^* - \mathbf{w}_\mu)^{\mathrm{T}}\mathbf{V}_\mu^{-1}(\mathbf{w}_\mu^* - \mathbf{w}_\mu) + \boldsymbol{\lambda}^{\mathrm{T}}\mathbf{g}(\mathbf{w}_\mu^*, \boldsymbol{\theta}^*)$$

Differentiating with respect to $\mathbf{w}_\mu^*$

$$\mathbf{V}_\mu^{-1}(\mathbf{w}_\mu^* - \mathbf{w}_\mu) + \mathbf{A}_\mu^{\mathrm{T}}\boldsymbol{\lambda} = \mathbf{0} \tag{7-14-6}$$

whence

$$\mathbf{w}_\mu^* - \mathbf{w}_\mu = -\mathbf{V}_\mu \mathbf{A}_\mu^T\boldsymbol{\lambda} \tag{7-14-7}$$

so that Eq. (5) becomes

$$-\mathbf{A}_\mu\boldsymbol{\varepsilon}_\mu - \mathbf{A}_\mu \mathbf{V}_\mu\mathbf{A}_\mu^{\mathrm{T}}\boldsymbol{\lambda} = \mathbf{0} \tag{7-14-8}$$

and

$$\boldsymbol{\lambda} = -(\mathbf{A}_\mu \mathbf{V}_\mu \mathbf{A}_\mu^{\mathrm{T}})^{-1}\mathbf{A}_\mu\boldsymbol{\varepsilon}_\mu \tag{7-14-9}$$

Now Eq. (1), Eq. (7) and Eq. (9) combine to form

$$\mathbf{e}_\mu^* = \mathbf{V}_\mu \mathbf{A}_\mu^{\mathrm{T}}(\mathbf{A}_\mu \mathbf{V}_\mu \mathbf{A}_\mu^{\mathrm{T}})^{-1}\mathbf{A}_\mu\boldsymbol{\varepsilon}_\mu \tag{7-14-10}$$

and remembering that $E(\boldsymbol{\varepsilon}_\mu\boldsymbol{\varepsilon}_\mu^{\mathrm{T}}) = \mathbf{V}_\mu$, we obtain

$$E(\mathbf{V}_\mu^*) \equiv E(\mathbf{e}_\mu^*\mathbf{e}_\mu^{*\mathrm{T}}) = \mathbf{V}_\mu \mathbf{A}_\mu^{\mathrm{T}}(\mathbf{A}_\mu \mathbf{V}_\mu \mathbf{A}_\mu^{\mathrm{T}})^{-1}\mathbf{A}_\mu \mathbf{V}_\mu \tag{7-14-11}$$

Assuming once more that $\mathbf{V}_\mu^*$ is proportional to $\mathbf{V}_\mu$, e.g., $\mathbf{V}_\mu^* = \rho\mathbf{V}_\mu$, Eq. (11) reduces to

$$\rho\mathbf{I}_r = \mathbf{V}_\mu \mathbf{A}_\mu^{\mathrm{T}}(\mathbf{A}_\mu \mathbf{V}_\mu \mathbf{A}_\mu^{\mathrm{T}})^{-1}\mathbf{A}_\mu \tag{7-14-12}$$

Taking traces

$$\begin{aligned}\rho r &= \mathrm{Tr}[\mathbf{V}_\mu \mathbf{A}_\mu^{\mathrm{T}}(\mathbf{A}_\mu \mathbf{V}_\mu \mathbf{A}_\mu^{\mathrm{T}})^{-1}\mathbf{A}_\mu] = \mathrm{Tr}[(\mathbf{A}_\mu \mathbf{V}_\mu \mathbf{A}_\mu^{\mathrm{T}})(\mathbf{A}_\mu \mathbf{V}_\mu \mathbf{A}_\mu^{\mathrm{T}})^{-1}] \\ &= \mathrm{Tr}\,\mathbf{I}_m = m\end{aligned} \tag{7-14-13}$$

so that

$$\rho = m/r \tag{7-14-14}$$

If $m = r$, i.e., there is only one inaccurately measured variable per equation, $\rho = 1$. There is no asymptotic bias and $\mathbf{V}_\mu^*$ is a consistent estimate of $\mathbf{V}_\mu$. This corresponds to the situation discussed in Section 7-13. In the problem of Section 6-13, however, we had one equation with x_1, x_2, and y subject to error. Here $\rho = \frac{1}{3}$, and we expect the residual covariance to be only one-third the true covariance, no matter how many experiments are performed.

It is clear that if we assume all $\mathbf{V}_\mu$ equal, then the matrix

$$\mathbf{V}^* = (1/n)\sum_{\mu=1}^{n} \mathbf{e}_\mu^*\mathbf{e}_\mu^{*\mathrm{T}} \tag{7-14-15}$$

has the same bias factor ρ. If, in addition, we correct for the bias caused by the l parameters $\boldsymbol{\theta}$, we arrive at the estimate

$$\tilde{\mathbf{V}} = [r/m(n - l/m)]\sum_{\mu=1}^{n} \mathbf{e}_\mu^*\mathbf{e}_\mu^{*\mathrm{T}} \tag{7-14-16}$$

7-15. Goodness of Fit

The crucial question that arises after the estimates have been obtained is whether or not our model fits the data. The question can be answered in the affirmative if the residuals of the fitted model can be explained as errors in the observations. On the other hand, if the residuals are so large, or of such a nonrandom nature, that they cannot be ascribed to random observation errors, then we say that the model does not fit the data. We stress that whereas a lack of fit constitutes strong grounds for rejecting, or at least amending the model, a good fit does not prove that the model is correct. A good fit merely establishes the fact that there is no reason to reject the model on the basis of the data at hand. In fact, no amount of data can ever prove a model; all we can hope is that it does not disprove it.

Our least squares or maximum likelihood estimates were usually based on the assumption that the errors $\boldsymbol{\varepsilon}_\mu$ in each experiment were realizations of a random variable with mean $\mathbf{0}$ and covariance matrix $\mathbf{V}$. After estimating the parameters we have a set of residuals $\mathbf{e}_\mu{}^*$ from which we compute an adjusted covariance matrix $\tilde{\mathbf{V}}$ (see, e.g., Eq. (7-13-22)). To establish the goodness of fit is to test the hypothesis that, with certain reservations,‡ the residuals form a sample from the distribution that we have postulated for the errors, corrected for the bias discussed in Sections 7-13–7-14.

To test a statistical hypothesis we usually compute a certain relevant statistic λ from the sample. We compare λ to a certain reference value λ_0, and reject the hypothesis if $\lambda \geqslant \lambda_0$. In doing so we incur the danger of erring in one of the following ways:

1. Error of the first kind; we reject the hypothesis although it is true.
2. Error of the second kind; we accept the hypothesis although it is false.

On the assumption that the model is correct, the distribution of the statistic λ may be determined, and hence we can find the value λ_0 such that

$$\Pr(\lambda \geqslant \lambda_0) = \alpha \qquad (7\text{-}15\text{-}1)$$

where α is a suitably chosen small number, e.g., 0.05 or 0.01. If we reject the model when $\lambda \geqslant \lambda_0$, the probability of committing an error of the first kind is α. The probability of committing an error of the second kind depends on what the true model actually is, and we shall not consider this question here.

The statistics that we shall use are valid for a wide class of error distributions. The distributions of the statistics, however, are known and tabulated primarily for the case when the distribution of the errors is normal. Only

‡ E.g., the residuals are serially correlated even when the errors are independent.

in this case is it easy to find the test value λ_0 associated with a given probability α. On the other hand, it makes no difference here whether or not the model equations are linear. Residuals should be attributable to errors, no matter what kind of model they were computed from.

When the distribution of λ is unknown and cannot be derived by analytic means, it is still easy to estimate the critical λ_0 by Monte Carlo techniques. Error samples with the proper distribution are generated on the computer, the statistic λ is computed for each sample, and λ_0 is chosen so that λ exceeds it in a fraction α of the samples.

Various commonly used statistics λ are described in the next section.

7-16. Tests on Residuals

The tests we wish to perform on the residuals relate to their mean and covariance. In many cases questions concerning the mean of the residuals are of no significance, since the estimation process guarantees that (except for the effect of rounding errors) the mean is zero. For instance, whenever a model of the form

$$y = \theta_1 + \phi(\mathbf{x}, \theta_2, \theta_3, \ldots, \theta_l) \tag{7-16-1}$$

is estimated by least squares, the average residual is zero. For, let the objective function be given by

$$\Phi = \sum_{\mu=1}^{n} e_\mu^2 = \sum_{\mu=1}^{n} (y_\mu - \theta_1 - \phi_\mu)^2 \tag{7-16-2}$$

To minimize Φ we form as one of the normal equations

$$\partial\Phi/\partial\theta_1 = -2\sum_{\mu=1}^{n} (y_\mu - \theta_1 - \phi_\mu) = -2\sum_{\mu} e_\mu = 0 \tag{7-16-3}$$

Thus the sum of the residuals is zero, and so is their average. Suppose, however, that we have a model that does not guarantee zero average residuals. Now if the errors in each experiment have covariance matrix $\mathbf{V}$, we expect the residuals to have covariance matrix $(1 - l/mn)\mathbf{V}$ [see Eq. (7-13-22)]. The mean residual vector

$$\bar{\mathbf{e}} \equiv (1/n) \sum_{\mu=1}^{n} \mathbf{e}_\mu^* \tag{7-16-4}$$

should have covariance matrix $(1/n)(1 - l/mn)\mathbf{V}$, since the variance of the mean of n independent observations is $1/n$ times the variance of the observations (we neglect the fact that the residuals are correlated even when the observations are not). If the errors $\boldsymbol{\varepsilon}_\mu$ are assumed to be $N_m(\mathbf{0}, \mathbf{V})$, then the average residuals $\bar{\mathbf{e}}$ should be $N_m(\mathbf{0}, (1/n)(1 - l/mn)\mathbf{V})$, and we may easily construct confidence regions for $\bar{\mathbf{e}}$, as we have done for $\boldsymbol{\theta}$ in Section 7-10. In particular, the statistic

$$\lambda \equiv [n/(1 - l/mn)]\bar{\mathbf{e}}^{\mathrm{T}}\mathbf{V}^{-1}\bar{\mathbf{e}} \tag{7-16-5}$$

is distributed as $\chi_m{}^2$.

If $\mathbf{V}$ is not known, we must introduce the matrix

$$\mathbf{S} \equiv [1/(n-1)] \sum_{\mu=1}^{n} (\mathbf{e}_\mu{}^* - \bar{\mathbf{e}})(\mathbf{e}_\mu{}^* - \bar{\mathbf{e}})^{\mathrm{T}} \tag{7-16-6}$$

If our hypothesis is true, then the statistic

$$\lambda = [(n-m)n/(n-1)m]\bar{\mathbf{e}}^{\mathrm{T}}\mathbf{S}^{-1}\bar{\mathbf{e}} \tag{7-16-7}$$

is distributed as $F_{m,n-m}$. The quantity $[(n-1)m/(n-m)]\lambda$ is sometimes known as T^2 [see Anderson (1958, Ch. 5)]. For a single equation model $m = 1$ and $T^2 = \lambda$. In this case, the quantity $\lambda^{1/2}$ is known as t, and the associated test is the well-known t-test.

If the zero-mean hypothesis is accepted, we may wish to test the hypothesis that the errors in each experiment possess a given covariance matrix $\mathbf{V}$. That is, we wish to compare the covariance of the residuals $\tilde{\mathbf{V}}$ given by Eq. (7-13-22) with $\mathbf{V}$. An appropriate statistic is given by

$$\lambda = n[\log \det(\mathbf{V}\tilde{\mathbf{V}}^{-1}) - m + \mathrm{Tr}(\tilde{\mathbf{V}}\mathbf{V}^{-1})] \tag{7-16-8}$$

Its distribution in the normal case is tabulated by Korin (1968).

Another frequently tested hypothesis is that two sets of residuals are uncorrelated. The need for such a test arises when we base our estimates on an assumed diagonal covariance matrix of the errors. We then obtain a covariance matrix $\tilde{\mathbf{V}}$ of the residuals, and wish to find out whether $\tilde{V}_{ab}$ $(a \neq b)$ differs significantly from zero.

We compute the correlation coefficient

$$r_{ab} = \tilde{V}_{ab}/(\tilde{V}_{aa}\tilde{V}_{bb})^{1/2} \tag{7-16-9}$$

If r_{ab} is the correlation coefficient computed from a sample of n^* independent pairs of mutually independent normal deviates, then the quantity

$$\lambda = r_{ab}[(n^* - 2)/(1 - r_{ab}^2)]^{1/2} \tag{7-16-10}$$

has the t-distribution with $n^* - 2$ degrees of freedom (Anderson, 1958, p. 65). For our purposes, we should probably take $n^* = n - l/m$, but with $n^* > 10$ the t-distribution is quite insensitive to the number of degrees of freedom.

Suppose $n^* = 20$ and $r_{ab} = -0.25$; we have $\lambda = -1.095$. According to Table 4 (Cramér, 1946), the probability that $|t_{18}|$ exceeds 1.095 is a sizable 29%, so we cannot reject the hypothesis that $V_{ab} = 0$. We could be 99% sure that $V_{ab} \neq 0$ if we have $|t_{18}| \geqslant 2.878$, corresponding to $|r_{ij}| \geqslant 0.561$. Further examples appear in Section 7-24.

7-17. Runs and Outliers

Residuals that have passed the tests of the previous section may still be unsatisfactory. Though of reasonable magnitude, they may display trends and other departures from randomness that call for modifications in the model. The reader is referred to Acton (1959, Ch. 3) or to Draper and Smith (1966) for an excellent treatment of this problem on a practical level. Briefly, the residuals should be plotted against the various variables that are included in the model, and also against the time at which the observations were taken. Linear, quadratic, or periodic trends may reveal themselves and will call for the inclusion of appropriate additional terms in the model. Trends in the variance of the errors may also be detected, and may shed some light on the measuring process. Finally we may test for randomness by counting the number of "runs" in the residuals, a run being a sequence of residuals of equal sign. If the number of runs is much lower than expected the randomness of the residuals is suspect.

If n_1 and n_2 are the numbers of negative and positive residuals respectively, then the expected number of runs (on the assumption of complete randomness) is

$$\mu = 2n_1 n_2/(n_1 + n_2) + 1 \tag{7-17-1}$$

and the variance of the number of runs is

$$\sigma^2 = 2n_1 n_2(2n_1 n_2 - n_1 - n_2)/[(n_1 + n_2)^2(n_1 + n_2 - 1)] \tag{7-17-2}$$

The actual distribution of the number of runs was derived and tabulated by Swed and Eisenhart (1943). A table also appears in Draper and Smith (1966, p. 98). When both n_1 and n_2 exceed 10, then the quantity

$$z \equiv (r - \mu + \tfrac{1}{2})/\sigma \tag{7-17-3}$$

(r = number of actual runs) is distributed approximately as $N_1(0, 1)$. A numerical example appears in Section 7-24.

We stress that failure to pass the number of runs test is no reason for outright rejection of the model. Usually it is merely an indication that some possibly minor effects have been neglected. Particularly in cases where the data are very accurate, neglected effects outweigh random errors in measurement. Consequently, nonrandomness of residuals is the rule, rather than the exception, when models are fitted to good data.

Many tests on residuals are best accomplished by graphical means. If the probability distribution of the errors is to be investigated, a histogram or a cumulative frequency plot is called for. Suppose the residuals are renumbered so that e_1 is the smallest (algebraically) and e_n is the largest. Let $P_i \equiv (i - \frac{1}{2})/n$; then P_i is an estimate of the quantity $\Pr(e \leqslant e_i)$. A plot of P_i versus e_i then approximates the cumulative distribution function of the errors. When this plot is made on normal probability paper, the result should be a straight line if the error distribution is normal. If all the points are reasonably close to a straight line except for a few at the low and high ends, then the presence of outliers (see below) is suspected. If the points seem to fall into a few clusters rather than follow a smooth curve, then one may conclude that different sources of error were operative in different subsets of the observations.

It may happen that some gross error is committed in the conduct or recording of some experiment. Naturally, the erroneous observations give rise to unusually large residuals, called *outliers*. More seriously, such erroneous values can gravely distort the parameter estimates. Therefore, one wishes to eliminate such observations from the analysis, and the easiest way to spot them is by examination of the residuals. If there is a clear-cut differentiation between the "regular" residuals which fall on the smooth part of the probability plot, and the "outliers", then we should not hesitate to remove the latter and recompute the estimates in their absence. However, if the distinction is blurred, then the problem of diagnosing outliers is a difficult one. A procedure often adopted in practice is to remove all residuals whose magnitude exceeds the standard deviation (either known or estimated using all residuals) by a fixed factor, say 2.5 or 3. When setting such a threshhold one should take into account the probability of residuals of such magnitude occuring by chance in a population of size n. For instance, with a normal distribution and 100 observations the probability of finding a residual exceeding 3σ is 23.7%, giving one little reason for rejecting such a residual out-of-hand. For a more systematic approach, see Anscombe (1960).

7-18. Causes of Failure

If the parameters turn out well-determined and the residuals are acceptable, then our estimation problem is solved. Only too often, however, we run into one of the following less satisfactory situations:

(a) Parameters ill-determined, residuals large but acceptable, since the measurement errors are known to be large. Barring the possibility of reducing measurement errors, we can improve our estimates only by conducting many more experiments. As a rule, the standard deviation of estimates decreases roughly as $n^{-1/2}$ so that a tenfold improvement in the estimate requires a hundredfold increase in the number of experiments.

(b) Parameters (or some linear combinations of them) ill-determined, residuals small. This may be due to a degeneracy in the model. For example, in the model

$$y = (\theta_1 + \theta_2)x \tag{7-18-1}$$

it is obviously impossible to estimate θ_1 and θ_2 separately. The degeneracy is not always quite so obvious. Consider for instance our falling sphere model Eq. (2-14-5). If we write out that equation in full we find that the distance s travelled by the sphere in time t is

$$s = [g(m - m_0)/6\pi r\mu]t - [gm(m - m_0)/36\pi^2 r^2 \mu^2]\,(1 - e^{-(6\pi r\mu/m)t}) \tag{7-18-2}$$

Some study is required to see that among the parameters g, m, m_0, r, and μ appearing in the equation, only two can be estimated independently.

An even more common source of degeneracy are the data themselves. For example, suppose that for the model

$$y = \theta_1 x_1 + \theta_2 x_2 \tag{7-18-3}$$

we have made many observations of y at different values of x_1 and x_2, but by chance in each experiment x_1 turned out to be approximately equal to x_2. For these data, model Eq. (3) is indistinguishable from

$$y = (\theta_1 + \theta_2)x_1 \tag{7-18-4}$$

in which θ_1 and θ_2 cannot be estimated independently. The above case appears trivial, but similar conditions obtain, perhaps in more subtle form, in many experimental situations. The only solution is to plan the experiments properly, as indicated in Chapter X.

(c) Parameters ill-determined and residuals unacceptable. The model must be rejected, or at least amended to include those effects that were observed in the residuals.

One of the hypotheses underlying a model is that the unknown parameters are constants that do not depend on the model variables. It is clearly desirable to test this hypothesis, and this can be done if the data are sufficiently rich. To test, for instance, the hypothesis that $\boldsymbol{\theta}$ is independent of some variable z_1, we break up the data into subsets each corresponding to a single value, or a narrow range of values, of z_1. We estimate the parameters separately from each data subset, and employ the usual statistical techniques to test whether the estimates obtained from the subsets are significantly different from the estimate obtained from the whole sample, or the subset estimates show any trends or other functional relationships with z_1. If such relationships exist, they may be used to amend the original model. This technique has been described by Hunter and Mezaki (1964) and Box and Hunter (1965).

It is not always possible to apply this method directly. For instance, let the model be

$$y = \theta_1 + \theta_2 x_1 + \theta_2 x_2 \tag{7-18-5}$$

It is impossible to estimate the parameters θ_1 and θ_2 separately if we restrict the data to a single value of x_1. However, we may still break up the entire range of x_1 values into a few fairly wide intervals, and obtain a separate estimate for each range.

7-19. Prediction

Perhaps the most important object of mathematical modeling of physical situations is that of predicting future responses to given conditions. The estimation procedures provide values for the parameters to be inserted into the prediction equations. These equations need not be the same as the model equations used for estimation, nor need the variables to be predicted coincide with the dependent variables of the model equations. For instance, we observe the time a liquid takes to flow through a capillary tube in order to estimate viscosity; we use the viscosity to predict damping factors for standing waves in a pool. At any rate, let us say that we wish to predict the value of some vector $\boldsymbol{\eta}$, based on the value of a vector of independent variables $\boldsymbol{\xi}$ and a vector of parameters $\boldsymbol{\theta}$. The prediction is to be made on the basis of the model

$$\boldsymbol{\eta}_p = \boldsymbol{\phi}(\boldsymbol{\xi}, \boldsymbol{\theta}^*) \tag{7-19-1}$$

where the subscript p stands for the predicted value.

Assuming the model itself is correct, there are three possible sources of inaccuracy in the predictions: errors in the estimated $\boldsymbol{\theta}^*$, errors in the setting of $\boldsymbol{\xi}$, and errors in the measurement of $\boldsymbol{\eta}$. All three sources contribute to the difference between the predicted $\boldsymbol{\eta}_p$ and the eventually observed $\boldsymbol{\eta}$. Usually (except in purely linear models) there will be some bias in the predicted $\boldsymbol{\eta}_p$, but there is little that we can say about it. Assuming, however, that this bias is small compared to the other errors involved, and that the errors from all three sources are statistically independent, then we can obtain an approximation to the covariance matrix of the prediction errors.

Suppose we denote the three errors by $\delta\boldsymbol{\theta}$, $\delta\boldsymbol{\xi}$ and $\delta\boldsymbol{\eta}$, respectively. The observed value of $\boldsymbol{\eta}$ will be given by

$$\boldsymbol{\eta}_0 = \boldsymbol{\phi}(\boldsymbol{\xi} + \delta\boldsymbol{\xi}, \boldsymbol{\theta}^* + \delta\boldsymbol{\theta}) + \delta\boldsymbol{\eta} \tag{7-19-2}$$

A Taylor series expansion up to linear terms yields

$$\boldsymbol{\eta}_0 - \boldsymbol{\eta}_p = (\partial\boldsymbol{\phi}/\partial\boldsymbol{\xi})\, \delta\boldsymbol{\xi} + (\partial\boldsymbol{\phi}/\partial\boldsymbol{\theta})\, \delta\boldsymbol{\theta} + \delta\,\boldsymbol{\eta} \tag{7-19-3}$$

The covariance matrix of the prediction errors is given by

$$\begin{aligned}\mathbf{V}_{\eta} &\equiv E(\boldsymbol{\eta}_0 - \boldsymbol{\eta}_p)(\boldsymbol{\eta}_0 - \boldsymbol{\eta}_p)^{\mathrm{T}} \\ &= (\partial\boldsymbol{\phi}/\partial\boldsymbol{\xi})\mathbf{V}_{\xi}\,(\partial\boldsymbol{\phi}/\partial\boldsymbol{\xi})^{\mathrm{T}} + (\partial\boldsymbol{\phi}/\partial\boldsymbol{\theta})\mathbf{V}_{\theta}\,(\partial\boldsymbol{\phi}/\partial\boldsymbol{\theta})^{\mathrm{T}} + \mathbf{V}_{\eta}\end{aligned} \tag{7-19-4}$$

where $\mathbf{V}_{\xi}$, $\mathbf{V}_{\theta}$ and $\mathbf{V}_{\eta}$ are, respectively, the covariance matrices of $\delta\boldsymbol{\xi}$, $\delta\boldsymbol{\theta}$, and $\delta\boldsymbol{\eta}$. The first term on the right hand side of Eq. (4) may be omitted if $\boldsymbol{\xi}$ can be set (or is known) precisely. The matrix $\mathbf{V}_{\theta}$ is obtained in the process of estimating the parameters, as shown in Section 7-5. If $\boldsymbol{\eta}$ coincides with the $\mathbf{y}$ in the model equations, then $\mathbf{V}_{\eta}$ is estimated (if not known previously) from the residuals, as in Section 7-13.

7-20. Parameter Transformation

It is frequently convenient to perform the estimation not in terms of the original parameters of the model, but in terms of transformed variables which simplify the mathematical form of the model equations. Examples of this have been given in Section 4-19, in connection with linearizing transformations, and the point is also illustrated in the problem of Section 5-23.

Let us assume, then, that we have estimated a vector $\boldsymbol{\theta}$ of l parameters which are functions $\boldsymbol{\theta}(\mathbf{c})$ of the original model parameters $\mathbf{c}$. Let $\boldsymbol{\theta}^*$ and $\mathbf{V}_{\theta}$ be the estimates for $\boldsymbol{\theta}$ and its covariance matrix, respectively. If the transformation from $\mathbf{c}$ to $\boldsymbol{\theta}$ is reversible around $\boldsymbol{\theta} = \boldsymbol{\theta}^*$, i.e., if in the neighborhood of $\boldsymbol{\theta}^*$ there exist functions $\boldsymbol{\gamma}(\boldsymbol{\theta})$ such that $\mathbf{c} = \boldsymbol{\gamma}(\boldsymbol{\theta})$ is a unique solution to the equations $\boldsymbol{\theta} = \boldsymbol{\theta}(\mathbf{c})$, then $\mathbf{J} \equiv \partial\boldsymbol{\gamma}/\partial\boldsymbol{\theta}$ is nonsingular at $\boldsymbol{\theta} = \boldsymbol{\theta}^*$. A first-order Taylor series expansion of $\boldsymbol{\gamma}$ has the form

$$\mathbf{c} = \boldsymbol{\gamma}(\boldsymbol{\theta}^*) + (\partial\boldsymbol{\gamma}/\partial\boldsymbol{\theta})(\boldsymbol{\theta} - \boldsymbol{\theta}^*) = \mathbf{c}^* + \mathbf{J}^*\,\delta\boldsymbol{\theta} \tag{7-20-1}$$

where $\mathbf{J}^* \equiv \mathbf{J}_{\boldsymbol{\theta}=\boldsymbol{\theta}^*}$. Hence, approximately

$$\mathbf{V}_{c} \equiv E(\mathbf{c} - \mathbf{c}^*)(\mathbf{c} - \mathbf{c}^*)^{\mathrm{T}} \approx E(\mathbf{J}^*\,\delta\boldsymbol{\theta}\,\delta\boldsymbol{\theta}^{\mathrm{T}}\,\mathbf{J}^{*\mathrm{T}}) = \mathbf{J}^*\mathbf{V}_{\theta}\,\mathbf{J}^{*\mathrm{T}} \tag{7-20-2}$$

Eq. (2) may be regarded as a special case of Eq. (7-19-4), where the $\boldsymbol{\eta}$ to be predicted are simply the $\mathbf{c}$. See Section 7-24 for a numerical example.

If the equations $\boldsymbol{\theta}^* = \boldsymbol{\theta}(\mathbf{c}^*)$ cannot be solved explicitly for $\mathbf{c}^*$, we have to resort to a numerical solution. In this case, we can still calculate $\mathbf{J}^* = (\partial\boldsymbol{\theta}/\partial\mathbf{c})^{-1}_{\mathbf{c}=\mathbf{c}^*}$. If we use the Newton method to solve for $\mathbf{c}^*$, we obtain $\mathbf{J}^*$ as a by-product.

7-21. Single-Equation Least Squares Problem

We shall now interpret the results of Section 5-21 in the light of the techniques described in this chapter. We recall that the least squares solution to the model Eq. (5-21-5) with the data of Table 5-2 is given by:

$$\boldsymbol{\theta}^* = \begin{bmatrix} 813.4583 \\ 960.9063 \end{bmatrix}, \qquad \Phi^* = 0.03980599$$

$$\mathbf{H}^* \approx \mathbf{N}^* = \begin{bmatrix} 0.271890 \times 10^{-5} & -0.957336 \times 10^{-5} \\ -0.957336 \times 10^{-5} & 3.50371 \times 10^{-5} \end{bmatrix}$$

According to Eq. (7-2-4) we may represent $\Phi(\boldsymbol{\theta})$ approximately by means of the equation

$$\begin{aligned} \Phi(\boldsymbol{\theta}) \approx 0.03980599 + \tfrac{1}{2}10^{-5}[&0.271890\,(\theta_1 - 813.4583)^2 \\ &- 1.914672\,(\theta_1 - 813.4583)(\theta_2 - 960.9063) \\ &+ 3.50371\,(\theta_2 - 960.9063)^2] \end{aligned} \tag{7-21-1}$$

How good is this approximation? In Fig. 7-3 we compare the contours of the true objective function to those of the approximation Eq. (1). We also show the boundary of the region in which the approximate value of Φ

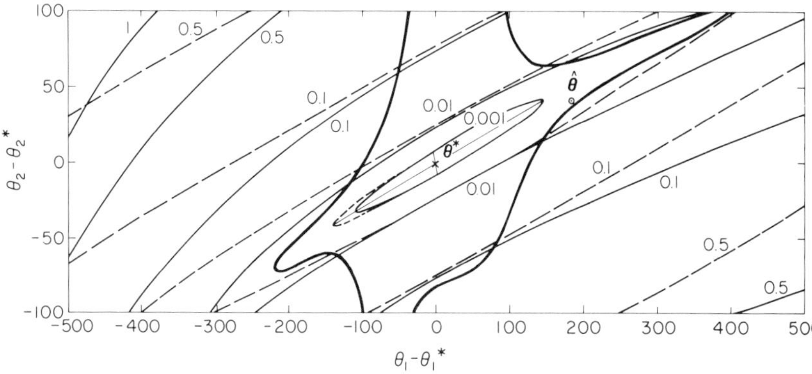

Fig. 7-3. Contours of objective function. Contours of $\Phi - \Phi^*$: —, true; ---, quadratic approximation; —, limits of 5% error region.

is in error by no more than 5%. We find that there is excellent agreement between the true and approximate values within the region $|\Phi(\boldsymbol{\theta}) - \Phi^*| \leqslant 0.005$, and in some areas the agreement extends far beyond this region.

The eigenvalues and vectors of $\mathbf{N}^*$ are:

$$\lambda_1 = 3.7660 \times 10^{-5}, \qquad \lambda_2 = 0.0096 \times 10^{-5}$$

$$\mathbf{u}_1 = \begin{bmatrix} 0.2642 \\ -0.9645 \end{bmatrix}, \qquad \mathbf{u}_2 = \begin{bmatrix} 0.9645 \\ 0.2642 \end{bmatrix}$$

Accordingly, Eq. (1) can be rewritten in canonical form as

$$\Phi - 0.03980599 = \tfrac{1}{2}10^{-5}(3.7660\,\psi_1^2 + 0.0096\,\psi_2^2) \tag{7-21-2}$$

where

$$\begin{aligned} \psi_1 &= 0.2642\,(\theta_1 - 813.4583) - 0.9645\,(\theta_2 - 960.9063) \\ \psi_2 &= 0.9645\,(\theta_1 - 813.4583) + 0.2642\,(\theta_2 - 960.9063) \end{aligned} \tag{7-21-3}$$

If we choose $\varepsilon = 0.005$, then the indifference region $|\Phi - \Phi^*| \leqslant 0.005$ is defined (approximately) by

$$3.7660\,\psi_1^2 + 0.0096\,\psi_2^2 \leqslant 2 \times 0.005/10^{-5} = 1000 \tag{7-21-4}$$

With ψ_2 held constant at zero, this corresponds to

$$|\psi_1| \leqslant (1000/3.7660)^{1/2} = 16.3 \qquad (\psi_2 = 0)$$

and with ψ_1 held constant at zero

$$|\psi_2| \leqslant (1000/0.0096)^{1/2} = 323 \qquad (\psi_1 = 0)$$

Thus ψ_1 (the short axis of the ellipse Eq. (4)) is relatively well-determined, but ψ_2 (the long axis of the ellipse) is poorly-determined. In Fig. 7-3, the ψ_1 and ψ_2 axes do not appear perpendicular to each other, because θ_1 and θ_2 are drawn to different scales.

To estimate the covariance matrix of the estimate, we use Eq. (7-5-15), but we must estimate σ^2 first. The residual sum of squares is $\Phi^* = 0.03980599$, and:

$$\sigma^2 = [1/(15-2)]0.03980599 = 0.00306200, \qquad \sigma = 0.05533533$$

$$\mathbf{V}_{\boldsymbol{\theta}} \approx 2 \times 0.00306200\,\mathbf{N}^{*-1} = \begin{bmatrix} 60561.7 & 16547.6 \\ 16547.6 & 4696.17 \end{bmatrix} \tag{7-21-5}$$

The standard deviations of the individual parameter estimates are

$$\sigma_1 = 60561.7^{1/2} = 246.093, \qquad \sigma_2 = 4696.17^{1/2} = 68.5286$$

The correlation between the estimates is

$$\rho_{1,2} = (16547.6/246.093 \times 68.5286) = 0.981214$$

The principal components, of course, coincide with ψ_1 and ψ_2. Their variances are given by

$$\pi_1 = 2\sigma^2/\lambda_1 = 162.61, \qquad \pi_2 = 2\sigma^2/\lambda_2 = 65095.3$$

with standard deviations

$$\sigma_1 = 12.752, \qquad \sigma_2 = 255.14$$

Again, we see that ψ_1 is well-determined, ψ_2 less so.

To compute the scaled principal components, we scale each parameter to have unit standard deviation, i.e., we define

$$v_1 = \theta_1/246.093, \qquad v_2 = \theta_2/68.5286$$

The covariance matrix of $\mathbf{v}$ is simply the correlation matrix of $\boldsymbol{\theta}$, i.e,.

$$\mathbf{V}_v = \begin{bmatrix} 1 & 0.981214 \\ 0.981214 & 1 \end{bmatrix}$$

whose eigenvalues and vectors (in the $\mathbf{v}$ coordinates) are:

$$\mu_1 = 1.981214, \qquad \mu_2 = 0.018786$$

$$\mathbf{p}_1 = \begin{bmatrix} 1/\sqrt{2} \\ 1/\sqrt{2} \end{bmatrix}, \qquad \mathbf{p}_2 = \begin{bmatrix} 1/\sqrt{2} \\ -1/\sqrt{2} \end{bmatrix}$$

To express $\mathbf{p}_1$ and $\mathbf{p}_2$ in terms of $\boldsymbol{\theta}$, we have to unscale, i.e.,

$$\mathbf{p}_1 = \begin{bmatrix} 1/(\sqrt{2} \times 246.093) \\ 1/(\sqrt{2} \times 68.5286) \end{bmatrix} = \begin{bmatrix} 0.00287333 \\ 0.0103184 \end{bmatrix}, \qquad \mathbf{p}_2 = \begin{bmatrix} 0.00287333 \\ -0.0103184 \end{bmatrix}$$

Thus the quantity

$$p_1 = 0.00287333(\theta_1 - \theta_1^*) + 0.0103184(\theta_2 - \theta_2^*)$$

has variance 1.981214, the quantity

$$p_2 = 0.00287333(\theta_1 - \theta_1^*) - 0.0103184(\theta_2 - \theta_2^*)$$

has variance 0.018786, and the two are uncorrelated.

To obtain a 95% confidence region for $\boldsymbol{\theta}$. we use the statistic of Eq. (7-11-5)

$$F_{2,13} \geqslant \frac{13}{2 \times 0.03980599} \tfrac{1}{2}10^{-5}\, \delta\boldsymbol{\theta}^{\mathrm{T}} \begin{bmatrix} 0.271890 & -0.957336 \\ -0.957336 & 3.50371 \end{bmatrix} \delta\boldsymbol{\theta}$$

$$= 163.292 \times \tfrac{1}{2}10^{-5}(0.271890\, \delta\theta_1{}^2 - 1.914672\, \delta\theta_1\, \delta\theta_2 + 3.50371\, \delta\theta_2{}^2) \tag{7-21-6}$$

The upper 0.05 point of F with 2 and 13 degrees of freedom is, according to the tables, 3.81. Our confidence region thus has the equation

$$\tfrac{1}{2}10^{-5}(0.271890\, \delta\theta_1{}^2 - 1.914672\, \delta\theta_1\, \delta\theta_2 + 3.50371\, \delta\theta_2{}^2) \leqslant 3.81/163.292 = 0.023332$$

Comparison with Eq. (1) indicates that this region is bounded by the contour

$$\Phi - \Phi^* = 0.023332$$

According to Fig. 7-3, this contour is partly outside the region where Eq. (1) is a reliable approximation. The fact that the exact contour is inside the approximate contour, suggests, however, that the latter should be a conservative estimate of the confidence region.

Finally, we examine the residuals, given in Table 7-1 and Fig. 7-4. A

Table 7-1

Residuals at $\boldsymbol{\theta}^* = \begin{bmatrix} 813.4583 \\ 960.9063 \end{bmatrix}$

μ	$x_{\mu 1} = t$	$x_{\mu 2} = T$	$e_\mu{}^* = y_\mu - f_\mu(\boldsymbol{\theta}^*, \mathbf{x}_\mu)$
1	0.1	100	−0.0145552
2	0.2	100	−0.00613993
3	0.3	100	−0.0287542
4	0.4	100	0.000602186
5	0.5	100	0.0199295
6	0.05	200	−0.0906165
7	0.1	200	0.0304608
8	0.15	200	0.0869893
9	0.2	200	−0.0387225
10	0.25	200	−0.0219878
11	0.02	300	0.0497515
12	0.04	300	0.0504873
13	0.06	300	−0.103587
14	0.08	300	−0.0550289
15	0.1	300	0.0293314

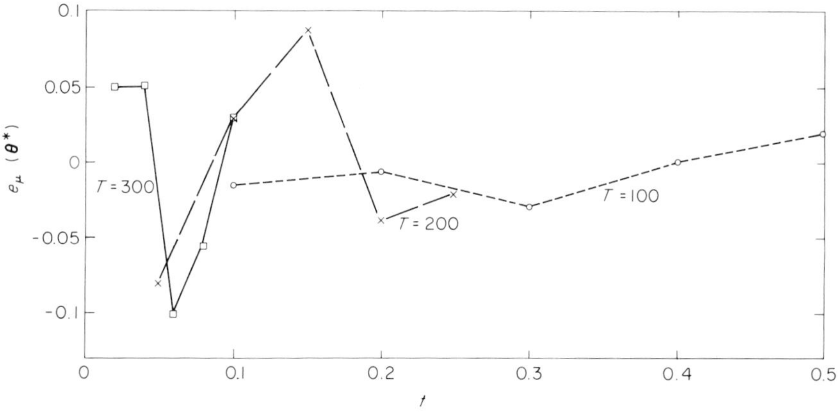

Fig. 7-4. Residuals (least squares problem).

glance at the latter suggests that the residuals at $T = 100$ are considerably smaller than those at $T = 200$ and $T = 300$. An F-test on the ratio of the sum of squares of the last ten residuals to that of the first five residuals indicates a significant difference even at the 99.5% confidence level. We shall deal with this problem further in Chapter IX, although this small body of data probably does not merit further analysis.

7-22. A Monte Carlo Study

To investigate the reliability of statistics obtained in the previous section, we used the simulation technique suggested in Section 3-3. We assumed that model Eq. (5-21-5) was correct, with $\theta_1 = \theta_2 = 1000$. We used this to compute y_μ for the fifteen data points, and added as "experimental error" a pseudorandom number drawn from an appropriate distribution. Six distributions were studied: normal and uniform distributions, each with $\sigma = 0.01, 0.03, 0.05$. The estimation procedures were always carried out, however, as though the errors were thought to be normally distributed.

For each one of the six cases, 100 replications (samples) of fifteen observations were generated. The parameters were estimated in each one of these samples, and

$$\text{bias} = \boldsymbol{\theta}^* - \begin{bmatrix} 1000 \\ 1000 \end{bmatrix}$$

was calculated, as well as the estimated covariance matrix. These were averaged over all samples. We also calculated the actual covariance of the estimates around their means [Eq. (3-3-2)]. The results appear in Table 7-2.

The following conclusions can be drawn from the table:

1. The average bias in all cases is small compared to the standard deviations of the estimates.
2. The estimated covariance matrix is, on the average, an acceptable estimate of the true covariance, particularly at small values of experimental error. Even at $\sigma = 0.05$, the estimates are not unreasonable, particularly when one takes square roots to obtain the standard deviations of the estimates.
3. The estimates are reasonably robust, at least as far as the difference between normal and uniform distributions is concerned.
4. The supposition that for this model the estimated variance is conservative (too large) is confirmed.
5. The standard deviations of the residuals (corrected for bias) provide, on the average, excellent estimates of the experimental error.

While these conclusions hold for the average of many replications, the results for individual replications vary quite sharply. In fact, the specific

Table 7-2
Monte Carlo Study of Single-Equation Least Squares Model[a]

		Experimental error					
		$\sigma = 0.01$		$\sigma = 0.03$		$\sigma = 0.05$	
		θ_1	θ_2	θ_1	θ_2	θ_1	θ_2
Normal distribution	Average bias	6.10	0.79	29.22	2.67	66.68	5.05
	True covariance Eq. (3-3-2)	$\begin{bmatrix} 3264 & 751 \\ 751 & 179 \end{bmatrix}$		$\begin{bmatrix} 28167 & 6468 \\ 6468 & 1560 \end{bmatrix}$		$\begin{bmatrix} 79169 & 17566 \\ 17566 & 4193 \end{bmatrix}$	
	Average Estimated covariance Eq. (7-5-17)	$\begin{bmatrix} 3315 & 725 \\ 725 & 165 \end{bmatrix}$		$\begin{bmatrix} 32752 & 6766 \\ 6766 & 1487 \end{bmatrix}$		$\begin{bmatrix} 105175 & 19828 \\ 19828 & 4131 \end{bmatrix}$	
	Standard deviation of residuals	0.01002497		0.03008677		0.05016445	
Uniform distribution	Average bias	−3.07	−1.01	0.83	−2.67	17.96	−3.82
	True covariance Eq. (3-3-2)	$\begin{bmatrix} 2743 & 622 \\ 622 & 146 \end{bmatrix}$		$\begin{bmatrix} 23911 & 5447 \\ 5447 & 1300 \end{bmatrix}$		$\begin{bmatrix} 67208 & 15061 \\ 15061 & 3597 \end{bmatrix}$	
	Average estimated covariance Eq. (7-5-17)	$\begin{bmatrix} 3328 & 735 \\ 735 & 169 \end{bmatrix}$		$\begin{bmatrix} 31396 & 6695 \\ 6695 & 1514 \end{bmatrix}$		$\begin{bmatrix} 96441 & 19173 \\ 19173 & 4191 \end{bmatrix}$	
	Standard deviation of residuals	0.01015331		0.03046174		0.05076649	

[a] Averages are Over 100 Replications.

problem that we have solved in Sections 5-21 and 7-21 is one of the replications (with y_μ rounded to three decimal places) of the normal distribution with $\sigma = 0.05$. The bias on this particular replication is

$$\begin{bmatrix} 813.4583 - 1000 \\ 960.9063 - 1000 \end{bmatrix} = \begin{bmatrix} -186.5 \\ -\ \ 39.1 \end{bmatrix}$$

The true value $\boldsymbol{\theta} = (1000, 1000)^{\mathrm{T}}$ is marked on Fig. 7-3. It lies just within the region of good approximation, and corresponds to an $F_{2,13}$ value given by Eq. (7-21-6)

$$F_{2,13} = 163.292 \times \tfrac{1}{2}10^{-5}(0.271890 \times 186.5^2 - 1.914672 \times 186.5 \times 39.1 + 3.50371 \times 39.1^2) = 0.695$$

Though this value is far from excessive, the bias in this replication is much larger than the average bias of (66.68, 5.05) in Table 7-2. On the other hand, the covariance estimate Eq. (7-21-5) is quite a lot closer to the true covariance than is the average estimate of Table 7-2. In some other replications this estimate is much worse. In one case, for instance (still with $\sigma = 0.05$), the estimated covariance is

$$\mathbf{V}_{\boldsymbol{\theta}} = \begin{bmatrix} 149984 & 33191.4 \\ 33191.4 & 7616.09 \end{bmatrix}$$

which is off approximately by a factor of two (still not very significant in an F-test). Oddly enough, this replication yielded the almost unbiased estimate

$$\boldsymbol{\theta}^* = \begin{bmatrix} 1000.512 \\ 1000.942 \end{bmatrix}$$

We conclude that in this particular problem our estimates for $\boldsymbol{\theta}$, $\mathbf{V}_{\boldsymbol{\theta}}$, and the confidence regions are quite reasonable.

7-23. Independent Variables Subject to Error

We shall now interpret the estimates obtained in Section 6-13 for the same model, but with all variables subject to error. According to Eq. (7-6-4) and using Eq. (6-13-5), we have

$$\mathbf{V}_{\boldsymbol{\theta}} \approx \mathbf{D}^{-1} = \begin{bmatrix} 93021.94 & 16298.55 \\ 16298.55 & 2912.917 \end{bmatrix}$$

This is not very different from what we found previously in Section 7-21 under different assumptions. We shall try to determine whether the assumptions underlying the estimate of Section 6-13 are validated by the data.

From the matrix $\mathbf{M}$ [Eq. (6-13-6)] we obtain, using Eq. (7-14-16), the following estimate for the covariance matrix of the residuals

$$[r/m(n - l/m)]\mathbf{M} = [3/1(15 - 2/1)]\mathbf{M} = (3/13)\mathbf{M}$$

$$= \begin{bmatrix} 0.000411 & 0.00224 & 0.000452 \\ 0.00224 & 0.0162 & 0.00278 \\ 0.000452 & 0.00278 & 0.000957 \end{bmatrix}$$

On the other hand, to obtain our estimate we had assumed an error covariance of

$$\mathbf{V}_{\mathbf{w}} = \begin{bmatrix} 0.0001 & 0 & 0 \\ 0 & 0.25 & 0 \\ 0 & 0 & v_3 \end{bmatrix}$$

The quantities $13 \times 0.000411/0.0001$ and $13 \times 0.0162/0.25$ should both be samples from a χ^2 distribution with 13 degrees of freedom. A glance at the tables shows the first to be much too large, the second too small to be acceptable; the odds for rejecting each are greater than 99:1. Even summing the two quantities (this is equivalent to evaluating $\mathrm{Tr}(\mathbf{V}^{-1}\mathbf{V}^*)$ as in Section 7-14), we obtain a number too large for χ^2_{26}. Our assumed covariance matrix is contradicted by the data.

7-24. Two-Equation Maximum Likelihood Problem

We now treat the estimates obtained in Section 5-23. Let us examine case (a), unknown **V**. The estimate $\boldsymbol{\theta}^*$ is given in the first row of Table 5-8, and the corresponding value $\mathbf{c}^*$ is found in Table 5-9. Since all calculations were performed in terms of $\boldsymbol{\theta}$, the inverse approximate Hessian with respect to $\boldsymbol{\theta}$ is found to be

$$\mathbf{V}_{\boldsymbol{\theta}} \approx (\mathbf{N}^*)^{-1} =$$

$$\begin{bmatrix} 0.834966-1 & -0.699289-4 & 0.121944-2 & -0.761264-1 & -0.269914-1 \\ & 0.121575 & -0.360968-4 & 0.517330-4 & 0.181755-4 \\ & & 0.122197-2 & -0.704885-3 & -0.247651-3 \\ & & & 0.698556-1 & 0.246071-1 \\ \text{symmetric} & & & & 0.875678-2 \end{bmatrix}$$

(The notation $0.834966-1$ is used to represent 0.834966×10^{-1}.) We write below the values of $\theta_\alpha{}^*$ along with their standard deviations, the latter being the square roots of the diagonal elements of $\mathbf{V}_{\boldsymbol{\theta}}$

$$\boldsymbol{\theta}^* = \begin{bmatrix} -0.0758463 \pm 0.288958 \\ -0.0115747 \pm 0.0011026 \\ 0.790686 \pm 0.034957 \\ 1.00224 \pm 0.26430 \\ 0.859255 \pm 0.093578 \end{bmatrix}$$

We are interested, however, in **c**, not in $\boldsymbol{\theta}$. According to Section 7-20 we need therefore to calculate the matrix

$$\mathbf{J}^* = \partial\mathbf{c}/\partial\boldsymbol{\theta})_{\boldsymbol{\theta}=\boldsymbol{\theta}^*}$$

This can be readily obtained from Eq. (5-23-8) by differentiation

$$\mathbf{J}^* =$$

$$\begin{bmatrix} -0.6905560 & 0 & 0.4178637 & -0.4263207 & -0.6835837 \\ 0 & -0.5492625 & -0.008040566 & 0 & 0 \\ 0 & 0 & -1.599525 & 0 & 0 \\ 0 & 0 & 0 & -0.9955402 & 0 \\ 0 & 0 & 0 & -0.0368951 & -0.5681071 \end{bmatrix}$$

Following Eq. (7-20-2) we now compute

$$\mathbf{V_c} = \mathbf{J^* V_\theta J^{*T}} =$$

$$\begin{bmatrix} 0.543786-3 & -0.412978-6 & 0.221255-3 & 0.564779-2 & 0.137911-2 \\ & 0.126945-6 & -0.159973-4 & 0.226459-4 & 0.537960-5 \\ & & 0.312639-2 & 0.112245-2 & 0.266644-3 \\ & & & 0.692338-1 & 0.164833-1 \\ \text{Symmetric} & & & & 0.395299-2 \end{bmatrix}$$

and the estimate $\mathbf{c}^*$ is represented as

$$\mathbf{c}^* = \begin{bmatrix} 0.5460134 & \pm 0.0233192 \\ 0.006357569 & \pm 0.000356293 \\ 1.264724 & \pm 0.0559141 \\ 0.9977676 & \pm 0.263123 \\ 0.8592555 & \pm 0.0628728 \end{bmatrix}$$

All the parameters are fairly well-determined, with c_4 less so than the others.

The residuals corresponding to our estimate are listed in Table 7-3.

Table 7-3
Residuals $\mathbf{e}_\mu{}^* = \mathbf{y}_\mu - \mathbf{f}(\mathbf{x}_\mu, \boldsymbol{\theta}^*)$ for Case (a)

μ	$e^*_{\mu 1}$	$e^*_{\mu 2}$	μ	$e^*_{\mu 1}$	$e^*_{\mu 2}$
1	−0.22944	−0.01629	22	0.42388	0.04043
2	−0.18880	0.00559	23	0.24983	0.02265
3	−0.19394	−0.01330	24	0.37242	−0.00994
4	−0.17473	−0.00069	25	0.24696	0.01022
5	−0.19199	−0.01578	26	0.16855	−0.00204
6	−0.21667	−0.01114	27	0.11696	−0.00205
7	−0.10269	0.00038	28	0.07203	−0.01195
8	−0.05086	−0.00752	29	0.08727	0.02173
9	0.00012	0.01701	30	0.02814	0.00542
10	−0.13722	0.00483	31	0.01613	0.00927
11	−0.06465	−0.02522	32	0.00542	0.02753
12	−0.00414	0.03791	33	0.05353	−0.04208
13	−0.01195	−0.03430	34	0.07066	−0.00772
14	−0.08990	−0.01499	35	−0.01496	−0.00765
15	−0.02357	−0.00057	36	−0.17103	−0.04543
16	−0.02433	−0.00699	37	−0.26740	−0.04499
17	0.00547	−0.01582	38	−0.19278	0.01363
18	0.06089	0.01065	39	−0.04582	0.03801
19	0.10571	0.02486	40	−0.00165	0.03415
20	0.23525	0.02979	41	−0.00726	0.01637
21	0.25371	0.03832			

Their moment matrix is

$$\mathbf{M}^* = \begin{bmatrix} 1.066369 & 0.06834212 \\ 0.06834212 & 0.02096695 \end{bmatrix}$$

and the estimated covariance matrix of the errors is

$$\mathbf{V}^* = 1/(41 - 5/2)\mathbf{M}^* = \begin{bmatrix} 0.0276979 & 0.00177512 \\ 0.00177512 & 0.000544596 \end{bmatrix}$$

corresponding to standard deviations of 0.166427 and 0.0233366 of the y_1 and y_2 errors, respectively.

We do not know enough about econometrics to decide whether errors of this magnitude are reasonable, and whether they can be ascribed to measurement errors alone. A glance at Table 7-3, however, reveals at once that at least the y_1 residuals are not random. They have been plotted in Fig. 7-5. It appears that Eq. (5-23-2) fails to account for certain strong

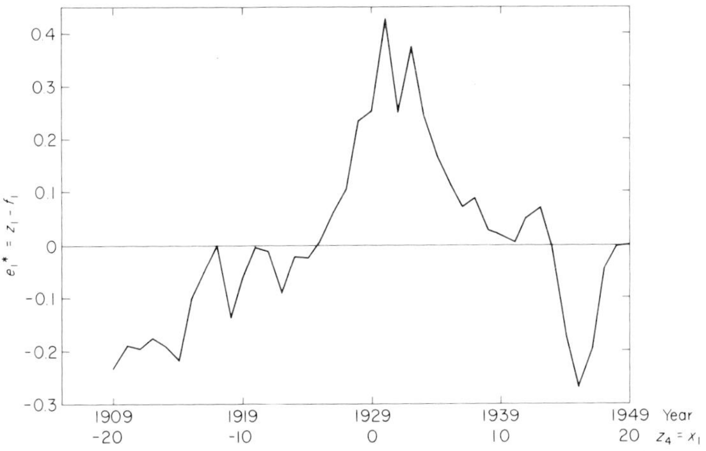

Fig. 7-5. First equation residuals, production model.

variations of z_1 with respect to time. The equation for z_2 seems somewhat more satisfactory. However, there are 21 negative residuals, 20 positive ones (both numbers exceed 10) and 16 runs. From Eqs. (7-17-1), (7-17-2), and (7-17-3) we have:

$$\mu = 2 \times 20 \times 21/(20 + 21) + 1 = 21.488$$
$$= \text{expected number of runs if residuals were random}$$

$$\sigma^2 = 2 \times 20 \times 21(2 \times 20 \times 21 - 20 - 21)/[(20 + 21)^2(20 + 21 - 1)] = 9.982$$
$$z = (16 - 21.488 + 0.5)/(9.982)^{1/2} = -1.579$$

where z is approximately a standard normal deviate.

The probability of finding 16 or fewer runs is approximately $P(z \leqslant -1.579)$, which according to tables of the normal distribution is only about 6%. Hence there is strong, though not conclusive evidence to indictae that the z_2 residuals are also not random.

In case (b) we assumed that $\mathbf{V}$ is diagonal. The estimated covariance of the residuals was

$$\mathbf{V}^* = \begin{bmatrix} 0.024722 & 0.00131260 \\ 0.00131260 & 0.000571524 \end{bmatrix}$$

The correlation between the residuals of z_1 and z_2 is

$$r_{12} = 0.00131260/(0.0247422 \times 0.000571524)^{1/2} = 0.349$$

Letting $n^* = 41 - 5/2 = 38.5$, we have from Eq. (7-16-10)

$$\lambda = (36.5)^{1/2} 0.349/(1 - 0.349^2)^{1/2} = 2.31$$

According to tables of the t-distribution, the chance of encountering a value of $|\lambda|$ as large or larger than 2.31 with 36.5 degrees of freedom is only about 3%. We reject therefore the hypothesis that $\mathbf{V}$ is diagonal.

In case (c) we assumed $\mathbf{V}$ proportional to $\mathbf{Q} = \text{diag}\,(4, 1)$. The covariance of the residuals turns out to be

$$\mathbf{V}^* = \begin{bmatrix} 0.0156715 & 0.000679944 \\ 0.000679944 & 0.00120482 \end{bmatrix}$$

The correlation $r_{12} = 0.157$ leads to $\lambda = 0.943$, which is not incompatible with the supposition that $V_{12} = 0$. On the other hand, if V_{11}^* is an estimate of a variance four times as large as the variance estimated by V_{22}^* (each based on 38.5 degrees of freedom), then

$$\lambda = 0.0156715/(4 \times 0.00120482) = 3.25$$

would be an $F_{38.5,\,38.5}$ variate. The probability of encountering such a value is less than 0.5%, so hypothesis (c) stands refuted.

In case (d) we made no assumptions concerning the value of $\mathbf{V}$. The residuals of $\log z_1$, however, behave no better than those of z_1 (Fig. 7-5) The same is true of the residuals in cases (e) and (f). At this time, on the basis of the data alone, we have no reason to prefer any of the models (a), (d), (e), and (f), and none of them account sufficiently for variations in z_1.

7-25. Problems

1. Verify Eqs. (7-12-6)–(7-12-7).

2. Show that Eq. (7-5-16) holds when the covariance matrix is proportional to a known matrix $\mathbf{Q}$ [see Eq. (4-21-2) and row 2 of Table 5-1].

3. Suppose **P** in Problem 8 of Section 4-21 is an unknown matrix. Show that its MLE is given by

$$(1/n)\mathbf{M}(\boldsymbol{\theta}^*) - (\mathbf{B}^{\mathrm{T}}\mathbf{Q}^{-1}\mathbf{B})^{-1}$$

where

$$\mathbf{M}(\boldsymbol{\theta}^*) = \sum_{\mu=1}^{n} [\hat{\mathbf{s}}_\mu - \mathbf{f}(\boldsymbol{\theta}^*, \mathbf{x}_\mu)][\hat{\mathbf{s}}_\mu - \mathbf{f}(\boldsymbol{\theta}^*, \mathbf{x}_\mu)]^{\mathrm{T}}$$

Derive an expression for **P** applicable to Problem 9 of Section 4-21.

4. Using the Monte Carlo technique, investigate the robustness of the test for correlation Eq. (7-16-10) for nonnormal distributions.

5. Suppose observations y_μ, $\mathbf{x}_\mu (\mu = 1, 2, \ldots n)$ are to be fitted by the model

$$y = \theta_0 + \theta_1 x_{\mu 1} + \theta_2 x_{\mu 2} + \cdots \qquad (7\text{-}25\text{-}1)$$

Let the error distribution be such as to justify estimation of $\boldsymbol{\theta}$ by least squares. Suppose Eq. (1) is to be used for predicting y at given values of $\mathbf{x}$. Show that the prediction error variance is minimum at the centroid of the observations used for estimating $\boldsymbol{\theta}$, i.e., at $\mathbf{x} = \sum_{\mu=1}^{n} \mathbf{x}_\mu/n$.

Chapter

VIII

Dynamic Models

8-1. Models Involving Differential Equations

Models are often formulated in terms of differential equations. That is, the model equations contain not only dependent and independent variables, but also derivatives of the former with respect to the latter. The model equations thus take the form

$$\mathbf{g}(\mathbf{x}, \mathbf{y}, \partial \mathbf{y}/\partial \mathbf{x}, \partial^2 \mathbf{y}/\partial \mathbf{x}\, \partial \mathbf{x}, \ldots, \boldsymbol{\theta}) = \mathbf{0} \tag{8-1-1}$$

When experiments are conducted, we measure values of $\mathbf{y}$ for given values of $\mathbf{x}$, but we do not usually directly measure the values of the derivatives. Hence the model equations cannot be used directly for the estimation of the parameters $\boldsymbol{\theta}$. However, this difficulty may be overcome in one of the following ways:

(a) Differentiation of Data. Approximate values of the derivatives appearing in the equation can be calculated by differencing adjacent data values. If $\mathbf{x}_\mu$ and $\mathbf{x}_{\mu+1}$ are neighboring points differing only in the ith coordinate, then $(\mathbf{y}_{\mu+1} - \mathbf{y}_\mu)/(x_{\mu+1,\,i} - x_{\mu,\,i})$ is an approximation to $\partial \mathbf{y}/\partial x_i$ in that region. Even though more accurate approximations are available in some cases, the maximum accuracy attainable with this method is severely limited, and its errors difficult to assess. The main advantage of this method (in those cases where it is feasible) is that the estimation is performed using Eqs. (1) directly, and these are usually much simpler than the integrated equations. Therefore, the computation can usually be performed much faster than in the method to be described next. We feel, however, that this advantage does not outweigh the disadvantage of limited accuracy. The method cannot be used at all if the separation between data points is large, but may be useful when experiments are specially planned for it, e.g., by the use of differential reactors.

(b) Integration of Equations. In principle, the differential Eqs. (1) may be integrated to yield expressions of the form

$$\mathbf{y} = \mathbf{f}(\mathbf{x}, \boldsymbol{\theta}) \tag{8-1-2}$$

which is identical to Eq. (2-4-2). Therefore, all standard estimation methods may be applied. If Eq. (1) can be solved analytically in closed form, we end up with explicit formulas for the functions **f** in Eq. (2), and their origin as solutions of differential equations need no longer concern us. The problem we shall deal with in the succeeding sections is that of estimating $\boldsymbol{\theta}$ when Eq. (1) must be integrated numerically, so that the functions **f** are only implicitly defined.

A special problem associated with method (b) is that of the initial or boundary values required for integrating the differential equations. These are frequently defined by the experimental conditions, in which case no further problem exists. When these conditions are entirely or partly unknown, they must be included in the problem as additional unknown parameters.

(c) Integration of Data. It is sometimes possible to integrate out all the derivatives appearing in the differential equations. The differential equations are, thereby, transformed into integral equations. If our observations cover the region of interest densely, we may integrate the data numerically to obtain the values of the integrals appearing in the equations, which can now be regarded as algebraic equations in $\boldsymbol{\theta}$.

This method, like (a), requires dense data, and gives rise to unassessable errors. In addition, it is applicable only in a limited number of cases. Its advantage over (a) is that numerical integration is generally more accurate than differentiation. Like (a), it is computationally faster than (b).

On the whole, we recommend method (b) whenever the required computations are not beyond the capability of available machinery. Method (a) or (c) may be used to obtain an initial guess for (b).

We illustrate these methods by means of a simple example, that of a system (e.g., a radioactive material) undergoing first-order decay. Here we have

$$dy/dx + \theta_1 y = 0 \tag{8-1-3}$$

as the model equation. Upon integration, this becomes

$$y = y_0 \exp(-\theta_1 x) \tag{8-1-4}$$

We have measured values y_μ at an ascending sequence of values x_μ ($\mu = 1, 2, \ldots, n$). If the initial value y_0 is known, we may use our data directly, in conjunction with Eq. (4), to estimate θ_1. If y_0 is unknown, we treat it as a parameter θ_2 and use

$$y = \theta_2 \exp(-\theta_1 x) \tag{8-1-5}$$

to estimate both θ_1 and θ_2. This constitutes method (b). Here we were able to integrate the equations analytically. The method applies equally well when the equations must be solved numerically.

To apply method (a), we could define

$$z_\mu \equiv (y_{\mu+1} - y_{\mu-1})/(x_{\mu+1} - x_{\mu-1}) \qquad (\mu = 2, 3, \ldots, n-1)$$

as an approximation to dy/dx at $x = x_\mu$. We then use

$$z + \theta_1 y = 0 \tag{8-1-6}$$

as the model equation, from which we estimate θ_1, by minimizing, say

$$\sum_{\mu=2}^{n-1} (z_\mu + \theta_1 y_\mu)^2$$

To apply method (c), we follow Himmelblau *et al.* (1967) and integrate Eq. (3) from $x = 0$ to $x = x_\mu$

$$y_\mu - y_0 + \theta_1 \int_0^{x_\mu} y(x)\,dx = 0 \tag{8-1-7}$$

If we have measured sufficiently many values of y between $x = 0$ and $x = x_\mu$, then we can obtain an approximate value I of the integral in Eq. (7), say by using the trapezoidal rule

$$I_\mu = \tfrac{1}{2} \sum_{\eta=1}^{\mu} (y_\eta + y_{\eta-1})(x_\eta - x_{\eta-1}) \tag{8-1-8}$$

Then θ_1 may be estimated from the linear model

$$y_\mu - y_0 + \theta_1 I_\mu = 0 \tag{8-1-9}$$

say, by least squares.

In an alternative data integration method, due to Shinbrot (1954), we multiply Eq. (3) by $\sin \alpha x$, and integrate the result over the range of x values, say from $x = 0$ to $x = A$, using the integration by parts technique

$$\begin{aligned} 0 &= \int_0^A (dy/dx) \sin \alpha x \, dx + \theta_1 \int_0^A y \sin \alpha x \, dx \\ &= [y \sin \alpha x]_0^A - \alpha \int_0^A y \cos \alpha x \, dx + \theta_1 \int_0^A y \sin \alpha x \, dx \\ &= y(A) \sin \alpha A + \int_0^A (\theta_1 \sin \alpha x - \alpha \cos \alpha x) y \, dx \end{aligned} \tag{8-1-10}$$

If we choose $\alpha = k\pi/A$, where k is any integer, then $y(A) \sin \alpha A$ vanishes. Hence we have

$$\theta_1 \int_0^A y \sin(k\pi x/A)\,dx = (k\pi/A) \int_0^A y \cos(k\pi x/A)\,dx \tag{8-1-11}$$

If y is known at a sufficiently dense set of points, then we can integrate both sides of Eq. (11) numerically for various values of k. This gives us several equations for the unknown θ_1, and we may choose that value of θ_1 which satisfies these equations in the least squares sense.

By choosing appropriate multiplier functions we can apply this method to problems involving higher derivatives, as well as to models involving partial differential equations [see Perdreauville and Goodson (1966)].

8-2. The Standard Dynamic Model

We do not propose to treat models represented by Eq. (8-1-1) in complete generality. Rather, we restrict ourselves to a subclass of models which are particularly tractable, yet at the same time extremely important in practice. These are the so-called *standard dynamic models*, which we define below by listing the variables included and the relations among them:

(a) A vector of *independent variables* $\mathbf{x}$.

(b) An additional independent variable t, usually referred to as *time*, although it need not represent the actual physical dimensions of time.

(c) A vector of *unknown parameters* $\boldsymbol{\theta}$.

(d) A vector of *state variables* $\mathbf{s}$, which are functions of t, $\mathbf{x}$, and $\boldsymbol{\theta}$. The functions are defined implicitly by means of

1. A set of simultaneous first-order ordinary differential equations

$$\dot{\mathbf{s}} \equiv d\mathbf{s}/dt = \mathbf{h}(t, \mathbf{x}, \mathbf{s}, \boldsymbol{\theta}) \tag{8-2-1}$$

where $\mathbf{h}$ is a vector of given functions, and

2. A set of initial conditions

$$\mathbf{s}(0) \equiv \mathbf{s}|_{t=0} = \mathbf{s}_0(\mathbf{x}, \boldsymbol{\theta}) \tag{8-2-2}$$

where $\mathbf{s}_0$ is a vector of given functions. Note that Eq. (2) includes the possibilities that some or all $\mathbf{s}(0)$ are given numbers (which are independent variables), or that they are themselves unknown parameters.

(e) A vector of *observed variables* $\mathbf{y}$, whose exact values $\hat{\mathbf{y}}$ are given functions of the state variables, and possibly of the other variables as well

$$\hat{\mathbf{y}} = \hat{\mathbf{y}}(\mathbf{s}, t, \mathbf{x}, \boldsymbol{\theta}) \tag{8-2-3}$$

A common special case is that in which the state variables are observed directly, i.e., $\hat{\mathbf{y}} = \mathbf{s}$. Note that of the set of parameters making up the vector $\boldsymbol{\theta}$, some may appear explicitly only in Eq. (1), others only in Eq. (2) or Eq. (3).

By solving (numerically, if necessary) the differential Eqs. (1) with the initial conditions Eq. (2), and substituting these solutions for **s** in Eq. (3), we bring Eq. (3) into the form Eq. (8-1-2), with **x** and t jointly playing the role of **x** in the latter expression. Hence, the model we have defined conforms to our general form, though in a somewhat roundabout fashion.

In essence, a dynamic system is characterized by a set of state variables which change with time (or some other independent variable) according to certain first-order differential equations. The initial conditions may or may not be fully known. The state of the system is observed at various points in time, but sometimes the state variables are not directly measurable, and we have to measure the related observed variables instead. Unknown parameters may appear in the initial conditions Eq. (2), in the differential equations (1), and in the observation equations (3). In the last case they usually represent unknown characteristics of the measuring devices, e.g., calibration constants.

Our main interest usually lies in estimating the parameters that appear in the differential equations, but we cannot escape estimating the others as well. Fortunately, good initial estimates for these are frequently available. Any inexact knowledge we have concerning the values of these parameters should be included in the form of a prior distribution.

We illustrate the concept of a dynamic system by means of a chemical reaction involving three species whose concentrations c_1, c_2, c_3, satisfy the following differential equations:

$$\begin{aligned} dc_1/dt &= -k_1 c_1{}^2 + k_2 c_2 c_3 \\ dc_2/dt &= k_1 c_1{}^2 - k_2 c_2 c_3 - k_3 c_2 \\ dc_3/dt &= k_1 c_1{}^2 - k_2 c_2 c_3 + k_3 c_2 \end{aligned} \tag{8-2-4}$$

The initial concentrations c_2 and c_3 are not known exactly, but all concentrations must add up to unity, so that we may write

$$c_1(0) = \alpha, \qquad c_2(0) = \beta, \qquad c_3(0) = 1 - \alpha - \beta \tag{8-2-5}$$

where α and β are respectively a known and unknown quantity.

At time t we withdraw three samples from our reactor. In two of these we determine c_1 directly by titration. The third sample is passed through an optical instrument, which measures the light absorptivity of the mixture. This is believed to be a linear function with unknown coefficients of c_1 and c_2. Denoting the results of the measurements on the three samples as y_1, y_2, and y_3, we may write

$$\hat{y}_1 = c_1, \qquad \hat{y}_2 = c_1, \qquad \hat{y}_3 = p + qc_1 + rc_2 \tag{8-2-6}$$

where p, q, and r are unknown quantities.

In this model, c_1, c_2, and c_3 are the state variables; y_1, y_2, and y_3 are the observed variables; α and t are the independent variables; β, k_1, k_2, k_3, p, q, and r are the unknown parameters. Eq. (4), Eq. (5), and Eq. (6) correspond to Eq. (1), Eq. (2), and Eq. (3), respectively. We are primarily concerned with estimating the reaction rate constants k_1, k_2, and k_3. Good initial guesses for β may be known from the manner in which the solution was made up, and for p, q, and r from previous experiments on the same apparatus.

An *experiment* performed on a dynamic system consists of measuring the values of the observed variables $\mathbf{y}$ for given values of the independent variables $\mathbf{x}$ and t. A group of experiments performed with identical values of $\mathbf{x}$ and identical initial conditions, and differing only in the values of t, constitute a *run*. For our purposes it does not matter whether all the experiments in a given run were actually performed as part of a physical run, or whether the apparatus was reset to the same conditions on separate occasions. If several runs each have unknown initial conditions, these constitute separate unknown parameters. In the above example, we may have unknown parameters β_1, $\beta_2, \ldots$ corresponding to distinct runs.

The covariance between the errors in different experiments may, however, depend on whether or not the experiments belong to the same physical run (see Problem 4 in Section 8-9).

8-3. Models Reducible to Standard Form

Our definition of a standard dynamic model is not as restrictive as might appear at first glance. Many problems not originally in this form may be recast to fit the definition. We show how this can be done in several cases.

(a) Suppose a model corresponds to our definition in all respects, except that it contains some second- or higher-order derivatives. If we can rearrange the differential equations in such a way that we have explicit equations for the highest-order derivative of each variable, then we can reformulate the model using the method illustrated by the following example.

Let a model be defined by means of the following two differential equations in the variables z_1 and z_2

$$\begin{aligned} \log d^2z_1/dt^2 + \theta_1\, d^2z_2/dt^2 + \theta_2(dz_1/dt)^3 + \theta_3\, dz_2/dt + z_1{}^2 &= 0 \\ d^3z_2/dt^3 + (d^2z_1/dt^2)^2 + \theta_4 \sin z_1z_2 &= 0 \end{aligned} \qquad (8\text{-}3\text{-}1)$$

The highest-order derivatives of each variable are d^2z_1/dt^2 and d^3z_2/dt^3. We may solve for these:

$$\begin{aligned} d^2z_1/dt^2 &= \exp\{-[\theta_1\, d^2z_2/dt^2 + \theta_2(dz_1/dt)^3 + \theta_3\, dz_2/dt + z_1{}^2]\} \\ d^3z_2/dt^3 &= -\exp\{-2[\theta_1\, d^2z_2/dt^2 + \theta_2(dz_1/dt)^3 + \theta_3\, dz_2/dt + z_1{}^2]\} \\ &\quad -\theta_4 \sin z_1z_2 \end{aligned} \qquad (8\text{-}3\text{-}2)$$

Let us introduce the following state variables

$$s_1 \equiv z_1, \quad s_2 \equiv dz_1/dt, \quad s_3 \equiv z_2, \quad s_4 \equiv dz_2/dt, \quad s_5 = d^2z_2/dt^2$$

whereupon Eq. (2) are equivalent to:

$$\dot{s}_1 = s_2, \qquad \dot{s}_2 = \exp-(\theta_1 s_5 + \theta_2 s_2^{\ 3} + \theta_3 s_4 + s_1^{\ 2}), \qquad \dot{s}_3 = s_4,$$
$$\dot{s}_4 = s_5, \qquad \dot{s}_5 = -\exp[-2(\theta_1 s_5 + \theta_2 s_2^{\ 3} + \theta_3 s_4 + s_1^{\ 2})] - \theta_4 \sin s_1 s_3 \tag{8-3-3}$$

which is in the desired form [Eq. (8-2-1)]. Initial conditions on z_1, z_2, and their derivatives are immediately translatable into conditions on the state variables.

One need not be able to solve the differential equations explicitly for the highest-order derivatives. It is sufficient that one have a numerical procedure for computing these derivatives if the values of all other quantities appearing in the equations are given.

We note that most computer programs for the numerical integration of ordinary differential equations require that the problem be formulated as a system of first-order equations.

(b) Some partial differential equations, particularly of the parabolic type, may be approximated by a dynamic model. For instance, consider the diffusion or heat conduction equation

$$\partial s/\partial t = \alpha \nabla^2 s \tag{8-3-4}$$

where α is a constant and ∇^2 is the Laplace operator

$$\nabla^2 s = \sum_{i=1}^{k} \partial^2 s/\partial x_i^{\ 2}$$

and where $k = 1, 2$, or 3 depending on the number of dimensions of the object we are considering. Suppose we select a grid of points within the object, and let $s_j(t)$ be the value of $s(t)$ at the jth grid point. Furthermore, let $\delta^2 s_j$ be some finite difference approximation to $\nabla^2 s_j$. For instance, in the one-dimensional case, $x_{1,j} = j\,\Delta x_1$, if we take $\delta^2 s_j = (s_{j+1} - 2s_j + s_{j-1})/(\Delta x_1)^2$ [better approximations are discussed by Hicks and Wei (1967)], then

$$\dot{\mathbf{s}}_j = \alpha\, \delta^2 s_j \tag{8-3-5}$$

has the desired form. Again, this is the way in which parabolic equations are often formulated for numerical solution (Rosenbrock and Storey, 1966, Ch. 7). Unfortunately, an excessive number of state variables may be required.

(c) A large variety of problems which are already in the desired form arises from the theory of process control. Linear control theory usually deals with models whose state variables satisfy the differential equations

$$\dot{\mathbf{s}} = \mathbf{A}\mathbf{s} + \mathbf{B}\mathbf{u}(t) + \boldsymbol{\varepsilon}(t) \tag{8-3-6}$$

where $\mathbf{A}$ and $\mathbf{B}$ are matrices, $\mathbf{u}(t)$ is a known function (the *control* signal), and

$\boldsymbol{\varepsilon}(t)$ is an unknown function (the *noise*) possessing certain statistical properties. The observed variables, in turn, are given by

$$\mathbf{y} = \mathbf{C}\mathbf{s} + \boldsymbol{\delta}(t)$$

where $\boldsymbol{\delta}(t)$ is another noise function, and $\mathbf{C}$ is a matrix. Generalization to nonlinear systems is obvious. When $\boldsymbol{\varepsilon}(t) = \mathbf{0}$, we have a dynamic system that conforms to our definition.

Commonly arising problems are those of *identification*, in which unknown elements of $\mathbf{A}$, $\mathbf{B}$, $\mathbf{C}$ are to be determined, and of *tracking*, in which $\mathbf{s}(t)$ is to be estimated from the measured values of $\mathbf{y}(\tau)$ $(\tau \leqslant t)$. The former problem belongs directly to the class of parameter estimation problems that we are considering here. The tracking problem is essentially one of filtering, and the methods for dealing with it, mostly due to Kalman (1960), are discussed extensively in the literature (for lucid expositions with many additional references, see Deutsch (1965) and Sorenson (1966)). Here we only wish to point out that if $\boldsymbol{\varepsilon}(t) = \mathbf{0}$, then the initial conditions completely determine the values of $\mathbf{s}(t)$ at any time. Once the initial conditions and the matrices $\mathbf{A}$ and $\mathbf{B}$ are known, $\mathbf{s}(t)$ can be obtained by straightforward integration. Hence, the tracking problem is equivalent to the problem of estimating unknown initial conditions and elements of $\mathbf{A}$ and $\mathbf{B}$, which is a special case of the parameter estimation problems that we shall treat.

The central problem of control theory is the determination of control functions $\mathbf{u}(t)$ that will cause the state variables to behave in a desired way. Even this problem may sometimes be treated within the parameter estimation framework. For practical reasons, one must usually restrict oneself to functions $\mathbf{u}(t)$ which depend on a finite number of parameters (e.g., polynomials with coefficients to be determined, or piecewise-constant functions). We then wish to determine the optimal values of these parameters, i.e., those values that maximize some performance index of the system. This is entirely analogous to a parameter estimation problem in which the performance index plays the role of the objective function. The general problem of determining $\mathbf{u}(t)$ can also be reduced to a two point boundary value problem by using the maximum principle (Pontryagin *et al.*, 1962). This problem can now be formulated as a parameter estimation problem, in which the missing initial conditions are the unknown parameters and the available final conditions act as the observations. In this form the problem can be solved using, say, the Gauss method.

8-4. Computation of the Objective Function and Its Gradient

In order to proceed with the estimation of the model parameters $\boldsymbol{\theta}$ of a dynamic system, we must be able to calculate the value of the objective function Φ for any given feasible values of the parameters. Now, once the parameter values have been prescribed, the initial conditions are determined by means of Eq. (8-2-2). The differential equation Eq. (8-2-1) can now be integrated, numerically if necessary, from $t = 0$ to $t = t_\mu$ (the time of the μth experiment) for $\mu = 1, 2, \ldots, n$. This determines $\mathbf{s}_\mu$, the predicted values of the state variables at the μth experiment. Now we are in position to determine the $\hat{\mathbf{y}}_\mu$ from Eq. (8-2-3), which in turn are used to compute the residuals $\mathbf{e}_\mu = \mathbf{y}_\mu - \hat{\mathbf{y}}_\mu$. From these, most objective functions (sum of squares, likelihood, etc.) can be calculated directly.

If we wish to use a gradient method (Chapter V) for the estimation of parameters in a dynamic system, we must compute not only the objective function $\Phi(\boldsymbol{\theta})$, but also its derivatives $\mathbf{q} \equiv \partial\Phi/\partial\boldsymbol{\theta}$. As we have stated before, gradient methods are the most efficient among currently available methods. The incentive to use an efficient (in terms of total number of function evaluations) method is particularly great in the case of dynamic systems, where each function evaluation is itself a complex procedure requiring the solution of a set of differential equations. We detail below several ways for calculating the required derivatives.

(a) Finite Differences. Finite difference methods, discussed in Section 5-18, are applicable to dynamic systems. As usual, we must face the problem of balancing the truncation error‡ (increasing with $\Delta\theta$) against the rounding error in differencing (decreasing with $\Delta\theta$). There is, however, an additional difficulty associated with dynamic models. Taking small $\Delta\theta$ and avoiding the concomitant rounding errors by using multiple-precision arithmetic is ineffective in itself, since the accuracy of Φ is limited not only by rounding errors, but primarily by the truncation errors of the integration method. Increased precision in Φ can be acquired only by combining multiple-precision arithmetic with decreased integration steps, or by using a higher-order integration method. Both solutions are costly in computer time. The finite difference method in its raw form works satisfactorily in many problems. In many others, however, it fails to provide the accurate derivative values that are required for convergence of the gradient method.

(b) Sensitivity Equations. Several methods, variously referred to as quasilinearization, sensitivity analysis, perturbations, etc. (Howland and Vaillancourt, 1961; Tomović, 1963; McGhee, 1963; Bellman *et al.*, 1967; Rosenbrock and Storey, 1966, Ch. 8), are based (at least implicitly) on the fact that the required derivatives must satisfy certain linear differential equations. These may be integrated along with the model Eq. (8-2-1) to yield the desired

‡ We are talking here of the truncation error incurred by representing $\partial\Phi/\partial\boldsymbol{\theta}$ as $\Delta\Phi/\Delta\boldsymbol{\theta}$. This is quite different from the truncation error of the integration method, which affects the accuracy of Φ itself.

gradient. In this way, the gradient can be computed with essentially the same degree of accuracy as the function itself without undue difficulties.

In order to apply the method, we must trace step-by-step the dependency of the objective function on the various model variables and parameters. We only list those dependencies which are relevant to our purposes:

1. Φ depends on $\mathbf{e}_\mu = \mathbf{y}_\mu - \hat{\mathbf{y}}_\mu$. The $\mathbf{y}_\mu$ are measured ($\mu = 1, 2, \ldots, n$). Φ may also depend directly on $\boldsymbol{\theta}$, e.g., when there is a prior density function. This requires addition of the appropriate terms to Eq. (1).

2. $\hat{\mathbf{y}}_\mu$ depends on $\mathbf{s}_\mu = \mathbf{s}(t_\mu)$ and $\boldsymbol{\theta}$ [Eq. (8-2-3)].

3. $\mathbf{s}(t_\mu)$ depends on $\mathbf{s}_0$ for the run containing the μth experiment, and on $\boldsymbol{\theta}$ [through integration of Eq. (8-2-1)].

4. $\mathbf{s}_0$ depends on $\boldsymbol{\theta}$ [Eq. (8-2-2)].

Using the chain rule of differentiation we find that

$$\mathbf{q} = \partial\Phi/\partial\boldsymbol{\theta} = \sum_\mu (\partial\Phi/\partial\mathbf{e}_\mu)(\partial\mathbf{e}_\mu/\partial\boldsymbol{\theta}) = -\sum_\mu (\partial\Phi/\partial\mathbf{e}_\mu) D\hat{\mathbf{y}}_\mu/D\boldsymbol{\theta} \tag{8-4-1}$$

where we have used $D\hat{\mathbf{y}}_\mu/D\boldsymbol{\theta}$ to indicate the total derivative of $\hat{\mathbf{y}}_\mu$ with respect to $\boldsymbol{\theta}$, given by

$$D\hat{\mathbf{y}}_\mu/D\boldsymbol{\theta} = \partial\hat{\mathbf{y}}_\mu/\partial\boldsymbol{\theta} + (\partial\hat{\mathbf{y}}_\mu/\partial\mathbf{s}_\mu)(\partial\mathbf{s}_\mu/\partial\boldsymbol{\theta}) \tag{8-4-2}$$

so that altogether

$$\mathbf{q} = -\sum_\mu (\partial\Phi/\partial\mathbf{e}_\mu)(\partial\hat{\mathbf{y}}_\mu/\partial\boldsymbol{\theta} + (\partial\hat{\mathbf{y}}_\mu/\partial\mathbf{s}_\mu)(\partial\mathbf{s}_\mu/\partial\boldsymbol{\theta})) \tag{8-4-3}$$

The quantities $\partial\Phi/\partial\mathbf{e}_\mu$, $\partial\hat{\mathbf{y}}_\mu/\partial\boldsymbol{\theta}$, and $\partial\hat{\mathbf{y}}_\mu/\partial\mathbf{s}_\mu$ are easily computed, the latter two by differentiation of Eq. (8-2-3). That leaves us the problem of determining $(\partial\mathbf{s}_\mu/\partial\boldsymbol{\theta})$.

Let us write down the original differential equation Eq. (8-2-1)

$$d\mathbf{s}/dt = \mathbf{h}(t, \mathbf{x}, \mathbf{s}, \boldsymbol{\theta}) \tag{8-4-4}$$

Differentiating both sides with respect to $\boldsymbol{\theta}$, and employing the chain rule, we find

$$\frac{\partial}{\partial\boldsymbol{\theta}}\left(\frac{d\mathbf{s}}{dt}\right) = \frac{D\mathbf{h}}{D\boldsymbol{\theta}} = \frac{\partial\mathbf{h}}{\partial\boldsymbol{\theta}} + \frac{\partial\mathbf{h}}{\partial\mathbf{s}}\frac{\partial\mathbf{s}}{\partial\boldsymbol{\theta}} \tag{8-4-5}$$

Interchanging the order of differentiations on the left-hand side of Eq. (5)

$$\frac{d}{dt}\left(\frac{\partial\mathbf{s}}{\partial\boldsymbol{\theta}}\right) = \frac{\partial\mathbf{h}}{\partial\boldsymbol{\theta}} + \frac{\partial\mathbf{h}}{\partial\mathbf{s}}\frac{\partial\mathbf{s}}{\partial\boldsymbol{\theta}} \tag{8-4-6}$$

The quantities $\partial\mathbf{h}/\partial\boldsymbol{\theta}$ and $\partial\mathbf{h}/\partial\mathbf{s}$ are easily determined by differentiation. We have, then, in Eq. (6) a set of simultaneous linear first-order ordinary differential equations in the unknown functions $\partial\mathbf{s}/\partial\boldsymbol{\theta}$. These are called the *sensitivity equations*, since their solutions indicate how sensitive the state variables are to changes in the parameters. The functions $\partial \mathbf{s}/\partial \boldsymbol{\theta}$ themselves are called *sensitivity coefficients*. To find $\partial \mathbf{s}_\mu/\partial \boldsymbol{\theta}$ we must integrate these equations, jointly with Eq. (4), from $t = 0$ to $t = t_\mu$. To do this, we need initial values, i.e., $(\partial \mathbf{s}/\partial \boldsymbol{\theta})_{t=0}$. These are obtained simply by differentiating the initial conditions Eq. (8-2-2), i.e.,

$$\left(\frac{\partial \mathbf{s}}{\partial \boldsymbol{\theta}}\right)_{t=0} = \frac{\partial \mathbf{s}_0}{\partial \boldsymbol{\theta}} \tag{8-4-7}$$

If p is the number of state variables and l_1 the number of parameters appearing either in the initial conditions or in the differential equations, then the number of quantities $\partial \mathbf{s}/\partial \boldsymbol{\theta}$ is pl_1, and the total number of equations to be integrated [Eqs. (4) and (6)] is $p(1 + l_1)$. On the other hand, if we were to use one-sided differences to estimate $\partial \Phi/\partial \boldsymbol{\theta}$ we would need to integrate the p Eqs. (4) for $1 + l_1$ different values of $\boldsymbol{\theta}$, again resulting in a total of $p(1 + l_1)$ equations. The computational effort involved in the two methods is roughly the same, but the accuracy attainable in the sensitivity-equations method is much higher, and more easily controlled. Admittedly, a greater effort is required to prepare a problem for treatment by the sensitivity equation method.

In the simplest case, the initial conditions are known and the state variables are observed directly. The initial conditions Eq. (7) then read simply

$$\partial \mathbf{s}/\partial \boldsymbol{\theta})_{t=0} = \mathbf{0}$$

and Eq. (2) reduces to

$$D\hat{\mathbf{y}}_\mu/D\boldsymbol{\theta} = \partial \mathbf{s}_\mu/\partial \boldsymbol{\theta}$$

Let us examine a simple example. There is one state variable, with known initial condition

$$ds/dt = -\theta s, \qquad s(0) = 1 \tag{8-4-8}$$

We know the solution to be $s = e^{-\theta t}$, so that

$$\partial s/\partial \theta = -te^{-\theta t}$$

However, let us form the sensitivity equation

$$\frac{d}{dt}\left(\frac{\partial s}{\partial \theta}\right) = \frac{\partial h}{\partial \theta} + \frac{\partial h}{\partial s}\frac{\partial s}{\partial \theta} = -s - \theta\frac{\partial s}{\partial \theta} \tag{8-4-9}$$

Substituting for s its value, we have to solve for $\partial s/\partial \theta$ the differential equation

$$\frac{d}{dt}\left(\frac{\partial s}{\partial \theta}\right) + \theta\left(\frac{\partial s}{\partial \theta}\right) = -e^{-\theta t}, \qquad \left.\frac{\partial s}{\partial \theta}\right)_{t=0} = 0 \tag{8-4-10}$$

The solution is $\partial s/\partial \theta = -te^{-\theta t}$, in agreement with the previous result.

The required steps in a more general case are illustrated on the example of Section 8-2. The unknown parameters involved in the initial conditions and differential equations are β, k_1, k_2, and k_3. From Eq. (8-2-5) we have

$$\frac{\partial c_1}{\partial \beta} = 0, \quad \frac{\partial c_2}{\partial \beta} = 1, \quad \frac{\partial c_3}{\partial \beta} = -1, \quad \frac{\partial c_i}{\partial k_j} = 0 \quad (i, j = 1, 2, 3), \quad t = 0 \qquad (8\text{-}4\text{-}11)$$

And from Eq. (8-2-4):

$$\begin{aligned}
\frac{d}{dt}\left(\frac{\partial c_1}{\partial \beta}\right) &= -2k_1c_1\left(\frac{\partial c_1}{\partial \beta}\right) + k_2c_3\left(\frac{\partial c_2}{\partial \beta}\right) + k_2c_2\left(\frac{\partial c_3}{\partial \beta}\right)\\
\frac{d}{dt}\left(\frac{\partial c_2}{\partial \beta}\right) &= 2k_1c_1\left(\frac{\partial c_1}{\partial \beta}\right) - (k_2c_3 + k_3)\left(\frac{\partial c_2}{\partial \beta}\right) - k_2c_2\left(\frac{\partial c_3}{\partial \beta}\right)\\
\frac{d}{dt}\left(\frac{\partial c_3}{\partial \beta}\right) &= 2k_1c_1\left(\frac{\partial c_1}{\partial \beta}\right) - (k_2c_3 - k_3)\left(\frac{\partial c_2}{\partial \beta}\right) - k_2c_2\left(\frac{\partial c_3}{\partial \beta}\right)\\
\frac{d}{dt}\left(\frac{\partial c_1}{\partial k_1}\right) &= -c_1^2 - 2k_1c_1\left(\frac{\partial c_1}{\partial k_1}\right) + k_2c_3\left(\frac{\partial c_2}{\partial k_1}\right) + k_2c_2\left(\frac{\partial c_3}{\partial k_1}\right)\\
\frac{d}{dt}\left(\frac{\partial c_2}{\partial k_1}\right) &= c_1^2 + 2k_1c_1\left(\frac{\partial c_1}{\partial k_1}\right) - (k_2c_3 + k_3)\left(\frac{\partial c_2}{\partial k_1}\right) - k_2c_2\left(\frac{\partial c_3}{\partial k_1}\right)\\
\frac{d}{dt}\left(\frac{\partial c_3}{\partial k_1}\right) &= c_1^2 + 2k_1c_1\left(\frac{\partial c_1}{\partial k_1}\right) - (k_2c_3 - k_3)\left(\frac{\partial c_2}{\partial k_1}\right) - k_2c_2\left(\frac{\partial c_3}{\partial k_1}\right) \qquad (8\text{-}4\text{-}12)\\
\frac{d}{dt}\left(\frac{\partial c_1}{\partial k_2}\right) &= c_2c_3 - 2k_1c_1\left(\frac{\partial c_1}{\partial k_2}\right) + k_2c_3\left(\frac{\partial c_2}{\partial k_2}\right) + k_2c_2\left(\frac{\partial c_3}{\partial k_2}\right)\\
\frac{d}{dt}\left(\frac{\partial c_2}{\partial k_2}\right) &= -c_2c_3 + 2k_1c_1\left(\frac{\partial c_1}{\partial k_2}\right) - (k_2c_3 + k_3)\left(\frac{\partial c_2}{\partial k_2}\right) - k_2c_2\left(\frac{\partial c_3}{\partial k_2}\right)\\
\frac{d}{dt}\left(\frac{\partial c_3}{\partial k_2}\right) &= -c_2c_3 + 2k_1c_1\left(\frac{\partial c_1}{\partial k_2}\right) - (k_2c_3 - k_3)\left(\frac{\partial c_2}{\partial k_2}\right) - k_2c_2\left(\frac{\partial c_3}{\partial k_2}\right)\\
\frac{d}{dt}\left(\frac{\partial c_1}{\partial k_3}\right) &= -2k_1c_1\left(\frac{\partial c_1}{\partial k_3}\right) + k_2c_3\left(\frac{\partial c_2}{\partial k_3}\right) + k_2c_2\left(\frac{\partial c_3}{\partial k_3}\right)\\
\frac{d}{dt}\left(\frac{\partial c_2}{\partial k_3}\right) &= -c_2 + 2k_1c_1\left(\frac{\partial c_1}{\partial k_3}\right) - (k_2c_3 + k_3)\left(\frac{\partial c_2}{\partial k_3}\right) - k_2c_2\left(\frac{\partial c_3}{\partial k_3}\right)\\
\frac{d}{dt}\left(\frac{\partial c_3}{\partial k_3}\right) &= c_2 + 2k_1c_1\left(\frac{\partial c_1}{\partial k_3}\right) - (k_2c_3 - k_3)\left(\frac{\partial c_2}{\partial k_3}\right) - k_2c_2\left(\frac{\partial c_3}{\partial k_3}\right)
\end{aligned}$$

Eq. (8-2-5) and (11) provide initial conditions to the differential equations Eq. (8-2-4) and Eq. (12), which may be integrated simultaneously from $t = 0$

to $t = t_\mu$. A separate integration is required for each run, each integration going up to the largest t_μ belonging to the run.

Setting up the Eq. (12) is a rather tedious task. The computer can perform this job, using Eq. (6), provided it is given subroutines that compute $\partial\mathbf{h}/\partial\boldsymbol{\theta}$ and $\partial\mathbf{h}/\partial\mathbf{s}$.

Once $\partial c_i/\partial\beta$ and $\partial c_i/\partial k_j$ $(i, j = 1, 2, 3)$ have been computed for $t = t_\mu$, we derive from Eq. (8-2-6):

$$\begin{aligned} D\hat{y}_1/D\beta &= D\hat{y}_2/D\beta = \partial c_1/\partial\beta \\ D\hat{y}_1/Dk_j &= D\hat{y}_2/Dk_j = \partial c_1/\partial k_j \qquad (j = 1, 2, 3) \\ D\hat{y}_3/D\beta &= q\,\partial c_1/\partial\beta + r\,\partial c_2/\partial\beta \\ D\hat{y}_3/Dk_j &= q\,\partial c_1/\partial k_j + r\,\partial c_2/\partial k_j \qquad (j = 1, 2, 3) \end{aligned} \tag{8-4-13}$$

and also, for the additional parameters, p, q, and r:

$$\begin{aligned} &D\hat{y}_a/Dp = D\hat{y}_a/Dq = D\hat{y}_a/Dr = 0 \qquad (a = 1, 2) \\ &D\hat{y}_3/Dp = 1, \qquad D\hat{y}_3/Dq = c_1, \qquad D\hat{y}_3/Dr = c_2 \end{aligned} \tag{8-4-14}$$

This gives us all the quantities needed to evaluate $\mathbf{q}$, using Eq. (1).

The quantities $D\hat{y}_\mu/D\boldsymbol{\theta}$ are also used to generate $\mathbf{N}$, the Gauss approximation to the Hessian. A numerical illustration of this method appears in Section 8-7.

It is possible to formulate the problem in such a way that only unknown initial conditions need be determined. This is done by replacing each parameter θ appearing in the differential equations by a new state variable s_θ subject to

$$\dot{s}_\theta = 0, \qquad s_\theta(0) = \theta \tag{8-4-15}$$

This procedure has been advocated by several authors, (e.g., Bellman *et al.* (1967) within the framework of quasilinearization), but it serves no purpose other than to increase the number of differential equations that must be integrated.

8-5. Numerical Integration

The methods described in the preceding section require the numerical integration of a set of simultaneous first-order ordinary differential equations. Methods for performing this task are described in textbooks on the subject, to which the reader is referred. Routines for evaluating the integrals are available at most computer installations. The following are some remarks pertinent to the choice of integration method in parameter estimation problems.

Most integration methods are either of the fixed- or the variable-step type. The former methods (e.g., Runge–Kutta) are easier to implement, but the latter provide better control over the truncation errors incurred in the calculations. Intelligent adjustment of step size can save a great deal of computer time. On the other hand, if we use a variable-step method, we must observe precautions. In such methods, the step length at any time is governed by the behavior of the equations. Two slightly different values of $\boldsymbol{\theta}$ may give rise to different sequences of step lengths, resulting in slight discontinuities in the computed functions. These may cause severe errors in derivatives obtained by differencing. It is suggested, therefore, that all $p(l_1 + 1)$ equations required for obtaining a complete set of differences be integrated *simultaneously*, all using the same integration step sizes.

In the algorithm for minimizing the objective function there occur some points (the main iterates) at which both the function and its derivatives are required, while at other points only the function is required. It is essential that the same *function* value should be obtained at a point whether or not derivatives are also required. Hence, regardless of the method used for computing derivatives, the integration step size should be determined by the behavior of the state equations at the point $\boldsymbol{\theta}$ alone. The sensitivity equations, or the state equations at the perturbed points $\boldsymbol{\theta} + \Delta\boldsymbol{\theta}_\alpha$, should have no effect on the integration step size. While this runs a certain risk of getting wrong values of the derivatives, in practice the alternative of computing a non-reproducible objective function has been found to give more trouble.

8-6. Some Difficulties Associated with Dynamic Systems

Solutions of differential equations behave in a variety of ways: some are stable and converge to a steady state; some are unstable and diverge to infinity; others oscillate or enter into limit cycles. What concerns us here is the fact that the nature of the solutions to a given set of equations may change drastically when one changes the values of the parameters. For instance, the solution of $ds/dt + \theta s = 0$ is stable when θ is positive or zero, and unstable when θ is negative. For this reason, we may find it difficult to estimate parameters if the initial guesses or any subsequent iterates give rise to solutions of the wrong type.

In a few cases, we can overcome this problem easily enough. If the system described by $ds/dt + \theta s = 0$ is known from physical considerations to be stable, all we need do is impose the constraint $\theta \geqslant 0$. Again, if a system is described by the set of equations:

$$\begin{aligned} ds_1/dt + h_{11}(\boldsymbol{\theta})s_1 + h_{12}(\boldsymbol{\theta})s_2 &= 0 \\ ds_2/dt + h_{21}(\boldsymbol{\theta})s_1 + h_{22}(\boldsymbol{\theta})s_2 &= 0 \end{aligned} \tag{8-6-1}$$

where $h_{ij}(\boldsymbol{\theta})$ are known functions, then the constraints

$$h_{11}(\boldsymbol{\theta}) > 0, \qquad h_{22}(\boldsymbol{\theta}) > 0, \qquad h_{11}(\boldsymbol{\theta})h_{22}(\boldsymbol{\theta}) - h_{12}(\boldsymbol{\theta})h_{21}(\boldsymbol{\theta}) > 0 \quad (8\text{-}6\text{-}2)$$

guarantee stability by making the matrix of coefficients positive definite. Unfortunately, in most practical situations such conditions turn out to be unwieldy. Besides, unless the equations are of the linear time-invariant type it is difficult to formulate stability conditions which hold at all times. We don't even have any reason to believe that the solutions must be locally stable at all times, since although appearing unstable at one time, they may eventually pass into a stable region.

In addition to unstable solutions in which the state variables increase rapidly beyond bounds, we may be troubled by solutions which are too stable, i.e., in which the state variables rapidly reach steady state values which are independent at least of some of the parameters.

Take, for instance, the system:

$$\begin{aligned} ds_1/dt &= -k_1 s_1 + k_2 s_2, \qquad (s_1(0) = c_1) \\ ds_2/dt &= k_1 s_1 - k_2 s_2, \qquad (s_2(0) = c_2) \end{aligned} \quad (8\text{-}6\text{-}3)$$

the solution of which is:

$$\begin{aligned} s_1 &= [(c_1 - Kc_2)/(1 + K)]\exp[-(k_1 + k_2)t] + K(c_1 + c_2)/(1 + K) \\ s_2 &= [(Kc_2 - c_1)/(1 + K)]\exp[-(k_1 + k_2)t] + (c_1 + c_2)/(1 + K) \end{aligned} \quad (8\text{-}6\text{-}4)$$

where $K \equiv k_2/k_1$. Suppose we have assigned to k_1 and k_2 initial guesses that are much too large, so that exponential terms are already negligible for the smallest t_μ at which measurements of $\mathbf{s}$ are available. Then

$$s_1 = K(c_1 + c_2)/(1 + K), \qquad s_2 = (c_1 + c_2)/(1 + K) \quad (8\text{-}6\text{-}5)$$

Clearly, we have lost all information pertaining to k_1 and k_2 individually, and we can hope to determine only their ratio K. In other words, the values $k_1 = 10{,}000$, $k_2 = 20{,}000$ will fit the data just as well (or just as poorly) as $k_1 = 100{,}000$, $k_2 = 200{,}000$, and the estimation procedure would have no incentive to reduce the values of k_1 and k_2, but only to adjust their ratio.

It seems clear then that we are most likely to avoid both instability and overstability if we start out with very small values of any unknown parameters which are rate coefficients. This gives us the best chance of obtaining solutions whose magnitude remains reasonable throughout the time intervals for which observations are available, and which are sensitive to the values of the parameters. In many cases it pays to place reasonable bounds on the magnitudes of the state variables. Should these be exceeded in the course of an integration, we reject the current value of $\boldsymbol{\theta}$ as infeasible. If we already have a feasible $\boldsymbol{\theta}$ from the previous iteration, then we can interpolate, i.e., return to a value of $\boldsymbol{\theta}$ halfway between the current and previous values. If necessary, this procedure may be repeated several times. If the infeasibility occurs in the course of the first iteration, simply reducing the magnitudes of all parameters by successive halvings often produces feasible values.

Alternatively, we may temporarily assign fictitious observed values zero to the state variables at time t equal to the value at which these variables exceed their bounds, and ignore for the present iteration observations taken at later t. For example, suppose s_1 is observed at $t = 1, 2, \ldots, 10$, and $|s_1| \leqslant 1000$ is the bound. If we integrate the equations for current values of $\boldsymbol{\theta}$ and find that $s_1 = 1000$ at $t = 4.5$, then we act as though we only had the observations at $t = 1, 2, 3, 4$, and in addition we add an "observation" $s_1 = 0$ at $t = 4.5$. It may be profitable to attach a large weight to this last observation in forming the objective function.

Degeneracies of various types may arise when the differential equations are linearly dependent. In the chemical reaction scheme of Eq. (3) for instance, we have $ds_1/dt = -ds_2/dt$, hence $s_1 + s_2 = c_1 + c_2$ remains constant. Suppose all our observations were taken in runs for which the initial conditions c_1 and c_2 always added up to the same value γ. Suppose, further, that the observed variable y is a linear function of s_1 and s_2 with unknown coefficients b_0, b_1, and b_2. Thus

$$\begin{aligned} y &= b_0 + b_1 s_1 + b_2 s_2 = b_0 + b_1(s_1 + s_2) + (b_2 - b_1)s_2 \\ &= b_0 + b_1\gamma + (b_2 - b_1)s_2 = a_0 + a_2 s_2 \end{aligned} \tag{8-6-6}$$

Under these conditions, then, y appears to be a linear function of s_2 (or s_1) alone. Any attempt to determine three coefficients independently will fail unless new observations with different values of $c_1 + c_2$ are made. Additional problems associated with linearly dependent systems are discussed in Section 8-8.

8-7. A Chemical Kinetics Problem

The following example is somewhat artificially concocted, but it serves to illustrate many points.

We consider a heterogeneous catalytic reaction in which a molecule of species A is reversibly transformed into two molecules of species B

$$A \rightleftarrows 2B$$

If it were not for the catalyst, then the rate of the forward reaction $A \rightarrow 2B$ would be proportional to s_1, the concentration of A

$$R_F = k_F s_1$$

The rate of the reverse reaction would be proportional to the square of s_2, the concentration of B

$$R_R = k_R s_2^{\,2}$$

If the reaction reaches a state of equilibrium, no further changes in concentrations occur because $R_F = R_R$; hence

$$k_R/k_F = s_1^{\,E}/(s_2^{\,E})^2$$

where $s_1^{\,E}$ and $s_2^{\,E}$ are the concentrations at equilibirum, and $K \equiv k_R/k_F$ is called an equilibrium constant. It is determined by thermodynamic considerations alone, and is unaffected by the catalyst. The net forward reaction rate is given by

$$R = R_F - R_R = k_F s_1 - k_R s_2^{\,2} = k_F(s_1 - K s_2^{\,2})$$

The presence of the catalyst affects the value of R. The nature of the effect depends on the mechanism of the reaction. We shall adopt the following expression for the rate of the catalyzed reaction

$$R = k_F(s_1 - K s_2^{\,2})/(1 + M s_1)^2$$

The three constants k_F, K, and M are functions of the temperature T, usually assumed to be of the form:

$$k_F = \theta_1 \exp(-\theta_2/T)$$
$$K = \alpha \exp(-\beta/T)$$
$$M = \theta_3 \exp(-\theta_4/T)$$

We assume that K has been determined accurately from thermodynamic data as

$$K = \exp(-1000/T)$$

Species A is disappearing at a rate equal to R, and B appears at a rate of $2R$. Hence the differential equations governing the system are:

$$ds_1/dt = h_1(\mathbf{s}, \boldsymbol{\theta}) = -\theta_1 \exp(-\theta_2/T)(s_1 - e^{-1000/T} s_2^{\,2})/(1 + \theta_3 \exp(-\theta_4/T) s_1)^2$$
$$ds_2/dt = h_2(\mathbf{s}, \boldsymbol{\theta}) = 2\theta_1 \exp(-\theta_2/T)(s_1 - e^{-1000/T} s_2^{\,2})/(1 + \theta_3 \exp(-\theta_4/T) s_1)^2 \qquad (8\text{-}7\text{-}1)$$

To estimate θ_1, θ_2, θ_3, and θ_4 we conduct three runs, at temperatures $T = 200°$, $400°$, and $600°$. The second run is started with pure A, and the third run with pure B. Otherwise, the initial concentrations are known only approximately. In the first run

$$s_1(0) = \theta_5 = 1 \pm 0.05, \qquad s_2(0) = \theta_6 = 1 \pm 0.05$$

In the second run

$$s_1(0) = \theta_7 = 1 \pm 0.05, \qquad s_2(0) = 0$$

In the third run

$$s_1(0) = 0, \qquad s_2(0) = \theta_8 = 1 \pm 0.05$$

In the course of each run, samples are widthdrawn at ten different times (including initially, at $t = 0$), and analyzed in a densitometer.

The instrument's readings are linear in the concentrations of A and B, i.e.,

$$y = 1 + \theta_9 s_1 + \theta_{10} s_2 \qquad (8\text{-}7\text{-}2)$$

The coefficients θ_9 and θ_{10} are known approximately from past experience

$$\theta_9 = 1 \pm 0.05, \qquad \theta_{10} = 2 \pm 0.05$$

The data are given in Table 8-1.

Table 8-1
Data for Kinetics Problem

Run	T	Initial conditions		μ	t_μ	y_μ
1	200	θ_5	θ_6	1	0	3.988
				2	10	4.073
				3	20	4.153
				4	30	4.231
				5	40	4.309
				6	50	4.376
				7	60	4.457
				8	70	4.522
				9	80	4.615
				10	90	4.667
2	400	θ_7	0	11	0	1.997
				12	2	2.149
				13	4	2.320
				14	6	2.465
				15	8	2.611
				16	10	2.754
				17	12	2.896
				18	14	3.034
				19	16	3.166
				20	18	3.278
3	600	0	θ_8	21	0	3.012
				22	0.5	2.956
				23	1	2.926
				24	1.5	2.877
				25	2.	2.853
				26	2.5	2.823
				27	3	2.800
				28	3.5	2.776
				29	4	2.767
				30	4.5	2.760

Along with θ_1, θ_2, θ_3, θ_4, we must also estimate the values of the unknown initial concentrations θ_5, θ_6, θ_7, and θ_8, and of the unknown coefficients θ_9 and θ_{10}. To account for our partial knowledge of the latter values, we assign to the six last parameters independent normal prior distributions with means [1, 1, 1, 1, 1, 2] and standard deviations 0.05.

If we do not know the standard deviation of the measurement errors, we are led to the objective function

$$\Phi(\boldsymbol\theta) = (10/2) \log \sum_{\mu=1}^{30} e_\mu^2(\boldsymbol\theta)$$

$$+ (1/2 \times 0.05^2)[(\theta_5 - 1)^2 + (\theta_6 - 1)^2 + (\theta_7 - 1)^2 + (\theta_8 - 1)^2$$

$$+ (\theta_9 - 1)^2 + (\theta_{10} - 2)^2]$$

where

$$e_\mu(\boldsymbol\theta) = y_\mu - 1 - \theta_9 s_1(t_\mu, \boldsymbol\theta') - \theta_{10} s_2(t_\mu, \boldsymbol\theta')$$

Here $\boldsymbol\theta'$ denotes the vector consisting of the first eight elements of $\boldsymbol\theta$. The functions $\mathbf{s}(t_\mu, \boldsymbol\theta')$ are to be determined by integrating Eq. (1) from $t = 0$ to $t = t_\mu$, using initial conditions (θ_5, θ_6), $(\theta_7, 0)$, or $(0, \theta_8)$ depending on whether the μth experiment belongs to run 1, 2, or 3.

We shall estimate $\boldsymbol\theta$ by means of the method of sensitivity equations. That is, along with the two Eqs. (1) we integrate at each iteration the set of sixteen differential equations for the functions $\partial \mathbf{s}/\partial \boldsymbol\theta'$. The initial conditions for these equations are

$$\partial \mathbf{s}/\partial \boldsymbol\theta')_{t=0} =
\begin{array}{c}
\begin{array}{cccccccc} \theta_1 & \theta_2 & \theta_3 & \theta_4 & \theta_5 & \theta_6 & \theta_7 & \theta_8 \end{array} \\
\begin{array}{c} s_1 \\ s_2 \end{array}\begin{bmatrix} 0 & 0 & 0 & 0 & 1 & 0 & 0 & 0 \\ 0 & 0 & 0 & 0 & 0 & 1 & 0 & 0 \end{bmatrix} \text{(run 1)} \\[10pt]
\begin{array}{c} s_1 \\ s_2 \end{array}\begin{bmatrix} 0 & 0 & 0 & 0 & 0 & 0 & 1 & 0 \\ 0 & 0 & 0 & 0 & 0 & 0 & 0 & 0 \end{bmatrix} \text{(run 2)} \\[10pt]
\begin{array}{c} s_1 \\ s_2 \end{array}\begin{bmatrix} 0 & 0 & 0 & 0 & 0 & 0 & 0 & 0 \\ 0 & 0 & 0 & 0 & 0 & 0 & 0 & 1 \end{bmatrix} \text{(run 3)}
\end{array}$$

To form the differential Eqs. (8-4-6) we need the matrices $\partial \mathbf{h}/\partial \boldsymbol\theta'$ and $\partial \mathbf{h}/\partial \mathbf{s}$. The first row of each matrix is given by:

$$\frac{\partial h_1}{\partial \boldsymbol\theta'} = \left[\frac{h_1}{\theta_1}, -\frac{h_1}{T}, -\frac{2h_1 \exp(-\theta_4/T)s_1}{1 + \theta_3 \exp(-\theta_4/T)s_1}, \right.$$

$$\left. \frac{2h_1\theta_3 \exp(-\theta_4/T)s_1}{T(1 + \theta_3 \exp(-\theta_4/T)s_1)}, 0, 0, 0, 0 \right]$$

$$\frac{\partial h_1}{\partial \mathbf{s}} = \left[-\frac{\theta_1 \exp(-\theta_2/T) - 2h_1\theta_3 \exp(-\theta_4/T)}{1 + \theta_3 \exp(-\theta_4/T)s_1}, \frac{2\theta_1 \exp[-(\theta_2 + 1000)/T]s_2}{1 + \theta_3 \exp(-\theta_4/T)s_1} \right]$$

To obtain the second row in each case we multiply the first row by -2. The reader may write out in full the differential equations for the sixteen functions $\partial\mathbf{h}/\partial\boldsymbol{\theta}'$.

We use the initial guesses [2, 500, 0.5, 50] for the first four parameters. The guesses [1, 1, 1, 1, 1, 2] for the remaining parameters are obvious.

In Table 8-2 we give the results of integrating the equations in **s** and

Table 8-2
Integrated First Run Data[a]. Using Initial Guess θ

	$t=0$		$t=10$	
	s_1	s_2	s_1	s_2
$\mathbf{s}$	1	1	0.3623495	2.275272
$\partial\mathbf{s}/\partial\theta_1$	0	0	−0.2064785	0.4129581
$\partial\mathbf{s}/\partial\theta_2$	0	0	0.002064790	−0.004129570
$\partial\mathbf{s}/\partial\theta_3$	0	0	0.3270462	−0.6540943
$\partial\mathbf{s}/\partial\theta_4$	0	0	−0.0008176151	0.001635234
$\partial\mathbf{s}/\partial\theta_5$	1	0	0.5249249	0.950146
$\partial\mathbf{s}/\partial\theta_6$	0	1	0.01804298	0.9639123

[a]Accurate to About Five Decimal Places.

$\partial\mathbf{s}/\partial\boldsymbol{\theta}'$ for the first run from $t = 0$ to $t = 10$, using the initial guess values for $\boldsymbol{\theta}$. The values of $\partial\mathbf{s}/\partial\theta_7$ and $\partial\mathbf{s}/\partial\theta_8$ were omitted, being all zero for the first run. From these values it is easy to compute the residuals and their derivatives for the first two observations. The residuals can be found from Eq. (2).

First, at $\mu = 1$ $(t = 0)$:

$$e_1 = y_1 - (1 + \theta_9 s_1 + \theta_{10} s_2) = 3.988 - 1 - 1 \times 1 - 2 \times 1 = -0.012$$

$$\partial f_1/\partial\boldsymbol{\theta} = -\partial e_1/\partial\boldsymbol{\theta} = \begin{bmatrix} \theta_9\,\partial s_1/\partial\boldsymbol{\theta}' + \theta_{10}\,\partial s_2/\partial\boldsymbol{\theta}' \\ s_1 \\ s_2 \end{bmatrix} = \begin{bmatrix} 0 \\ 0 \\ 0 \\ 0 \\ 1 \\ 2 \\ 0 \\ 0 \\ 1 \\ 1 \end{bmatrix}$$

and, at $\mu = 2$ ($t = 10$):

$$e_2 = 4.073 - 1 - 1 \times 0.3623495 - 2 \times 2.275272 = -0.83892$$

$$\partial f_2/\partial \boldsymbol{\theta} = \begin{bmatrix} 1 \times (-0.2064785) + 2 \times (0.4129581) \\ 1 \times (0.002064790) + 2 \times (-0.004129570) \\ \cdots \\ \cdots \\ \cdots \\ \cdots \\ 0 \\ 0 \\ 0.3623495 \\ 2.275272 \end{bmatrix} = \begin{bmatrix} 0.6194376 \\ -0.006194349 \\ -0.9811422 \\ 0.002452854 \\ 2.45216 \\ 1.945867 \\ 0 \\ 0 \\ 0.3623495 \\ 2.275272 \end{bmatrix}$$

In similar manner, we can compute e_μ and $\partial f_\mu/\partial \boldsymbol{\theta}$ for $\mu = 3, 4, \ldots, 30$. From these, $\mathbf{q}$ and $\mathbf{N}$ can be computed and the Gauss method applied. The process does not converge unless penalty functions are used to maintain all parameters (or at least the first four) strictly positive.

The solution is

$$\boldsymbol{\theta}^* = \begin{bmatrix} 1.39266 \pm 0.20891 \\ 1140.01 \pm 75.16 \\ 1.82052 \pm 0.80081 \\ 366.524 \pm 194.213 \\ 1.00601 \pm 0.04502 \\ 0.998853 \pm 0.02988 \\ 0.986829 \pm 0.02606 \\ 1.01898 \pm 0.01677 \\ 1.01086 \pm 0.02620 \\ 1.97541 \pm 0.03198 \end{bmatrix}$$

One would judge all parameters well-determined except for θ_3 and θ_4. The amount of information concerning the values of θ_5 through θ_{10} that was gained from the data can be gaged by comparing the present standard deviations of these parameters with the prior standard deviations of 0.05. There is substantial improvement in all cases but θ_5.

8-8. Linearly Dependent Equations

We return to the kinetics problem of the previous section, but now the initial conditions for each run are known precisely (see Table 8-3). This time the concentrations of A and B are measured directly in each experiment, and appear under the headings $y_{\mu 1}$ and $y_{\mu 2}$ respectively in Table 8-4. We recall

Table 8-3
Run Data for Kinetics Problem

Run	Initial conditions $s_1(0)$	$s_2(0)$	$\alpha = 2s_1(0) + s_2(0)$	T
1	1.00830	0.99662	3.01322	200°
2	0.98862	0	1.97724	400°
3	0	1.01731	1.01731	600°

Table 8-4
Data for Kinetics Problem

Run	μ	t	$y_{\mu 1}$	$y_{\mu 2}$	$\bar{s}_{\mu 1} = \bar{y}_{\mu 1}$	$\bar{y}_{\mu 2}$
1	1	10	0.98040	1.05134	0.980832	1.051556
	2	20	0.95262	1.10796	0.952628	1.107964
	3	30	0.92703	1.16017	0.926626	1.159968
	4	40	0.90120	1.21002	0.901520	1.210180
	5	50	0.87706	1.26056	0.876476	1.260268
	6	60	0.84883	1.31534	0.848918	1.315384
	7	70	0.82480	1.36299	0.825052	1.363116
	8	80	0.80163	1.41035	0.801474	1.410272
	9	90	0.77726	1.45822	0.777452	1.458316
2	10	2	0.93445	0.10857	0.934358	0.108524
	11	4	0.88192	0.21395	0.881700	0.213840
	12	6	0.82997	0.31716	0.830026	0.317188
	13	8	0.78015	0.41627	0.780418	0.416404
	14	10	0.73047	0.51561	0.730746	0.515748
	15	12	0.68271	0.61219	0.682562	0.612116
	16	14	0.63816	0.70085	0.638188	0.700864
	17	16	0.59445	0.78925	0.594086	0.789068
	18	18	0.55375	0.86976	0.553742	0.869756
3	19	0.5	0.01688	0.98263	0.017248	0.982814
	20	1	0.03331	0.94991	0.033622	0.950066
	21	1.5	0.04726	0.92359	0.046940	0.923430
	22	2	0.05604	0.90485	0.056192	0.904926
	23	2.5	0.06330	0.89059	0.063348	0.890614
	24	3	0.07073	0.87613	0.070618	0.876074
	25	3.5	0.07811	0.86138	0.077994	0.861322
	26	4	0.08214	0.85298	0.082160	0.852990
	27	4.5	0.08782	0.84250	0.087488	0.842334

that for each molecule of A that disappears, two molecules of B are created. Hence the quantity $\alpha = 2s_1(t) + s_2(t)$ remains constant throughout any run. We can compute its value from the known initial conditions $\alpha = 2s_1(0) + s_2(0)$ (see Table 8-3), and it does not depend on what values of $\boldsymbol{\theta}$ we choose.

Suppose $\hat{\boldsymbol{\theta}}$ is the true value of $\boldsymbol{\theta}$. The observed concentrations are given by

$$y_{\mu 1} = s_1(t_\mu, \hat{\boldsymbol{\theta}}) + \varepsilon_{\mu 1}, \qquad y_{\mu 2} = s_2(t_\mu, \hat{\boldsymbol{\theta}}) + \varepsilon_{\mu 2}$$

where $\varepsilon_{\mu 1}$ and $\varepsilon_{\mu 2}$ are errors, hopefully small. Hence

$$2y_{\mu 1} + y_{\mu 2} = 2s_1 + s_2 + 2\varepsilon_{\mu 1} + \varepsilon_{\mu 2} = \alpha + 2\varepsilon_{\mu 1} + \varepsilon_{\mu 2} \tag{8-8-1}$$

But also, for any trial value $\boldsymbol{\theta}$

$$2s_1(t_\mu, \boldsymbol{\theta}) + s_2(t_\mu, \boldsymbol{\theta}) = \alpha \tag{8-8-2}$$

Subtracting Eq. (2) from Eq. (1) and remembering that

$$\mathbf{e}_\mu(\boldsymbol{\theta}) \equiv \mathbf{y}_\mu(\boldsymbol{\theta}) - \mathbf{s}(t_\mu, \boldsymbol{\theta})$$

we obtain

$$2e_{\mu 1}(\boldsymbol{\theta}) + e_{\mu 2}(\boldsymbol{\theta}) = 2\varepsilon_{\mu 1} + \varepsilon_{\mu 2} \tag{8-8-3}$$

Unless $\boldsymbol{\theta}$ is close to $\hat{\boldsymbol{\theta}}$, the residuals $\mathbf{e}_\mu(\boldsymbol{\theta})$ are large compared to the errors $\boldsymbol{\varepsilon}_\mu$, hence Eq. (3) takes the approximate form

$$e_{\mu 2}(\boldsymbol{\theta}) \approx -2e_{\mu 1}(\boldsymbol{\theta}) \qquad (\boldsymbol{\theta} \neq \hat{\boldsymbol{\theta}})$$

From this it follows that the moment matrix $\mathbf{M}(\boldsymbol{\theta})$ is nearly singular

$$\mathbf{M}(\boldsymbol{\theta}) \approx \begin{bmatrix} \sum_\mu e_{\mu 1}^2 & -2\sum_\mu e_{\mu 1}^2 \\ -2\sum_\mu e_{\mu 1}^2 & 4\sum_\mu e_{\mu 1}^2 \end{bmatrix} \qquad (\boldsymbol{\theta} \neq \hat{\boldsymbol{\theta}})$$

Indeed, an attempt to estimate $\boldsymbol{\theta} = [\theta_1, \theta_2, \theta_3, \theta_4]^{\mathrm{T}}$ by minimizing $\log \det \mathbf{M}(\boldsymbol{\theta})$ fails when one starts with $\boldsymbol{\theta}_1 = [2, 500, 0.5, 50]^{\mathrm{T}}$. One simply finds $\det \mathbf{M}(\boldsymbol{\theta}_1) = 0$, and no progress can be made. However, using the results of Problem 8, Section 4-21, let us take s_1 as our sole state variable (if we know s_1 we can always compute $s_2 = \alpha - 2s_1$). If we define $y_1' \equiv y_1$ and $y_2' \equiv \alpha - y_2$, then we have the representation

$$y_1' = s_1, \qquad y_2' = 2s_1$$

or

$$\mathbf{y} = \mathbf{B}s_1 \tag{8-8-4}$$

where $\mathbf{B} = [1, 2]^{\mathrm{T}}$. Let us assume, further, that the measurement errors of $y_{\mu 1}$ and $y_{\mu 2}$ are independent and have the same standard deviation σ. Since α is a known constant, the error in $y_{\mu 2}'$ likewise has standard deviation σ.

Hence, at each experiment we can obtain the least squares estimate of $\bar{\mathbf{s}}_\mu$ on the basis of the measured $y_{\mu 1}$ and $y_{\mu 2}$

$$\bar{\mathbf{s}}_\mu = (\mathbf{B}^T\mathbf{B})^{-1}\mathbf{B}^T\mathbf{y}_\mu \tag{8-8-5}$$

or

$$\bar{s}_{\mu 1} = \left([1, 2]\begin{bmatrix}1\\2\end{bmatrix}\right)^{-1}[1, 2]\begin{bmatrix}y'_{\mu 1}\\y'_{\mu 2}\end{bmatrix}$$

$$= \tfrac{1}{5}y'_{\mu 1} + \tfrac{2}{5}y'_{\mu 2} = \tfrac{1}{5}y_{\mu 1} + \tfrac{2}{5}(\alpha - y_{\mu 2})$$

The values of $\bar{s}_{\mu 1}$ appear in Table 8-4, as do the values of $\bar{\mathbf{y}}_\mu$ computed according to

$$\bar{y}_{\mu 1} = \bar{s}_{\mu 1}, \qquad \bar{y}_{\mu 2} = \alpha - 2\bar{s}_{\mu 1} \tag{8-8-6}$$

The standard deviation of the measurement errors may be estimated from

$$\sigma = \left\{[1/27(2-1)]\sum_{\mu=1}^{27}[(y_{\mu 1} - \bar{y}_{\mu 1})^2 + (y_{\mu 2} - \bar{y}_{\mu 2})^2]\right\}^{1/2} = 0.0002876$$

The values $\bar{s}_{\mu 1}$ can now be used as "data" for estimating $\boldsymbol{\theta}$. We do this by minimizing

$$\Phi(\boldsymbol{\theta}) = \sum_{\mu=1}^{27}[\bar{s}_{\mu 1} - s_1(t_\mu, \boldsymbol{\theta})]^2$$

We employ the Gauss method, and use penalty functions to keep all θ_α positive. Starting from $\boldsymbol{\theta}_1 = [2, 500, 0.5, 50]^T$ we arrive in 19 iterations at the solution

$$\boldsymbol{\theta}^* = \begin{bmatrix} 1.48393 \pm 0.0558515 \\ 1175.56 \pm 19.5129 \\ 2.29692 \pm 0.344311 \\ 471.100 \pm 66.5655 \end{bmatrix} \tag{8-8-7}$$

The estimates of all parameters are fairly well-determined. The minimum sum of squares is

$$\Phi(\boldsymbol{\theta}^*) = 0.1353722 \times 10^{-4}$$

The estimate Eq. (7) should be sufficiently close to $\hat{\boldsymbol{\theta}}$ so as to make $\mathbf{M}(\boldsymbol{\theta}^*)$ nonsingular. Hence we can use $\boldsymbol{\theta}^*$ as the initial guess for minimizing

$$\Phi(\boldsymbol{\theta}) = (27/2)\log\det\mathbf{M}(\boldsymbol{\theta}) \tag{8-8-8}$$

Indeed, three iterations bring us to

$$\boldsymbol{\theta}^{**} = \begin{bmatrix} 1.48555 \pm 0.0395848 \\ 1176.30 \pm 13.5234 \\ 2.31028 \pm 0.241886 \\ 473.880 \pm 46.4845 \end{bmatrix}$$

8-9. Problems

1. State variables are usually computed by solving finite difference equations which are approximations to the differential equations. It is suggested by Kelley and Denham (1969) that one ought to obtain exact derivatives of the approximate **s** (by differentiating the difference equations with respect to $\boldsymbol{\theta}$), rather than approximate derivatives of the exact **s** (by solving the sensitivity equations approximately with a finite difference method). Show that with the Runge-Kutta method both approaches lead to precisely the same results. Illustrate with the models of Eqs. (8-2-4) and (8-4-8).

2. Using the method suggested at the end of Section 8-3, solve the following two point boundary value problem:

Find $s_1(0)$ for the system

$$\dot{s}_1 = s_1 - 2ts_1/(1 + s_2), \qquad \dot{s}_2 = 2\log[s_1(1 + s_2)], \quad s_2(0) = 0, \quad s_2(1) = 1$$

Use the initial guess $s_1(0) = 1.5$.

3. Suppose the observed variables are measured continuously, so that a record exists of $\mathbf{y}(t)$ $(0 \leqslant t \leqslant T)$. Let the objective function be

$$\Phi(\boldsymbol{\theta}) = \int_0^T \mathbf{e}^T(t, \boldsymbol{\theta})\mathbf{Q}\mathbf{e}(t, \boldsymbol{\theta})\, dt$$

where $\mathbf{Q}$ is a given positive definite matrix. Derive the sensitivity equation, i.e., a differential equation for $\partial\Phi/\partial\boldsymbol{\theta}$. Apply a variable metric method to finding the minimum of Φ for an example given by Bellman, Kagiwada, *et al.* (1964): There are two parameters and one state variable

$$\dot{s} = -s + \theta_1 s^3, \qquad s(0) = \theta_2 .$$

The observed data are given as

$$y(t) = \tilde{s}(t) + 0.5\cos 60t$$

where $\tilde{s}(t)$ is the solution of the differential equation with $\theta_1 = 1/30$ and $\theta_2 = 1$. Use $T = 5$ and $\mathbf{Q} = \mathbf{I}$, i.e.,

$$\Phi(\boldsymbol{\theta}) = \int_0^5 [y(t) - s(t)]^2\, dt$$

4. The errors in measurements taken in the course of one physical run are generally not independent. Using the theory of power spectra, one can obtain expressions for the covariance between the errors in different measurements, provided the differential equations are linear. Specifically, suppose

$$ds/dt = \theta s + \varepsilon(t), \qquad s(0) = s_0$$

where $\varepsilon(t)$ is a random noise with given power spectrum. Let $\hat{s}$ satisfy

$$d\hat{s}/dt = \theta\hat{s}, \qquad \hat{s}(0) = s_0$$

and let $u(t) = s(t) - \hat{s}(t)$ be the error function. Derive expressions for the power spectrum of $u(t)$, and for the autocovariance function $V(t, \tau) = E[u(t)u(\tau)]$. Generalize to the case of a vector of state variables, i.e., $d\mathbf{s}/dt = \mathbf{A}(\boldsymbol{\theta})\mathbf{s} + \boldsymbol{\varepsilon}(t)$.

Chapter

IX

Some Special Problems

9-1. Missing Observations

It is not uncommon to find that one is missing one or more data items, i.e., elements of the matrix **W**. We distinguish between two situations, which are illustrated by the following cases:

(a) A single equation model, with the objective function $\sum_{\mu=1}^{n} e_\mu^2(\boldsymbol{\theta})$. The value of $x_{1,1}$ is missing. It is clear that we cannot do better than determine $\boldsymbol{\theta}^*$ so as to minimize $\sum_{\mu=2}^{n} e_\mu^2(\boldsymbol{\theta})$. The entire first experiment contributes nothing to the estimation of $\boldsymbol{\theta}$, and may be dropped from the objective function.

Another example in which a term may be dropped occurs when $y_{1,1}$ is missing and the objective function is

$$\sum_{\mu=1}^{n} \sum_{a=1}^{m} b_{\mu a}[y_{\mu a} - f_a(\mathbf{x}_\mu, \boldsymbol{\theta})]^2$$

(b) On the other hand, with $y_{1,1}$ missing, suppose the objective function has the form

$$\sum_{\mu=1}^{n} \sum_{a,b=1}^{m} B_{\mu ab}[y_{\mu a} - f_a(\mathbf{x}_\mu, \boldsymbol{\theta})][y_{\mu b} - f_b(\mathbf{x}_\mu, \boldsymbol{\theta})]$$

Now $y_{1,1}$ appears in several terms, which cannot all be dropped.

Similarly, with $x_{1,1}$ missing, let the model consist of two equations. The values of $y_{1,1}$ and $y_{1,2}$ jointly do contain some information on $\boldsymbol{\theta}$, since we cannot in general solve the equations

$$y_{1,1} - f_1(x_{1,1}, \ldots, \boldsymbol{\theta}^*) = 0, \qquad y_{1,2} - f_2(x_{1,1}, \ldots, \boldsymbol{\theta}^*) = 0 \quad (9\text{-}1\text{-}1)$$

simultaneously for $x_{1,1}$. The first experiment residuals should not be dropped from the objective function. Instead, the missing value $x_{1,1}$ or $y_{1,1}$ can be regarded as an additional unknown parameter, whose value is to be determined together with $\boldsymbol{\theta}$ so as to optimize the objective function.

When several noneliminable items are missing, they can all be treated as unknown parameters. However, from a practical computational point of view only a few such parameters can be handled in this manner.

The following is a systematic approach:

1. Write down the objective function, with all missing data items represented as unknown parameters.
2. Differentiate this expression with repsect to all the missing-data parameters, and equate the derivatives to zero.
3. Solve the equations thus formed for the missing data parameters.
4. Substitute the solutions in the objective function for those parameters where the substitution results in a simplification of the expression. Retain other unknown parameters in the objective function.

Examples:

1. $\Phi(\boldsymbol{\theta}) = \sum_{\mu=1}^{n} e_\mu{}^2(\boldsymbol{\theta}), \qquad x_{1,1}$ unknown.

$$\partial\Phi/\partial x_{1,1} = -2e_1(\boldsymbol{\theta})\,\partial f_1/\partial x_{1,1} = 0$$

$$\therefore\ e_1(\boldsymbol{\theta}) = 0$$

Substituting in $\Phi(\boldsymbol{\theta})$, we find $\Phi(\boldsymbol{\theta}) = \sum_{\mu=2}^{n} e_\mu{}^2(\boldsymbol{\theta})$, i.e., we drop the first term.

2. $\Phi(\boldsymbol{\theta}) = \sum_{\mu=1}^{n} \sum_{a,b=1}^{m} B_{\mu ab}\, e_{\mu a}\, e_{\mu b}, \qquad y_{1,1}$ missing.

$$\partial\Phi/\partial y_{1,1} = 2\sum_{a=1}^{m} B_{1a,1}\, e_{1,a} = 0$$

$$\therefore\ e_{1,1} = -B_{11,1}^{-1} \sum_{a=2}^{m} B_{1a,1} e_{1,a}$$

Substitution of this expression in $\Phi(\boldsymbol{\theta})$ only complicates matters, so we may as well retain $y_{1,1}$ as an unknown parameter. Incidentally, the same result would be obtained if $e_{1,1}$ is made an unknown parameter, and then $y_{1,1}$ can be computed from $y_{1,1} = e_{1,1}^* + f_1(\mathbf{x}_1, \boldsymbol{\theta}^*)$, where $e_{1,1}^*$ and $\boldsymbol{\theta}^*$ are the estimated values.

A survey of the literature on the problem of missing observations is given by Afifi and Elashoff (1966), but most of the reported results pertain only to multiple linear regression.

Numerical illustrations appear in Sections 9-6–9-7.

9-2. Inhomogeneous Covariance

Most of the estimation formulas were derived on the assumption that the covariance matrices $\mathbf{V}_\mu$ ($\mu = 1, 2, \ldots, n$) of the errors in the μth experiment were all equal to a fixed (though possibly unknown) matrix $\mathbf{V}$. The modifications required when the $\mathbf{V}_\mu$ differ from each other are trivial, provided the manner of the variation is known. If the $\mathbf{V}_\mu$ vary in an entirely unknown manner, nothing much can be done; we simply cannot estimate a variance from a single observation.

The following three cases can be treated easily:

(a) Suppose the following holds

$$\mathbf{V}_\mu = \mathbf{A}_\mu \mathbf{V} \mathbf{A}_\mu^{\mathrm{T}} \tag{9-2-1}$$

where $\mathbf{V}$ is an $m \times m$ positive definite known or unknown matrix, and the $\mathbf{A}_\mu$ are known $m \times m$ nonsingular matrices. This includes the case where the $\mathbf{V}_\mu$ are known matrices, since then $\mathbf{V} = \mathbf{I}$ and there exist $\mathbf{A}_\mu$ such that $\mathbf{V}_\mu = \mathbf{A}_\mu \mathbf{A}_\mu^{\mathrm{T}}$. In the single equation case, Eq. (1) amounts to

$$\sigma_\mu^2 = a_\mu^2 \sigma^2 \tag{9-2-2}$$

where the a_μ are known constants. For a normal distribution, the objective function takes the form

$$\Phi(\boldsymbol{\theta}, \mathbf{V}) = \sum_{\mu=1}^{n} \log \det \mathbf{A}_\mu + n/2 \log \det \mathbf{V} + \tfrac{1}{2} \sum_{\mu=1}^{n} \mathbf{e}_\mu^{\mathrm{T}} (\mathbf{A}_\mu^{\mathrm{T}})^{-1} \mathbf{V}^{-1} \mathbf{A}_\mu^{-1} \mathbf{e}_\mu \tag{9-2-3}$$

We may drop the first term on the right-hand side, which is a constant.

Let us redefine our model equations so that instead of $\mathbf{y}_\mu = \mathbf{f}(\mathbf{x}_\mu, \boldsymbol{\theta})$ we write $\mathbf{A}_\mu^{-1} \mathbf{y}_\mu = \mathbf{A}_\mu^{-1} \mathbf{f}(\mathbf{x}_\mu, \boldsymbol{\theta})$. We obtain new residuals

$$\tilde{\mathbf{e}}_\mu \equiv \mathbf{A}_\mu^{-1} \mathbf{e}_\mu = \mathbf{A}_\mu^{-1} \mathbf{y}_\mu - \mathbf{A}_\mu^{-1} \mathbf{f}(\mathbf{x}_\mu, \boldsymbol{\theta}) \tag{9-2-4}$$

The objective function now becomes

$$\Phi(\boldsymbol{\theta}, \mathbf{V}) = (n/2) \log \det \mathbf{V} + \tfrac{1}{2} \sum_{\mu=1}^{n} \tilde{\mathbf{e}}_\mu^{\mathrm{T}} \mathbf{V}^{-1} \tilde{\mathbf{e}}_\mu \tag{9-2-5}$$

which is identical to the expressions derived in Chapter IV.

Example Suppose, in a single-equation model, the standard deviation is proportional to the magnitude of the measurement, that is

$$V_\mu = y_\mu \sigma^2 y_\mu = \sigma^2 y_\mu^2 \tag{9-2-6}$$

The redefined residuals are

$$\tilde{e}_\mu = (1/y_\mu)[y_\mu - f(\mathbf{x}_\mu, \boldsymbol{\theta})] = 1 - (1/y_\mu) f(\mathbf{x}_\mu, \boldsymbol{\theta}) \tag{9-2-7}$$

These have variance σ^2. If instead the errors are assumed proportional to the true values of the measured variables, we should use

$$\tilde{e}_\mu = y_\mu / f(\mathbf{x}_\mu, \boldsymbol{\theta}) - 1 \tag{9-2-8}$$

Eq. (7) is easier to deal with, and the error committed in using it in place of Eq. (8) is likely to be small.

(b) Suppose we know that experiments $\mu = 1, 2, \ldots, n_1$ have the unknown covariance matrix $\mathbf{V}_1$, experiments $\mu = n_1 + 1, n_1 + 2, \ldots, n_1 + n_2$ the matrix $\mathbf{V}_2$, and so on. The objective function has the form

$$\begin{aligned}\Phi(\boldsymbol{\theta}, \mathbf{V}_1, \mathbf{V}_2, \ldots) = (n_1/2) \log \det \mathbf{V}_1 + (n_2/2) \log \det \mathbf{V}_2 + \cdots \\ + \tfrac{1}{2} \operatorname{Tr}(\mathbf{V}_1^{-1}\mathbf{M}_1) + \tfrac{1}{2} \operatorname{Tr}(\mathbf{V}_2^{-2}\mathbf{M}_2) + \cdots\end{aligned} \tag{9-2-9}$$

where $\mathbf{M}_i$ is the moment matrix of the residuals in the experiments with covariance $\mathbf{V}_i$. Proceeding as in Section 4-9, we differentiate with respect to $\mathbf{V}_i$ and obtain eventually

$$\tilde{\mathbf{V}}_i = (1/n_i)\mathbf{M}_i(\boldsymbol{\theta}) \tag{9-2-10}$$

and the concentrated objective function becomes

$$\Phi(\boldsymbol{\theta}) = \tfrac{1}{2} \sum_i n_i \log \det \mathbf{M}_i(\boldsymbol{\theta}) \tag{9-2-11}$$

Should there be some experiments with known covariances, the objective function would be

$$\Phi(\boldsymbol{\theta}) = \tfrac{1}{2} \sum_i n_i \log \det \mathbf{M}_i(\boldsymbol{\theta}) + \sum_\mu{}' \mathbf{e}_\mu^{\mathrm{T}}(\boldsymbol{\theta})\mathbf{V}_\mu^{-1}\mathbf{e}_\mu(\boldsymbol{\theta}) \tag{9-2-12}$$

where the summation $\sum_\mu'$ is extended only over those experiments.

The minimum number of required experiments given in Section 4-12 now applies to each n_i separately, since Eq. (11) cannot be used unless none of the $\mathbf{M}_i$ are singular. Hence we must usually have

$$n_i \geqslant \max(l + 1, m) \qquad (i = 1, 2, \ldots) \tag{9-2-13}$$

Another problem that can be solved by means of the maximum likelihood method is one in which the covariance matrix varies regularly as some function of the independent variables. These functions may depend on unknown parameters, which can be estimated. The case where both the model equation and the standard deviation of the erorrs are linear functions of the independent variables is treated by Rutemiller and Bowers (1968).

(c) Known Serial Correlations. Correlations between errors in different experiments are called *serial correlations*. Suppose, in a single-equation model, the covariance matrix of all errors is given by $\mathbf{R}$, where

$$R_{\mu\eta} = E(\varepsilon_\mu \varepsilon_\eta) \tag{9-2-14}$$

The likelihood takes the form

$$\log L(\boldsymbol{\theta}) = -(n/2)\, log\, 2\pi - \tfrac{1}{2} \log \det \mathbf{R} - \tfrac{1}{2} \sum_{\mu,\eta=1}^{n} [\mathbf{R}^{-1}]_{\mu\eta}\, e_\mu(\boldsymbol{\theta}) e_\eta(\boldsymbol{\theta}) \qquad (9\text{-}2\text{-}15)$$

If $\mathbf{R}$ is known, we can find a matrix $\mathbf{S}$ such that $\mathbf{S}\mathbf{S}^{\mathrm{T}} = \mathbf{R}$. Defining a new set of residuals

$$\tilde{e}_\eta(\boldsymbol{\theta}) \equiv \sum_{\mu=1}^{n} [\mathbf{S}^{-1}]_{\eta\mu}\, e_\mu(\boldsymbol{\theta}) \qquad (9\text{-}2\text{-}16)$$

we find that maximizing Eq. (15) is equivalent to minimizing the sum of squares of the $\tilde{e}_\eta(\boldsymbol{\theta})$.

The estimation of $\mathbf{R}$ when the serial correlations are unknown is relatively difficult, and we shall not consider this problem here. We observe, however, that residuals almost always show serial correlations even when the errors possess none. For instance, in the case of the linear model Eq. (4-4-2) with $\mathbf{V} = \sigma^2 \mathbf{I}$, we find that the covariance matrix of the residuals is given by $\mathbf{V}_r = \sigma^2[\mathbf{I} - \mathbf{B}(\mathbf{B}^{\mathrm{T}}\mathbf{B})^{-1}\mathbf{B}^{\mathrm{T}}]$. Thus while $\mathbf{V}$ is diagonal (no serial correlations), $\mathbf{V}_r$ is nondiagonal. Furthermore, $\mathbf{V}_r$ is singular since clearly $\mathbf{V}_r \mathbf{B} = \mathbf{0}$.

9-3. Sequential Reestimation

Suppose a series of experiments is being conducted, and we wish to reestimate the parameters as the results of each experiment come in.

Many of the objective functions that we have studied consist of a sum of terms, each containing the results of a single experiment. Examples are sums of squares, weighted sums of squares, and log-likelihood functions with known covariance matrices. In such a case, let us denote by $\Phi_\mu(\boldsymbol{\theta})$ the term corresponding to the μth experiment, and by $\Phi^{(n)}(\boldsymbol{\theta})$ the objective function for n experiments. Then

$$\Phi^{(n)}(\boldsymbol{\theta}) \equiv \sum_{\mu=1}^{n} \Phi_\mu(\boldsymbol{\theta}) \qquad (9\text{-}3\text{-}1)$$

If follows that

$$\Phi^{(n+1)}(\boldsymbol{\theta}) = \Phi^{(n)}(\boldsymbol{\theta}) + \Phi_{n+1}(\boldsymbol{\theta}) \qquad (9\text{-}3\text{-}2)$$

If we had estimated the parameters after the nth experiment, we would have found $\boldsymbol{\theta}^{(n)}$ which minimizes $\Phi^{(n)}$. We have also obtained the matrix $\mathbf{H}^{(n)}$ which is an approximation to the Hessian of $\Phi^{(n)}$ at $\boldsymbol{\theta} = \boldsymbol{\theta}^{(n)}$. A Taylor series approximation to $\Phi^{(n)}$ in the neighborhood of $\boldsymbol{\theta}^{(n)}$ is given by

$$\Phi^{(n)}(\boldsymbol{\theta}) \approx \Phi^{(n)}(\boldsymbol{\theta}^{(n)}) + \tfrac{1}{2}(\boldsymbol{\theta} - \boldsymbol{\theta}^{(n)})^{\mathrm{T}} \mathbf{H}^{(n)} (\boldsymbol{\theta} - \boldsymbol{\theta}^{(n)}) \qquad (9\text{-}3\text{-}3)$$

When the results of the $n+1$st experiment are available, we wish to find $\boldsymbol{\theta}^{(n+1)}$ which minimizes $\Phi^{(n+1)}$. It is reasonable to expect that $\boldsymbol{\theta}^{(n+1)}$ will not differ very much from $\boldsymbol{\theta}^{(n)}$, so that the approximation Eq. (3) is valid, and may be substituted in Eq. (2). Instead of minimizing $\Phi^{(n+1)}$ we may, instead, minimize

$$\tilde{\Phi}^{(n+1)}(\boldsymbol{\theta}) \equiv \tfrac{1}{2}(\boldsymbol{\theta} - \boldsymbol{\theta}^{(n)})^{\mathrm{T}}\mathbf{H}^{(n)}(\boldsymbol{\theta} - \boldsymbol{\theta}^{(n)}) + \Phi_{n+1}(\boldsymbol{\theta}) \tag{9-3-4}$$

The function $\tilde{\Phi}^{(n+1)}$ is much simpler and easier to calculate than $\Phi^{(n+1)}$, so that a great deal of computer time may be saved by this substitution. Of course, if it turns out that $\tilde{\Phi}^{(n+1)}$ is minimized at a point so far removed from $\boldsymbol{\theta}^{(n)}$ that Eq. (3) cannot be accepted, then we may have to revert to $\Phi^{(n+1)}$. We use $\boldsymbol{\theta}^{(n)}$ as the initial guess for the minimization of $\tilde{\Phi}^{(n+1)}$, and the result of that minimization as the initial guess for minimizing $\Phi^{(n+1)}$ if required. A single iteration may suffice for the latter.

If $\Phi_\mu(\boldsymbol{\theta})$ is the negative log-likelihood for the μth experiment, then Eq. (4) may be regarded as the logarithm of a posterior density function, in which $\tfrac{1}{2}(\boldsymbol{\theta} - \boldsymbol{\theta}^{(n)})^{\mathrm{T}}\mathbf{H}^{(n)}(\boldsymbol{\theta} - \boldsymbol{\theta}^{(n)})$ plays the part of a log prior density. It corresponds to a normal distribution of the parameters $\boldsymbol{\theta}$ with mean $\boldsymbol{\theta}^{(n)}$ and covariance matrix $(\mathbf{H}^{(n)})^{-1}$. This accords well with the fact (established in Chapter VII) that $(\mathbf{H}^{(n)})^{-1}$ is an approximation to the covariance matrix of the estimate $\boldsymbol{\theta}^{(n)}$, and that $\tfrac{1}{2}(\boldsymbol{\theta} - \boldsymbol{\theta}^{(n)})^{\mathrm{T}}\mathbf{H}^{(n)}(\boldsymbol{\theta} - \boldsymbol{\theta}^{(n)})$ is approximately (apart from an additive constant) the logarithm of the posterior distribution after n experiments. Minimizing $\Phi^{(n+1)}$ corresponds to finding the mode of the posterior distribution after $n+1$ experiments, where the posterior distribution after n experiments is taken as the prior distribution.

These results may be extended to the case where the parameters are to be reestimated only after ν additional experiments have been completed. Here

$$\Phi^{(n+\nu)}(\boldsymbol{\theta}) = \Phi^{(n)}(\boldsymbol{\theta}) + \sum_{\mu=1}^{\nu} \Phi_{n+\mu}(\boldsymbol{\theta}) \approx \tfrac{1}{2}(\boldsymbol{\theta} - \boldsymbol{\theta}^{(n)})^{\mathrm{T}}\mathbf{H}^{(n)}(\boldsymbol{\theta} - \boldsymbol{\theta}^{(n)}) + \sum_{\mu=1}^{\nu} \Phi_{n+\mu}(\boldsymbol{\theta}) \tag{9-3-5}$$

Sequential estimation procedures are of particular interest when the computer designs, controls, and analyzes the results of experiments on line (see Chapter X).

A numerical illustration appears in Section 9-8.

9-4. Computational Aspects

Let us consider a single new observation in the single equation least squares case. Here

$$\Phi^{(n)}(\boldsymbol{\theta}) \approx \Phi^{(n)}(\boldsymbol{\theta}^{(n)}) + (\boldsymbol{\theta} - \boldsymbol{\theta}^{(n)})^{\mathrm{T}}\mathbf{A}_n(\boldsymbol{\theta} - \boldsymbol{\theta}^{(n)}) \tag{9-4-1}$$

where

$$\mathbf{A}_n \equiv \sum_{\mu=1}^{n} \mathbf{b}_\mu \mathbf{b}_\mu{}^{\mathrm{T}}, \qquad \mathbf{b}_\mu \equiv \partial f_\mu / \partial \boldsymbol{\theta} \tag{9-4-2}$$

At the same time

$$\Phi_{n+1}(\boldsymbol{\theta}) = [y_{n+1} - f(\mathbf{x}_{n+1}, \boldsymbol{\theta})]^2 = e_{n+1}^2(\boldsymbol{\theta}) \tag{9-4-3}$$

Now, for $\boldsymbol{\theta}$ close to $\boldsymbol{\theta}^{(n)}$, we have approximately

$$f(\mathbf{x}, \boldsymbol{\theta}) = f(\mathbf{x}, \boldsymbol{\theta}^{(n)}) + \mathbf{b}_{n+1}^{\mathrm{T}}(\boldsymbol{\theta} - \boldsymbol{\theta}^{(n)})$$

so that the Gauss approximation to Eq. (3) is

$$\Phi_{n+1}(\boldsymbol{\theta}) \approx e_{n+1}^2(\boldsymbol{\theta}^{(n)}) - 2e_{n+1}(\boldsymbol{\theta}^{(n)})\mathbf{b}_{n+1}^{\mathrm{T}}(\boldsymbol{\theta} - \boldsymbol{\theta}^{(n)}) + [(\boldsymbol{\theta} - \boldsymbol{\theta}^{(n)})^{\mathrm{T}}\mathbf{b}_{n+1}]^2 \tag{9-4-4}$$

and Eq. (1) becomes, after dropping constant terms

$$\Phi^{(n)}(\boldsymbol{\theta}) \approx (\boldsymbol{\theta} - \boldsymbol{\theta}^{(n)})^{\mathrm{T}}\mathbf{A}_{n+1}(\boldsymbol{\theta} - \boldsymbol{\theta}^{(n)}) - 2e_{n+1}(\boldsymbol{\theta}^{(n)})\mathbf{b}_{n+1}^{\mathrm{T}}(\boldsymbol{\theta} - \boldsymbol{\theta}^{(n)}) \tag{9-4-5}$$

where

$$\mathbf{A}_{n+1} = \mathbf{A}_n + \mathbf{b}_{n+1}\mathbf{b}_{n+1}^{\mathrm{T}} \tag{9-4-6}$$

The minimum of Eq. (5) is easily seen to occur at

$$\boldsymbol{\theta}^{(n+1)} \equiv \boldsymbol{\theta}^{(n)} + \mathbf{A}_{n+1}^{-1}\mathbf{b}_{n+1}e_{n+1}(\boldsymbol{\theta}^{(n)}) \tag{9-4-7}$$

Having computed $\boldsymbol{\theta}^{(n)}$ and $\mathbf{A}_n$ after n experiments, the updating procedure after the $(n + 1)$st experiment may be summarized as follows:

1. Compute $e_{n+1}(\boldsymbol{\theta}^{(n)}) = y_{n+1} - f(\mathbf{x}_{n+1}, \boldsymbol{\theta}^{(n)})$.
2. Compute $\mathbf{b}_{n+1} = \partial f(\mathbf{x}_{n+1}, \boldsymbol{\theta})/\partial\boldsymbol{\theta})_{\boldsymbol{\theta}=\boldsymbol{\theta}^{(n)}}$.
3. Compute $\mathbf{A}_{n+1}$ (Eq. 6).
4. Compute $\mathbf{A}_{n+1}^{-1}$ (see below).
5. Use Eq. (7) to estimate $\boldsymbol{\theta}^{(n+1)}$.

Step 4 requires elaboration. One does not wish to invert an $l \times l$ matrix at each step. Fortunately, this is not necessary. Suppose that $\mathbf{A}_n^{-1}$ has already been computed. Define:

$$\mathbf{a}_{n+1} = \mathbf{A}_n^{-1}\mathbf{b}_{n+1} \tag{9-4-8}$$

$$\beta_{n+1} = \mathbf{b}_{n+1}^{\mathrm{T}}\mathbf{a}_{n+1} \tag{9-4-9}$$

Then, as may be verified by multiplying Eq. (6) by Eq. (10)

$$\mathbf{A}_{n+1}^{-1} = \mathbf{A}_n^{-1} - \mathbf{a}_{n+1}\mathbf{a}_{n+1}^{\mathrm{T}}/(1 + \beta_{n+1}) \tag{9-4-10}$$

Thus $\mathbf{A}_{n+1}^{-1}$ can be calculated without explicit reinversion. Somewhat more complicated inverse updating formulas are given by Powell (1969); these help reduce the accumulation of rounding errors.

9-5. Stochastic Approximation

When reestimations have to be performed at a very rapid pace, even the formulas of the preceding section may be too cumbersome. It is the aim of *stochastic approximation* methods to introduce further simplifications.

Equation (9-4-7) is a special case of the general stochastic approximation formula

$$\boldsymbol{\theta}^{(n+1)} = \boldsymbol{\theta}^{(n)} + \mathbf{c}_n e_{n+1}(\boldsymbol{\theta}^{(n)}) \tag{9-5-1}$$

where $\mathbf{c}_n$ is some suitably chosen vector. This formula states that the correction to be applied to $\boldsymbol{\theta}^{(n)}$ is proportional to $e_{n+1}(\boldsymbol{\theta}^{(n)})$, i.e., to the error incurred in predicting y_{n+1} given $\mathbf{x}_{n+1}$ and $\boldsymbol{\theta}^{(n)}$. In Eq. (9-4-7) we used

$$\mathbf{c}_n = \mathbf{A}_{n+1}^{-1}\mathbf{b}_{n+1}$$

Sometimes it is preferable to multiply this value by some positive constant less than one. The following variations represent progressive simplifications:

1. The procedure of Section 9-4 can only be started after enough observations have been accumulated to make $\mathbf{A}_n$ nonsingular (at least $n = l$). Instead, we can start with an arbitrarily chosen positive definite $\mathbf{A}_0$.
2. Use

$$\mathbf{c}_n = \mathbf{b}_{n+1} \Big/ \sum_{\mu=1}^{n+1} \mathbf{b}_\mu{}^{\mathrm{T}} \mathbf{b}_\mu \tag{9-5-2}$$

These methods, and some others, are discussed in detail by Albert and Gardner (1967).

9-6. A Missing Data Problem

We return to case (a) of the two equation maximum likelihood problem of Section 5-23. We assume, however, that measurements on $z_{1,1}$ and $z_{2,3}$ are missing. It is clear that relevant data still remain in data points $\mu = 1$ and 2, and these should not be discarded. Instead, we treat $z_{1,1}$ as an unknown parameter θ_6, and $z_{2,3}$ as an unknown parameter θ_7. The first equation of Eq. (5-23-6) now takes the form

$$0 = -\theta_6 + f_1(\mathbf{x}_\mu, \boldsymbol{\theta})$$

for $\mu = 1$ only, and Eq. (5-23-7) becomes

$$a = \theta_1 + \theta_2 x_{\mu 1} + \theta_3 \log \theta_7 + \theta_4 \log \{\theta_5 + \exp[(x_{\mu 3}/(1 + \theta_4)]\}$$

for $\mu = 2$ only. The model equations for $\mu = 3, 4, \ldots, 41$ remain unchanged. The matrices $\mathbf{B}_\mu = \partial \mathbf{f}_\mu / \partial \boldsymbol{\theta}$ have zero sixth and seventh columns for $\mu = 3, 4, \ldots, 41$. For $\mu = 1$ the sixth row is $[-1, 0]$ and the seventh row is zero; for $\mu = 2$, the sixth row is zero and the seventh row is $[(\theta_3/\theta_7) f_{2,1}, (\theta_3/\theta_7) f_{2,2}]$. As initial guesses for θ_6 and θ_7 we take 1.3 and 0.4 respectively, values which are reasonable in view of Table 5-6.

The estimate, obtained by means of the Marquardt method, is

$$\boldsymbol{\theta}^* = [-0.00551000, -0.0115201, 0.789967, 0.939189, \\ 0.835258, 1.51783, 0.417195]^{\mathrm{T}}$$

The estimate for θ_6, i.e., $z_{1,1}$ differs considerably from the known value 1.33135. However, this value leads to a residual of -0.22944 (Table 7-3), whereas the present estimate has a residual of only -0.0479336. The estimate θ_7 is quite close to the true value 0.4084 of $z_{2,3}$.

The parameters $\theta_1, \theta_2, \ldots, \theta_5$ can be converted into $\mathbf{c}$ as usual

$$\mathbf{c}^* = [0.5397926, 0.006333321, 1.265875, 1.064749, 0.5918664]^{\mathrm{T}}$$

The covariance of $\boldsymbol{\theta}^*$ can be obtained as usual; the marginal covariance of $\theta_1^*, \theta_2^*, \ldots, \theta_5^*$ consists of the 5×5 upper left hand corner of the 7×7 matrix $\mathbf{V}_{\boldsymbol{\theta}}$, and $\mathbf{V}_{\mathbf{c}}$ can be computed from it as usual.

We wish to see how much information has been lost due to our ignorance of $z_{1,1}$ and $z_{2,3}$ and also how much more would be lost if we dropped the first two observations completely. For this purpose we also obtained the estimate based on observations $\mu = 3, 4, \ldots, 41$. We computed det $\mathbf{V}_{\mathbf{c}}$ for all three cases; this quantity is proportional to the square of the volume of any confidence ellipsoid, and is also a measure of the uncertainty in the sampling distribution (see Section 10-2). The results are given in Table 9-1. We see that the more data are lost, the more uncertain are our estimates.

Table 9-1
Comparison of Information in Data

Case	det $\mathbf{V}_C$
41 full observations	0.246678×10^{-20}
41 observations, $z_{1,1}$ and $z_{2,3}$ missing	0.291663×10^{-20}
39 observations	0.384105×10^{-20}

9-7. Further Problem with Missing Data

Galambos and Cornell (1962) supply data on the observed proportions $y_{\mu 1}$ and $y_{\mu 2}$ of radioactive tracer in two human body compartments at times x_μ after injection. These data are presented in Table 9-2. The value $y_{1,1}$ is missing. The model equations are

$$y_1 = \theta_1 \exp(-\theta_2 x) + (1 - \theta_1)\exp(-\theta_3 x) \tag{9-7-1}$$

$$y_2 = 1 - \frac{\theta_1\theta_3}{\theta_1(\theta_3 - \theta_2) + \theta_2} \exp(-\theta_2 x) + \left[\frac{\theta_1\theta_3}{\theta_1(\theta_3 - \theta_2) + \theta_2} - 1\right]\exp(-\theta_3 x) \tag{9-7-2}$$

Table 9-2
Data for Radioactive Tracer Problem

		Proportion radioactive tracer	
μ	Time, x_μ (hr)	Compartment 1, $y_{\mu 1}$	Compartment 2, $y_{\mu 2}$
1	0.33	missing	0.03
2	2	0.84	0.10
3	3	0.79	0.14
4	5	0.64	0.21
5	8	0.55	0.30
6	12	0.44	0.40
7	24	0.27	0.54
8	46	0.12	0.66
9	72	0.06	0.71

Beauchamp and Cornell (1966) used the following method to estimate $\boldsymbol{\theta}$: First, least squares estimates were obtained for $\boldsymbol{\theta}$ using the y_1 data alone, giving

$$\boldsymbol{\theta}^{(1)} = \begin{bmatrix} 0.555524 & \pm 0.072741 \\ 0.0314238 & \pm 0.0038325 \\ 0.171109 & \pm 0.027847 \end{bmatrix}$$

Corresponding to this estimate is a minimum sum of squares of residuals $= 0.0009510273$, a variance of $0.0009510273/(8-3) = 0.0001902054$, and a standard deviation $\sigma_1 = 0.01379$. The same procedure was applied to the y_2 data alone, yielding

$$\boldsymbol{\theta}^{(2)} = \begin{bmatrix} 0.0606528 & \pm 0.0124113 \\ 0.00680764 & \pm 0.00107404 \\ 0.09316 & \pm 0.00593128 \end{bmatrix}$$

with residuals having a variance of $0.0002860781/(9-3) = 0.0000476797$ and standard deviation $\sigma_2 = 0.006905$. The residuals of each equation are close to the rounding error of the data. The two estimates of $\boldsymbol{\theta}$, however, are so far apart (measured by the scale of their standard deviations) as to cast doubt on the hypothesis that the same parameters apply to both equations.

Nevertheless, let us proceed with the joint fitting of the two equations. Beauchamp and Cornell compute the residuals from the two separate fits, and form their covariance matrix (neglecting the first observation on y_2) without compensating for degrees of freedom. They quote the matrix as being

$$\mathbf{V} = \begin{bmatrix} 0.1189 & 0.009753 \\ 0.009753 & 0.03179 \end{bmatrix} \times 10^{-3} \tag{9-7-3}$$

They now use the inverse of this matrix as a weight in the objective function

$$\sum_{\mu=2}^{9} \mathbf{e}_\mu^{\mathrm{T}}(\boldsymbol{\theta})\mathbf{V}^{-1}\mathbf{e}_\mu(\boldsymbol{\theta})$$

The minimum occurs at

$$\boldsymbol{\theta}^* = [0.06751, 0.00706, 0.08393]^{\mathrm{T}}$$

We shall now proceed to calculate an estimate based on the method used in the preceding section. We let θ_4 denote the missing value of $y_{1,1}$, and we assume $\mathbf{V}$ unknown. Our objective function then is

$$\Phi(\boldsymbol{\theta}) = (9/2) \log \det \sum_{\mu=1}^{9} \mathbf{e}_\mu(\boldsymbol{\theta})\mathbf{e}_\mu^{\mathrm{T}}(\boldsymbol{\theta})$$

where

$$e_{11}(\boldsymbol{\theta}) = \theta_4 - \theta_1 \exp(-\theta_2 x_1) - (1 - \theta_1)\exp(-\theta_3 x_1)$$

and all other residuals are defined as usual. Using Beauchamp and Cornell's initial guess (with a value for θ_4 appended)

$$\boldsymbol{\theta}_1 = (0.381, 0.21, 0.197, 1)^{\mathrm{T}}$$

The Gauss method (with nonnegativity constraints, using penalty functions) converged to

$$\boldsymbol{\theta}^* = [0.0782549, 0.00792904, 0.0975048, 1.04852]^{\mathrm{T}} \tag{9-7-4}$$

This result is unacceptable since it requires $y_{1,1} = 1.04852$, but no value of y

can exceed one. In fact, we must have $y_{\mu 1} + y_{\mu 2} \leqslant 1$; therefore, since $y_{12} = 0.03$, we impose the additional constraint $\theta_4 \leqslant 0.97$.‡ The result this time is

$$\boldsymbol{\theta}^* = \begin{bmatrix} 0.0358913 & \pm 0.0073400 \\ 0.00463630 & \pm 0.0007897 \\ 0.0812979 & \pm 0.0037664 \\ 0.910611 & \pm 0.023837 \end{bmatrix} \tag{9-7-5}$$

Note that this is an interior minimum (θ_4 is below its constraint). Curiously, the objective function attains a lower value at Eq. (5) than at Eq. (4), indicating that the latter is only a local minimum, and that Eq. (5) is the proper estimate even in the absence of the constraint on θ_4.

Our estimate is quite well determined, and significantly different from Beauchamp and Cornell's estimate. This may be accounted for by the fact that the estimated covariance of residuals corresponding to Eq. (5)

$$\mathbf{V} = \begin{bmatrix} 3.42423 & -0.565601 \\ -0.565601 & 0.118719 \end{bmatrix} \times 10^{-3}$$

is very different and much larger than Eq. (3). In other words, the combined fit attainable for both equations is much worse than the fit obtainable for each equation separately. The residuals found in fitting the individual equations are very poor measures for the errors in the simultaneous fit. If, however the residuals from the joint fit, which have standard deviations of $\sigma_1 = 0.0582$ and $\sigma_2 = 0.01090$, are still considered not in excess of experimental error, then we have no compelling reason for rejecting the joint model even though the separate models give much better fits.

9-8. A Sequential Reestimation Problem

In Section 5-21 we solved a single equation least squares problem. On the basis of fifteen observations we found:

$$\Phi^* = 0.03980599, \qquad \boldsymbol{\theta}^* = \begin{bmatrix} 813.4583 \\ 960.9063 \end{bmatrix}$$

$$\mathbf{N}^* = \begin{bmatrix} 0.271890 \times 10^{-5} & -0.957336 \times 10^{-5} \\ -0.957336 \times 10^{-5} & 3.50371 \times 10^{-5} \end{bmatrix}$$

‡ Since 0.03 is not an exact value, we really should have used the constraint

$$\theta_4 + 1 - \frac{\theta_1 \theta_3}{\theta_1(\theta_3 - \theta_2) + \theta_2} \exp(-\theta_2 x) + \left[\frac{\theta_1 \theta_3}{\theta_1(\theta_3 - \theta_2) + \theta_2} - 1\right] \exp(-\theta_3 x) \leqslant 1$$

but in practice, the simpler constraint suffices.

Therefore, the objective function Eq. (5-21-6) has the approximate representation Eq. (7-21-1). Suppose four additional observations were made. Our new objective function could be written approximately as

$$\Phi^{(19)}(\boldsymbol{\theta}) = \Phi^{(15)}(\boldsymbol{\theta}) + \sum_{\mu=16}^{19} e_\mu{}^2(\boldsymbol{\theta}) \approx 0.03980599$$
$$+ (10^{-5}/2)[0.271890(\theta_1 - 813.4583)^2 - 1.914672(\theta_1 - 813.4583)$$
$$\times (\theta_2 - 960.9063) + 3.50371(\theta_2 - 960.9063)^2 + \sum_{\mu=16}^{19} [y_\mu - f(\mathbf{x}_\mu, \boldsymbol{\theta})]^2$$

The data for the four new observations are given in Table 9-3. Starting

Table 9-3
Additional Good Data

μ	$x_{\mu 1}$	$x_{\mu 2}$	y_μ
16	0.1	150	0.851
17	0.1	250	0.176
18	0.2	150	0.825
19	0.2	250	0.011

with [813.4583, 960.9063] as the initial guess, a single Gauss iteration takes us to

$$\boldsymbol{\theta} = \begin{bmatrix} 891.4626 \\ 984.3818 \end{bmatrix}$$

A total of three iterations bring us to the minimum at

$$\boldsymbol{\theta} = \begin{bmatrix} 895.2656 \\ 985.1655 \end{bmatrix}$$

On the other hand, the true minimum of $\Phi^{(19)}(\boldsymbol{\theta}) = \sum_{\mu=1}^{19} e_\mu{}^2(\boldsymbol{\theta})$ occurs at

$$\boldsymbol{\theta} = \begin{bmatrix} 892.9341 \pm 213.702 \\ 983.7429 \pm 53.0124 \end{bmatrix}$$

We see then that the single Gauss iteration on the approximate objective function produced very acceptable results.

The new estimate is very close to the old one because the data of Table 9-3 were generated by the same model as the previous data. The data of Table 9-4, however, came from a different model. Nevertheless, when these

Table 9-4
Additional Poor Data

μ	$x_{\mu 1}$	$x_{\mu 2}$	y_μ
16	0.1	150	0.760
17	0.1	250	0.300
18	0.2	150	0.608
19	0.2	250	0.095

data are used in place of those of Table 9-3, we find after one Gauss iteration on the approximate objective function

$$\boldsymbol{\theta} = \begin{bmatrix} 462.8711 \\ 863.4094 \end{bmatrix}$$

The minimum of the approximate function is found in six iterations at

$$\boldsymbol{\theta} = \begin{bmatrix} 448.9961 \\ 852.5732 \end{bmatrix}$$

And the minimum of the exact function is at

$$\boldsymbol{\theta} = \begin{bmatrix} 484.7656 \pm 135.943 \\ 841.6172 \pm \ \ 61.2455 \end{bmatrix}$$

Even here, where the new estimate is very far from the old, we obtain an acceptable result in the single iteration.

9-9. Problems

Show that Eq. (9-4-7) can be generalized for a multiple equation model as follows

$$\boldsymbol{\theta}^{(n+1)} = \boldsymbol{\theta}^{(n)} + \mathbf{A}_{n+1}^{-1}\mathbf{B}_{n+1}^{\mathrm{T}}\mathbf{V}^{-1}\mathbf{e}_{n+1}(\boldsymbol{\theta}^{(n)})$$

where

$$\mathbf{B}_{n+1} = \partial \mathbf{e}_{n+1}/\partial \boldsymbol{\theta}, \qquad \mathbf{A}_n = \sum_{\mu=1}^{n} \mathbf{B}_\mu{}^{\mathrm{T}}\mathbf{V}^{-1}\mathbf{B}_\mu$$

Show that

$$\mathbf{A}_{n+1}^{-1} = \mathbf{A}_n^{-1} - \mathbf{A}_n^{-1}\mathbf{C}_{n+1}(\mathbf{I} + \mathbf{C}_{n+1}^{\mathrm{T}}\mathbf{A}_n^{-1}\mathbf{C}_{n+1})^{-1}\mathbf{C}_{n+1}^{\mathrm{T}}\mathbf{A}_n^{-1}$$

where $\mathbf{C}_\mu = \mathbf{B}_\mu{}^{\mathrm{T}}\mathbf{S}$, and $\mathbf{S}$ is a matrix such that $\mathbf{S}\mathbf{S}^{\mathrm{T}} = \mathbf{V}^{-1}$, e.g., the Cholesky decomposition of $\mathbf{V}^{-1}$.

Chapter

X

Design of Experiments

10-1. Introduction

Parameters are estimated on the basis of data obtained in experiments. It is natural to ask whether we can plan experiments so as to facilitate the task of estimating the parameters. The answer is generally in the affirmative, and this chapter is devoted to the study of suitable experimental strategies.

For our present purposes, we define an *experiment* as the act of observing the value of certain dependent variables $\mathbf{y}_\mu$ at given values of the independent variables $\mathbf{x}_\mu$. We *design an experiment* by choosing in some rational way the values of $\mathbf{x}$ at which $\mathbf{y}$ is to be observed. We shall use the phrase "the experiment $\mathbf{x}$" to denote the experiment whose independent variables take the value of $\mathbf{x}$. The values of the independent variables are referred to as the *experimental conditions*.

The design of the experiments and the estimation of the parameters form but two stages in a scientific investigation. What constitutes a "rational way" of choosing experimental conditions can be decided only on the basis of the overall aims of the investigation. A somewhat idealized scheme of a typical investigation is depicted in Fig. 10-1. In practice, investigators rarely adopt such a scheme explicitly, but nevertheless they adhere to it in a loose informal way.

This book is concerned with investigations in which parameter estimation plays a crucial role, forming the contents of box 3 in Fig. 10-1. Such investigations are naturally concerned with the development of mathematical models to represent physical situations. To devise a formal scheme for producing such a model in a general situation is, as yet, beyond our capabilities. Therefore, we place somewhat more modest goals into box 1 of our scheme. Typically, the goal may be one of the following:

(a) The estimation of the parameters in a given model to a specified degree of precision. For instance, we may wish to estimate the kinematic viscosity of a liquid, using Eq. (2-1-1) as the model.

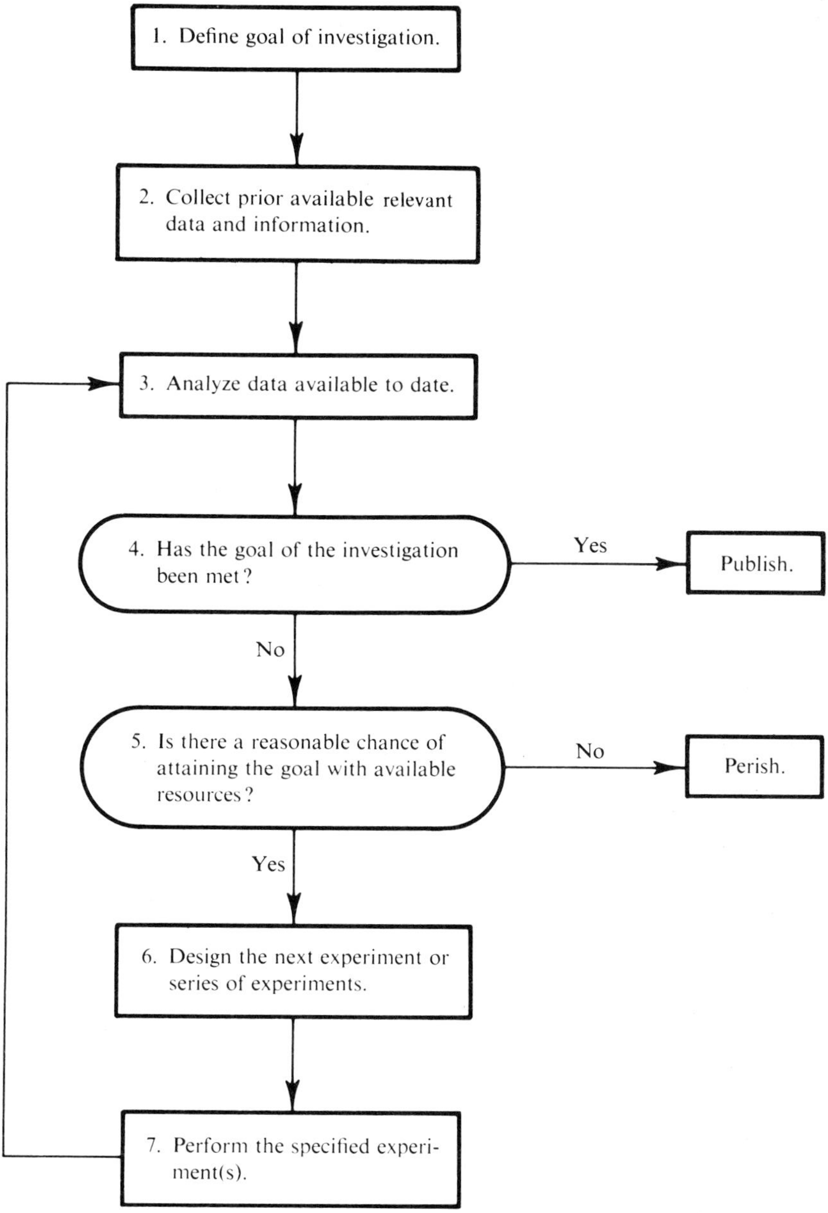

Fig. 10-1. A scheme for scientific investigations.

(b) The prediction of the values of certain variables which depend on some unknown parameters. For instance, we may wish to predict the power required to pump the liquid at a specified rate through a given pipe. To do this, we need to determine the liquid's viscosity.

(c) The selection of which one of several proposed models best accords with reality. Returning to the liquid and its viscosity, we may wish to determine whether the liquid is Newtonian (viscosity constant) or non-Newtonian (viscosity depending on shear rate, past history, or other factors).

(d) The determination of a course of action in a situation where the optimal action depends on what the correct model is and what the values of the parameters are. For instance, the proper design of a structure depends on the tensile strength of the materials used; the proper design of a chemical reactor depends on whether or not the reaction can be catalyzed; and the inventory required in a stockroom depends on the predicted demand, which in turn depends on the values of parameters appearing in an econometric model.

The method of selecting the experiments to be performed must be tailored to the goal of the investigation. A simple example suffices to illustrate this point:

Suppose we propose the model $y = \theta_1 + \theta_2 x$. For physical reasons, measurements are restricted to the range $-1 \leqslant x \leqslant 1$. It is intuitively obvious (and we shall later derive this fact rigorously) that best estimates for θ_1 and θ_2 will be obtained if all experiments are performed at the two extreme points of the range $x = -1$ and $x = 1$. On the other hand, if our main concern is to prove that the model is as given and not, say, $y = \theta_1 + \theta_2 x + \theta_3 x^2$, it becomes imperative to perform experiments with at least three distinct values of x. In fact, the best three experiments are at $x = -1, 0$, and 1. It is meaningful, then, to ask "what is the best experiment for the attainment of our goal?", but not simply "what is the best experiment?"

The classical methods of experimental design were devised by Fisher (1935), Davies and coworkers (1954), and others to satisfy goals different from those we are concerned with here. They referred to agricultural or industrial situations where no a priori mathematical models were available. Generally, one designed in advance a large number of experiments to be performed simultaneously (a necessary condition in agriculture, where an experiment takes months). In the scientific laboratory, on the other hand, an experiment usually takes only a short time, but requires expensive apparatus of which not many specimens are available. Experiments are perforce carried out in sequence, one (or at most several) at a time.

Wald (1947) has demonstrated that when experiments are carried out in sequence a smaller number of them is required, on the average, than when they are performed simultaneously. This is true even where no use is made of information gained in one experiment for planning the next one; The gain in this case accrues entirely from the ability to terminate the experimentation precisely at the point at which one's goal has been met. If, in addition, one is able to design each experiment in the light of the results of the previous ones, the gain in efficiency can be even more impressive. Informally, this strategy is adopted by most chemists, physicists, and other experimental scientists. What we are seeking here is the formalization of well-established intuitive procedures. The major contributions to the attainment of this goal are those of G.E.P. Box and his coworkers, starting with Box and Lucas (1959). Many more of their papers will be cited in the sequel.

10-2. Information and Uncertainty

It is the purpose of an experiment to gain relevant information. The best experiment is the one that is most informative. It is only natural that we should turn to information theory in our quest for a quantitative criterion for selecting the experiments to be performed.

Suppose ζ is a random vector. From the probability distribution of ζ we can gain a picture of the uncertainty associated with ζ; the more disperse the distribution of ζ, the more uncertain is the value any specific realization of ζ will assume. These intuitive notions of uncertainty have been formalized by Shannon (1948), who showed that the unique (except for a positive multiplicative factor) suitable measure of uncertainty associated with the probability density function $p(\zeta)$ is given by

$$H(p) \equiv -E(\log p) = -\int p(\zeta) \log p(\zeta)\, d\zeta \tag{10-2-1}$$

We gain information by reducing the uncertainty. Suppose $p_0(\zeta)$ and $p^*(\zeta)$ are, respectively, the prior density of ζ, and the posterior density after an experiment has been performed. According to Lindley (1956), the amount of information I that is gained by the experiment equals the reduction in uncertainty from the prior to the posterior distributions

$$I = H(p_0) - H(p^*) \tag{10-2-2}$$

Our aim is to find that experiment which maximizes I. Since $H(p_0)$ is unaffected by the experiment, we may equally well look for the experiment that minimizes $H(p^*)$.

When ζ is the vector of unknown parameters $\boldsymbol{\theta}$, p_0 and p^* may be the prior and posterior densities in the usual Bayesian sense. If we wish to eschew this interpretation, we may take p_0 and p^* to be the estimated sampling distribution densities before and after the experiment is conducted.‡ When the normal approximations are adopted, the two interpretations yield identical results.

We shall need to evaluate $H(p)$ for the multivariate normal distribution. Let $p(\zeta) = N_n(\mathbf{a}, \mathbf{V})$.

We have

$$\begin{aligned} H(p) &= -E(\log p) = -E[-(n/2)\log 2\pi - \tfrac{1}{2}\log\det\mathbf{V} - \tfrac{1}{2}(\zeta - \mathbf{a})^{\mathrm{T}}\mathbf{V}^{-1}(\zeta - \mathbf{a})] \\ &= (n/2)\log 2\pi + \tfrac{1}{2}\log\det\mathbf{V} + \tfrac{1}{2}\operatorname{Tr}[\mathbf{V}^{-1}E(\zeta - \mathbf{a})(\zeta - \mathbf{a})^{\mathrm{T}}] \\ &= (n/2)\log 2\pi + \tfrac{1}{2}\log\det\mathbf{V} + \tfrac{1}{2}\operatorname{Tr}\mathbf{V}^{-1}\mathbf{V} \\ &= (n/2)(1 + \log 2\pi) + \tfrac{1}{2}\log\det\mathbf{V} \end{aligned} \tag{10-2-3}$$

Discarding irrelevant constants, we can say that

$$H^*(p) \equiv \log\det\mathbf{V} \tag{10-2-4}$$

is a measure of the uncertainty in the distribution $N_n(\mathbf{a}, \mathbf{V})$.

We have remarked previously (Section 7-10) that for a normal distribution $\det^{1/2}\mathbf{V}$ is proportional to the volume of a confidence region in ζ space. Eq. (4) tells us that the uncertainty increases linearly with the logarithm of the volume of the confidence region. An experiment that seeks to minimize uncertainty also seeks to shrink the volume of the confidence region as much as possible.

10-3. Design Criterion for Parameter Estimation

Suppose our current state of knowledge concerning the value of the parameters $\boldsymbol{\theta}$ may be summarized in a normal prior distribution $N_l(\boldsymbol{\theta}_0, \mathbf{V}_0)$. Typically, this is the posterior (or sampling) distribution relative to experiments already performed. We are contemplating performing n additional experiments, in which $\mathbf{y}_\mu$ $(\mu = 1, 2, \ldots, n)$ are to be measured. Our task is to determine the values of $\mathbf{x}_\mu$ $(\mu = 1, 2, \ldots, n)$ at which these measurements are to be taken. We assume that the errors $\mathbf{y}_\mu - \mathbf{f}(\mathbf{x}_\mu, \boldsymbol{\theta})$ are distributed as $N_m(\mathbf{0}, \mathbf{V})$.

After these experiments are performed, we shall be able to construct a new posterior distribution. Let the normal approximation to that distribution be $N_l(\tilde{\boldsymbol{\theta}}, \tilde{\mathbf{V}})$. The vector $\tilde{\boldsymbol{\theta}}$ will be obtained as the mode of the posterior density; the posterior covariance $\tilde{\mathbf{V}}$ is given by Eq. (7-12-5)

$$\tilde{\mathbf{V}} = \left[\sum_{\mu=1}^{n} \mathbf{B}_\mu{}^{\mathrm{T}}\mathbf{V}^{-1}\mathbf{B}_\mu + \mathbf{V}_0^{-1}\right]^{-1} \tag{10-3-1}$$

‡ We assume that several experiments have already been performed, and p_0 estimated from the results.

where, as usual, $\mathbf{B}_\mu \equiv \partial \mathbf{f}_\mu / \partial \boldsymbol{\theta}$ evaluated at $\mathbf{x} = \mathbf{x}_\mu$, $\boldsymbol{\theta} = \tilde{\boldsymbol{\theta}}$. Since the experiments have not been performed yet, we cannot tell what value $\tilde{\boldsymbol{\theta}}$ will take; but in trying to calculate $\tilde{\mathbf{V}}$ we can use $\boldsymbol{\theta}_0$ in place of $\tilde{\boldsymbol{\theta}}$ when evaluating $\partial \mathbf{f}_\mu / \partial \boldsymbol{\theta}$.

Given any proposed set of experimental conditions $\mathbf{x}_1, \mathbf{x}_2, \ldots, \mathbf{x}_n$ we are thus able, by using Eq. (1), to estimate what the value of $\tilde{\mathbf{V}}$ will be after the proposed experiments are conducted. This is the same thing as saying that the estimated $\tilde{\mathbf{V}}$ is a function of $\mathbf{x}_1, \mathbf{x}_2, \ldots, \mathbf{x}_n$.

To maximize the amount of information gained by the experiments, we wish to select the $\mathbf{x}_\mu$ ($\mu = 1, 2, \ldots, n$) in such a way that the uncertainty is minimized, i.e., so that

$$\tilde{H} = \log \det \tilde{\mathbf{V}} \tag{10-3-2}$$

is minimized. This also minimizes the volume of the confidence region for the parameters. Clearly, minimizing $\log \det \tilde{\mathbf{V}}$ is equivalent to minimizing $\det \tilde{\mathbf{V}}$, or maximizing $\det(\tilde{\mathbf{V}})^{-1}$.

Let us reintroduce the notation

$$\mathbf{B} \equiv \begin{bmatrix} \mathbf{B}_1 \\ \mathbf{B}_2 \\ \vdots \\ \mathbf{B}_n \end{bmatrix}, \qquad \boldsymbol{\Pi} \equiv \begin{bmatrix} \mathbf{V} & \mathbf{0} & \cdots & \mathbf{0} \\ \mathbf{0} & \mathbf{V} & \cdots & \mathbf{0} \\ \vdots & & & \\ \mathbf{0} & \mathbf{0} & \cdots & \mathbf{V} \end{bmatrix} \tag{10-3-3}$$

The matrix $\boldsymbol{\Pi}$ is the joint covariance matrix‡ of the errors $\boldsymbol{\varepsilon}_1, \boldsymbol{\varepsilon}_2, \ldots, \boldsymbol{\varepsilon}_n$. Then, Eq. (1) becomes

$$\tilde{\mathbf{V}} = (\mathbf{B}^{\mathrm{T}} \boldsymbol{\Pi}^{-1} \mathbf{B} + \mathbf{V}_0^{-1})^{-1} \tag{10-3-4}$$

and

$$\det(\tilde{\mathbf{V}})^{-1} = \det(\mathbf{V}_0^{-1} + \mathbf{B}^{\mathrm{T}} \boldsymbol{\Pi}^{-1} \mathbf{B}) = \det \mathbf{V}_0^{-1} \det(\mathbf{I} + \mathbf{V}_0 \mathbf{B}^{\mathrm{T}} \boldsymbol{\Pi}^{-1} \mathbf{B}) \tag{10-3-5}$$

We recall now [see Eq. (A-1-33)] that $\det(\mathbf{I} + \mathbf{AB}) = \det(\mathbf{I} + \mathbf{BA})$ so that

$$\begin{aligned} \det(\tilde{\mathbf{V}})^{-1} &= \det \mathbf{V}_0^{-1} \det(\mathbf{I} + \mathbf{B} \mathbf{V}_0 \mathbf{B}^{\mathrm{T}} \boldsymbol{\Pi}^{-1}) \\ &= \det \mathbf{V}_0^{-1} \det \boldsymbol{\Pi}^{-1} \det(\boldsymbol{\Pi} + \mathbf{B} \mathbf{V}_0 \mathbf{B}^{\mathrm{T}}) \end{aligned} \tag{10-3-6}$$

Since $\det \mathbf{V}_0^{-1} \det \boldsymbol{\Pi}^{-1}$ is a positive constant, we may simply maximize the function

$$T(\mathbf{x}_1, \mathbf{x}_2, \ldots, \mathbf{x}_n) \equiv \det(\boldsymbol{\Pi} + \mathbf{B} \mathbf{V}_0 \mathbf{B}^{\mathrm{T}}) \tag{10-3-7}$$

Let us examine the matrix $\boldsymbol{\Pi} + \mathbf{B} \mathbf{V}_0 \mathbf{B}^{\mathrm{T}}$. As stated before, $\boldsymbol{\Pi}$ is the joint covariance matrix of the errors in all the measurements to be taken in the

‡ If the $\boldsymbol{\epsilon}_\mu$ are serially correlated, we introduce the proper off-diagonal elements into the definition of $\boldsymbol{\Pi}$.

course of the n proposed experiments, and $\mathbf{BV}_0\mathbf{B}^T$ is the covariance matrix of the errors incurred in computing the predicted outcomes $\mathbf{f}(\mathbf{x}_\mu, \boldsymbol{\theta})$ of the proposed experiments due to the current uncertainty in the values of the parameters. Therefore, $\boldsymbol{\Pi} + \mathbf{BV}_0\mathbf{B}^T$ is the total covariance of the predicted outcome (see Eq. (7-19-4), with $\mathbf{V}_0$ and $\boldsymbol{\Pi}$ playing the roles of $\mathbf{V}_{\boldsymbol{\theta}}$ and $\mathbf{V}_{\boldsymbol{\eta}}$, respectively; $\mathbf{V}_{\boldsymbol{\xi}}$ is assumed zero). Eq. (7) is then a measure of the joint uncertainty of the predicted outcomes.‡ We have shown that to obtain maximum information we must perform those experiments whose outcome is the most uncertain. This result is not surprising; experiments whose outcomes are most uncertain represent the greatest gaps in our knowledge of the system under consideration; to fill the gaps we must perform those experiments.

We have a choice of minimizing $\det(\mathbf{V}_0^{-1} + \mathbf{B}^T\boldsymbol{\Pi}^{-1}\mathbf{B})$ or, equivalently, maximizing $\det(\boldsymbol{\Pi} + \mathbf{BV}_0\mathbf{B}^T)$. Our choice should depend on the relative dimensions of the two matrices, which are $l \times l$ and $mn \times mn$, respectively. We should obviously choose the determinant of lower dimension. The case that is most favorable to the second formulation is one where a single experiment is to be conducted on a single equation model. Here $\boldsymbol{\Pi}$ reduces to a single number σ^2, and $\mathbf{B}$ is a row vector $\mathbf{b}^T = [\partial f/\partial\boldsymbol{\theta}]^T$. Hence Eq. (7) reduces to

$$T(\mathbf{x}_1) = \sigma^2 + \mathbf{b}^T\mathbf{V}_0\mathbf{b} \tag{10-3-8}$$

If σ^2 is a constant, we need only find the $\mathbf{x}_1$ which maximizes the error of prediction variance $\mathbf{b}^T\mathbf{V}_0\mathbf{b}$.

We cite the following simple example:

Suppose the model is linear

$$y = f(x, \boldsymbol{\theta}) = \theta_1 + \theta_2 x \tag{10-3-9}$$

We have $\mathbf{b}^T = [1, x]$. Let $\sigma^2 = 0.1$ and suppose the current estimates are $\theta_1 = 2$, $\theta_2 = 1$, with covariance matrix $\mathbf{V}_0 = \text{diag}(0.1, 0.5)$. The predicted outcome y of any experiment x is given by $2 + x$, and the variance of this prediction is, according to Eq. (8)

$$T(x) = 0.1 + [1, x]\begin{bmatrix} 0.1 & 0 \\ 0 & 0.5 \end{bmatrix}\begin{bmatrix} 1 \\ x \end{bmatrix} = 0.2 + 0.5x^2 \tag{10-3-10}$$

To improve the estimates of θ_1 and θ_2 we should perform an experiment maximizing $T(x)$, that is we should choose as large (in absolute value) an x as is practically feasible. The situation is illustrated in Fig. 10-2, where the predicted curve $2 + x$ is plotted surrounded by a confidence curve of width $\pm(0.2 + 0.5x^2)^{1/2}$. We choose for our experiment a value of x at which the confidence band is as wide as possible.

‡ The determinant of a covariance matrix is sometimes referred to as the *generalized variance*.

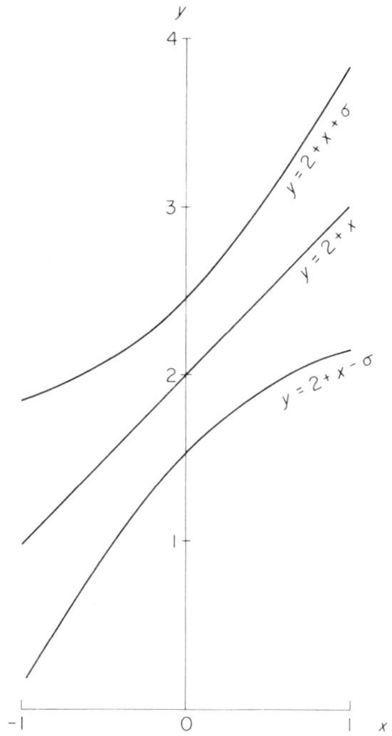

Fig. 10-2. Predicted y with confidence bands.

The design criterion that was described here has been arrived at from different points of view by Box and Lucas (1959) and Box and Hunter (1963), with further details supplied by Draper and Hunter (1966, 1967a, 1967b) and Atkinson and Hunter (1968). The use of the method on a computer-simulated chemical kinetics model is described by Kittrell, Hunter, and Watson (1966), and details for the estimation of polymerization parameters are worked out by Behnken (1964).

10-4. Design Criterion for Prediction

If $\boldsymbol{\theta}$ is to be estimated purely for the purpose of predicting certain quantities $\boldsymbol{\eta} = \boldsymbol{\phi}(\boldsymbol{\xi}, \boldsymbol{\theta})$, then the uncertainty of the prediction is given by det $\mathbf{V}_\mathbf{p}$, where $\mathbf{V}_\mathbf{p}$ is defined in Eq. (7-19-4) and $\tilde{\mathbf{V}}$ is used in place of $\mathbf{V}_{\boldsymbol{\theta}}$. Choose $\mathbf{x}_\mu$ $(\mu = 1, 2, \ldots, n)$ so as to minimize the uncertainty of the prediction

$$\tilde{H} = \det \mathbf{V}_\mathbf{p} = \det[(\partial\boldsymbol{\phi}/\partial\boldsymbol{\xi})\mathbf{V}_{\boldsymbol{\xi}}\,(\partial\boldsymbol{\phi}/\partial\boldsymbol{\xi})^\mathrm{T} + (\partial\boldsymbol{\phi}/\partial\boldsymbol{\theta})\tilde{\mathbf{V}}\,(\partial\boldsymbol{\phi}/\partial\boldsymbol{\theta})^\mathrm{T} + \mathbf{V}_{\boldsymbol{\eta}}] \quad (10\text{-}4\text{-}1)$$

Here $\tilde{\mathbf{V}}$, given by Eq. (10-3-1), is the only quantity that depends on the $\mathbf{x}_\mu$.

In the special case when the parameters $\boldsymbol{\theta}$ themselves are the $\boldsymbol{\eta}$ to be predicted, then Eq. (1) reduces to the criterion of Section 10-3. We may be interested only in a subset of the parameters, in which case we associate that subset with $\boldsymbol{\eta}$, and minimize the determinant of the matrix obtained from $\tilde{\mathbf{V}}$ by deleting the rows and columns corresponding to the unwanted parameters.

10-5. Design Criterion for Model Discrimination

Sometimes several alternative models are proposed for the same physical situation. We wish to conduct experiments that will enable us to select the "best" model, i.e., the one that best fits the data.

Each one of our models attempts to predict $\mathbf{y}$ as a function of $\mathbf{x}$ and $\boldsymbol{\theta}$. What varies from model to model is the mathematical form of the function, and the set of parameters involved (although some of the parameters appearing in different models may possess the same physical interpretation). We attach a superscript (i) to quantities pertaining to the ith model. The ith model equation reads

$$\mathbf{y} = \mathbf{f}^{(i)}(\mathbf{x}, \boldsymbol{\theta}^{(i)}) \tag{10-5-1}$$

Suppose that we already have estimates $\boldsymbol{\theta}_0^{(i)}$ for the parameters appearing in the ith model, and estimates $\mathbf{V}_0^{(i)}$ for the associated covariance matrices. Typically, these are obtained by fitting each model in turn to data from previously performed experiments. Using the parameter values $\boldsymbol{\theta}_0^{(i)}$ we can predict the outcome $\mathbf{y}^{(i)}$ of any proposed experiment $\mathbf{x}$, assuming the ith model is the correct one. This prediction is given by

$$\mathbf{y}^{(i)}(\mathbf{x}) \equiv \mathbf{f}^{(i)}(\mathbf{x}, \boldsymbol{\theta}_0^{(i)}) \tag{10-5-2}$$

Still assuming that the ith model is correct, we can compute the covariance of the prediction error in Eq. (2). Following Eq. (7-19-4), this is

$$\mathbf{V}_{\mathbf{y}}^{(i)}(\mathbf{x}) = \mathbf{V} + \mathbf{B}^{(i)}(\mathbf{x})\mathbf{V}_0^{(i)}\mathbf{B}^{(i)\mathrm{T}}(\mathbf{x}) \tag{10-5-3}$$

where $\mathbf{B}^{(i)} \equiv \partial\mathbf{f}^{(i)}/\partial\boldsymbol{\theta}^{(i)}$ and $\mathbf{V}$ is the covariance of the measurement errors of $\mathbf{y}$ (which may also be a function of $\mathbf{x}$).

The hypothesis that the ith model is correct leads us to regard the outcome of a proposed experiment $\mathbf{x}$ as a random variable $\boldsymbol{\eta}$ with pdf $p^{(i)}(\boldsymbol{\eta}\,|\,\mathbf{x})$ having mean and covariance given by Eq. (2) and Eq. (3), respectively. Suppose the experiment has actually yielded an outcome $\mathbf{y}$. Then we can compute the number $p^{(i)}(\mathbf{y}\,|\,\mathbf{x})$, which is the likelihood associated with the ith hypothesis.

For the moment we restrict ourselves to the case of two alternative models. The quantity

$$a_{12}(\mathbf{x}) \equiv \log[p^{(1)}(\mathbf{y}\,|\,\mathbf{x})/p^{(2)}(\mathbf{y}\,|\,\mathbf{x})]$$

is a measure of how much the observed $\mathbf{y}$ supports model 1 in preference to model 2 (it is related to the likelihood ratio, see Section 10-6). In advance of performing the experiment we do not know $\mathbf{y}$, so we cannot compute a_{12}, but we can compute its expected value under the assumption that model 1 is correct (the symbol $E^{(1)}$ denoting expectation under this assumption)

$$E^{(1)}[a_{12}(\mathbf{x})] = \int p^{(1)}(\mathbf{y}\,|\,\mathbf{x})\log[p^{(1)}(\mathbf{y}\,|\,\mathbf{x})/p^{(2)}(\mathbf{y}\,|\,\mathbf{x})]\,d\mathbf{y} \tag{10-5-4}$$

If indeed model 1 is correct, we wish to conduct an experiment $\mathbf{x}$ which is likely to confirm this, i.e., is expected to produce a large value of a_{12}. Conversely, if model 2 is correct, we wish our experiment to have a large value of the corresponding quantity $E^{(2)}[a_{21}(\mathbf{x})]$. Since we do not know which model is correct, we form the sum of these two quantities

$$\begin{aligned} J_{1,2}(\mathbf{x}) &\equiv E^{(1)}[a_{12}(\mathbf{x})] + E^{(2)}[a_{21}(\mathbf{x})] \\ &= \int [p^{(1)}(\mathbf{y}\,|\,\mathbf{x}) - p^{(2)}(\mathbf{y}\,|\,\mathbf{x})]\log[p^{(1)}(\mathbf{y}\,|\,\mathbf{x})/p^{(2)}(\mathbf{y}\,|\,\mathbf{x})]\,d\mathbf{y} \end{aligned} \tag{10-5-5}$$

The experiment to be selected is the one that maximizes $J_{1,2}(\mathbf{x})$; a large value of $J_{1,2}$ can only be obtained if $p^{(2)}$ is much larger than $p^{(1)}$, or vice versa. In either case, the outcome shows a strong preference for one model as opposed to the other.

The quantity $J_{1,2}$ is called the *divergence* or the *information for discrimination* (Kullback and Leibler, 1951; Kullback, 1959). Its similarity to Eq. (10-2-1) is evident.

If both models assume normal error distributions with covariance matrices $\mathbf{V}_\mathbf{y}^{(1)}$ and $\mathbf{V}_\mathbf{y}^{(2)}$, respectively, then it can be shown that

$$\begin{aligned} J_{1,2}(\mathbf{x}) = {} & -m + \tfrac{1}{2}\,\mathrm{Tr}(\mathbf{Q}^{(1)}\mathbf{V}_\mathbf{y}^{(2)} + \mathbf{Q}^{(2)}\mathbf{V}_\mathbf{y}^{(1)}) \\ & + \tfrac{1}{2}(\mathbf{y}^{(2)} - \mathbf{y}^{(1)})^\mathrm{T}(\mathbf{Q}^{(1)} + \mathbf{Q}^{(2)})(\mathbf{y}^{(2)} - \mathbf{y}^{(1)}) \end{aligned} \tag{10-5-6}$$

where $\mathbf{Q}^{(i)} \equiv (\mathbf{V}_\mathbf{y}^{(i)})^{-1}$. The dependence of $J_{1,2}$ on the experimental conditions $\mathbf{x}$ comes about through Eqs. (2) and (3). An important special case occurs when the models are of the single-equation type, with $m = 1$. Then $\mathbf{V}_\mathbf{y}^{(i)} = \sigma_i^2$, $\mathbf{Q}^{(i)} = \sigma_i^{-2}$ $(i = 1, 2)$, and

$$J_{1,2}(\mathbf{x}) = -1 + \tfrac{1}{2}[(\sigma_2/\sigma_1)^2 + (\sigma_1/\sigma_2)^2] + \tfrac{1}{2}[(1/\sigma_1^2) + (1/\sigma_2^2)](y^{(2)} - y^{(1)})^2 \tag{10-5-7}$$

This equation was derived by Box and Hill (1967). The analogues to Eqs. (2) and (3) are in this case:

$$y^{(i)}(\mathbf{x}) = f^{(i)}(\mathbf{x}, \boldsymbol{\theta}_0^{(i)}) \tag{10-5-8}$$

$$\sigma_i^2(\mathbf{x}) = \sigma^2 + \mathbf{b}^{(i)\mathrm{T}}(\mathbf{x})\mathbf{V}_0^{(i)}\mathbf{b}^{(i)}(\mathbf{x}) \tag{10-5-9}$$

where σ is the standard deviation of the measurements errors and $\mathbf{b}^{(i)}(\mathbf{x}) = \partial f^{(i)}/\partial \boldsymbol{\theta}^{(i)}$.

Equations (6) and (7) have a simple heuristic interpretation, particularly in the single-equation case. Let us plot the predicted values $y^{(1)}$ and $y^{(2)}$ as functions of $\mathbf{x}$; this is done for a hypothetical situation in Fig. 10-3 where $\mathbf{x}$

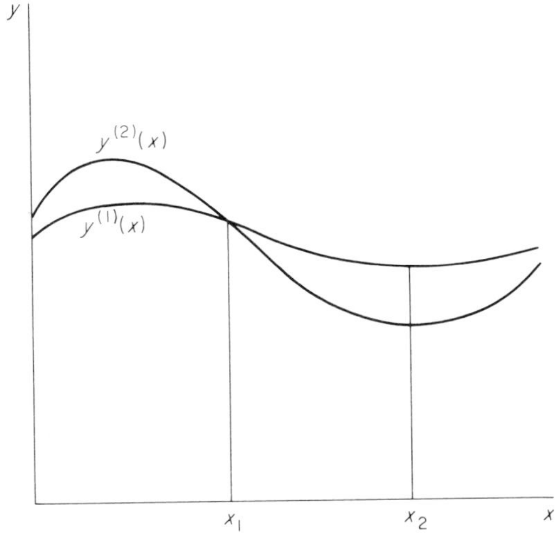

Fig. 10-3. Discrimination between two predicted responses.

is assumed one-dimensional. If we chose to perform the experiment x_1, where $y^{(1)}$ and $y^{(2)}$ coincide, the results of the experiment will tell us nothing about which prediction was the better one. On the other hand, the two predictions are most divergent at x_2, and the result of the experiment (unless it happens to fall exactly midway between the two predictions) is likely to confirm one or the other of the two models depending on which prediction it falls nearer to. It seems reasonable to select, then, the experiment $\mathbf{x}$ for which $(y^{(2)} - y^{(1)})^2$ is maximum. It may happen, however, that at that value of $\mathbf{x}$ (x_2 in Fig. 10-3) one or both of the predictions are particularly uncertain, possessing large values of σ_i. Performing this experiment then is likely to be inconclusive, and we may prefer another experiment for which $(y^{(2)} - y^{(1)})^2$ is somewhat smaller, but where the uncertainty is much smaller. Therefore we must attach to the term $(y^{(2)} - y^{(1)})^2$ a weight which is small when even one σ_i is large, and large when both σ_i are small. Eq. (7) provides the right weight, and the same is true of Eq. (6) in the multiresponse case.

It frequently turns out that the σ_i do not vary strongly with $\mathbf{x}$, so the weights are nearly the same for all values of $\mathbf{x}$. In this case we need only find the maximum of $(y^{(2)} - y^{(1)})^2$ or $|y^{(2)} - y^{(1)}|$.

Our results can be generalized in several directions. To design several experiments simultaneously, we maximize $J_{1,2}$ constructed with $\mathbf{y}^{(i)}$ and $\mathbf{V}^{(i)}$ augmented to include the responses from all the planned experiments; in Eq. (3) $\mathbf{B}^{(i)}$ takes the meaning defined in Eq. (10-3-3), and $\mathbf{\Pi}$ of Eq. (10-3-3) replaces $\mathbf{V}$.

There are several ways in which we may treat more than two models. After each experiment is performed, we can compute the likelihood $L^{(i)}$ associated with each model and the best current estimate of its parameters. We then design the next experiment so as to discriminate specifically between the two models with largest values of the likelihood. Or, following Box and Hill (1967), we may form a joint divergence as a linear combination of the pairwise divergences

$$J_{1,2,3,\ldots}(\mathbf{x}) \equiv \sum_{i \neq j} L^{(i)} L^{(j)} J_{i,j}(\mathbf{x}) \tag{10-5-10}$$

We have at this point no experience to guide us in the choice of the method to use, but it is obvious that the first one requires fewer calculations.

Our aim may be both to find the best among alternative models, and at the same time find good estimates for the parameters in the best model. A solution to this problem suggested by Hill *et al.* (1968) is to use as design criterion a weighted sum of Eq. (6) and Eq. (10-3-2), the latter quantity being evaluated for the currently best model. Initially, a relatively large weight is placed on Eq. (6), but as one model becomes increasingly preferred, the relative weight given to Eq. (10-3-2) is progressively increased.

10-6. Termination Criteria

We now turn our attention to box 4 of Fig. 10-1. How do we decide whether more experiments are needed? How and when do we decide that a given model is better than the alternatives?

We have advocated the use of the maximum likelihood method for estimating parameters. We preferred to assign to our parameters $\boldsymbol{\theta}$ the value $\boldsymbol{\theta}_1$ rather than $\boldsymbol{\theta}_2$, provided that the likelihood associated with $\boldsymbol{\theta}_1$ was greater than that associated with $\boldsymbol{\theta}_2$. The same idea applies to the choice of models; we prefer model 1 to model 2 if the maximum likelihood attainable with model 1 is greater than that attainable with model 2. These considerations lead to Wald's (Wald, 1947) *sequential probability ratio* (or *likelihood ratio*) test. Suppose our aim is to choose one of two alternative hypotheses, H_1 (model 1 is correct), or H_2 (model 2 is correct). Let $L^{(i)}(\mathbf{y}, \boldsymbol{\theta}_0^{(i)})$ be the likelihood (i.e., the value of the joint probability density function) associated with the data obtained to date, and with the current best estimate $\boldsymbol{\theta}_0^{(i)}$ for the parameters based on the ith hypothesis ($i = 1, 2$).

Let A and B be two constants satisfying

$$0 < B < 1 < A \tag{10-6-1}$$

Then the likelihood ratio test proceeds as follows:

1. If $L^{(1)}/L^{(2)} \leqslant B$ accept hypothesis 2.
2. If $L^{(1)}/L^{(2)} \geqslant A$ accept hypothesis 1.
3. If $B < L^{(1)}/L^{(2)} < A$ continue experimentation.

The choice of the constants A and B is determined by what confidence we desire to place on the results. Let α be the probability that H_1 is accepted when H_2 is true, and β the probability that H_2 is accepted when H_1 is true. It was shown by Wald (1947) that the following relations hold approximately (the last two being consequences of the first two)

$$\begin{aligned} A &\approx (1-\beta)/\alpha, & B &\approx \beta/(1-\alpha), \\ \alpha &\approx (1-B)/(A-B), & \beta &\approx B(A-1)/(A-B) \end{aligned} \tag{10-6-2}$$

If we want, say, to be 95% certain that we accept H_1 only if H_1 is true, and 90% certain that we accept H_2 only if H_2 is true, then $\alpha = 0.05$ and $\beta = 0.1$ so that $A = 0.9/0.05 = 18$ and $B = 0.1/0.95 = 0.105$. Conversely, suppose we choose $A = 10$, $B = 0.1$. This is tantamount to accepting error probabilities $\alpha = 0.9/9.9 = 0.0909$ and $\beta = 0.1 \times 9/9.9 = 0.0909$. The choice $\alpha = \beta$ leads to $B = 1/A$, and hence $\alpha = \beta = 1/(1 + A)$.

When more than two alternatives are present, we need only apply the test to the two currently most likely models.

It is instructive to derive an expression for the likelihood ratio after n experiments in the single equation case. Assuming normal distributions, we have

$$L^{(i)} = (2\pi)^{-(n/2)}\sigma^{-n} \exp\left\{-(1/2\sigma^2)\sum_{\mu=1}^{n} [y_\mu - f_\mu^{(i)}(\boldsymbol{\theta}^{(i)})]^2\right\} \tag{10-6-3}$$

For the ith model, L is maximized if we estimate σ to be

$$\sigma^{(i)} = \left\{(1/n)\sum_{\mu=1}^{n} [y_\mu - f_\mu^{(i)}(\boldsymbol{\theta}^{(i)})]^2\right\}^{1/2} \tag{10-6-4}$$

Hence

$$L^{(i)} = (2\pi)^{-(n/2)}(\sigma^{(i)})^{-n} \exp(-n/2) \tag{10-6-5}$$

and the likelihood ratio is

$$\frac{L^{(1)}}{L^{(2)}} = \left(\frac{\sigma^{(2)}}{\sigma^{(1)}}\right)^n = \left\{\frac{\sum_{\mu=1}^{n}[y_\mu - f_\mu^{(2)}(\boldsymbol{\theta}^{(2)})]^2}{\sum_{\mu=1}^{n}[y_\mu - f_\mu^{(1)}(\boldsymbol{\theta}^{(1)})]^2}\right\}^{n/2} \tag{10-6-6}$$

If after n experiments $\sigma^{(2)} > \sigma^{(1)}$, we expect to find ultimately that model 1 is to be preferred. If $[\sigma^{(2)}/\sigma^{(1)}]^n < A$, we must defer final conclusions until some more experiments are performed. Having no reason to expect the estimates of $\sigma^{(2)}$ or $\sigma^{(1)}$ to be changed much by the results of future experiments, we can predict that $L^{(1)}/L^{(2)}$ will exceed A after we conduct n_0 additional experiments with $[\sigma^{(2)}/\sigma^{(1)}]^{n+n_0} \geqslant A$

$$n_0 \geqslant (\log A)/(\log \sigma^{(2)}/\sigma^{(1)}) - n \tag{10-6-7}$$

The smallest integer n_0 satisfying Eq. (7) is an estimate of the number of additional experiments required to reach the conclusion that model 1 is the better one.

If $\sigma^{(1)} < \sigma^{(2)}$, then

$$n_0 \geqslant -(\log B)/(\log \sigma^{(1)}/\sigma^{(2)}) - n \tag{10-6-8}$$

provides an estimate for the number of additional experiments required to establish a preference for model 2. The reliability of these estimates, which is very small when $n \ll n_0$, increases steadily as n_0 approaches zero. For further discussion of the expected number of experiments, the reader is referred to Wald (1947).

When experiments are being conducted for the purpose of estimating parameters in a single model, the termination criterion is usually formulated in terms of the variance of the estimates. One demands that det $\mathbf{V}_\theta$ fall below a specified value, or that the individual parameter variances $V_{\theta ii}$ $(i = 1, 2, \ldots, l)$ all fall below specified levels σ_i^2. The number of additional experiments required at any stage may be estimated easily from the fact that the elements of $\mathbf{V}_\theta$ are roughly proportional to $(n - l)^{-1}$. If det $\mathbf{V}_\theta = a$ after n experiments, and the number of additional experiments n_0 required to reach det $\mathbf{V}_\theta = b < a$ is to be determined, then we must solve the equation

$$(n + n_0 - l)^l b = (n - l)^l a \tag{10-6-9}$$

for n_0.

10-7. Some Practical Considerations

We have derived several experimental design criteria, given by Eqs. (10-3-2), (10-3-7), (10-4-1), (10-5-6), and (10-5-7) for the various cases that may arise. Let $D(\mathbf{x})$ denote the criterion adopted in a given situation. The experimental conditions $\mathbf{x}$ are to be chosen so as to maximize $D(\mathbf{x})$. We discuss here some of the problems associated with finding these experimental conditions.

In the first place, we must realize that the choice of experimental conditions is generally not unrestricted. Mole fractions can only range from zero to one, the temperature of a liquid is constrained between its freezing and boiling points, and the pressure in a vessel is limited by the strength of its walls. Therefore, searching for the maximum of $D(\mathbf{x})$ involves constrained optimization, with the variables (experimental conditions) confined to a bounded feasible region. Experience has shown that the maximum usually falls on the boundary of the feasible region (Atkinson and Hunter, 1968) have derived conditions under which this must be so). The experimenter must apply the design criterion with caution; the extreme values of the experimental conditions prescribed by the criterion may be far removed from the region of interest, and it may be well to impose stricter bounds on the variables than is required by physical or technical limitations. There is also the danger that the properties (i.e., the model equations or parameter values) of the system under investigation are not the same at the boundary as in the center of the feasible region. We recommend therefore that occasional experiments be chosen in the interior of the region, even when not prescribed by the design criterion.

The reader will have noticed that the design criterion cannot be computed unless initial estimates $\boldsymbol{\theta}_0$ and $\mathbf{V}_0$ are given for the parameters and their covariance matrix. At the start of the investigation such estimates may not be available, and some initial experiments must be performed to get things going. The number of such experiments must exceed somewhat the number of unknown parameters, so that the estimates $\boldsymbol{\theta}_0$ and $\mathbf{V}_0$ can be obtained. The initial experiments may be selected by standard methods such as factorial, fractional factorial, or rotatable designs covering the feasible range of the experimental conditions.

An experimenter using these designs must remember that he cannot expect to get more out of the procedure than he has put into it. He cannot expect to obtain clear-cut preference for one model or one value of the parameters, if major effects have been neglected. For example, suppose a compound A is converted into a product C according to the consecutive reaction scheme

$$\mathrm{A} \rightarrow \mathrm{B}, \qquad \mathrm{B} \rightarrow \mathrm{C} \tag{10-7-1}$$

However, the experimenter has set down models involving only the reaction

$$\mathrm{A} \rightarrow \mathrm{C} \tag{10-7-2}$$

He should not be disappointed then if the design criterion does not tell him to run experiments with varying initial concentrations of B.

In our derivations, expected information was the sole criterion for selecting experiments. In practice, considerations of economics and convenience in experimental setup must also play a role. In many situations, particularly those involving dynamic systems, experiments are conducted in runs; several observations are made at different times on a process starting from given initial conditions. In such cases one should design whole runs, rather than single observations. We must select, then, a set of initial conditions $\mathbf{s}_0$, and times $t_1, t_2, \ldots, t_n$ at which observations are to be made. Computing the total information obtainable in each possible run is a formidable task because of the high correlation between the predicted values of successive observations of a run. It is, however, easy to calculate the expected information in any single observation taken at times t_μ with initial conditions $\mathbf{s}_0$. If we plot the expected information I as a function of t for given $\mathbf{s}_0$ we usually find that there is a definite time $t_M(\mathbf{s}_0)$ at which the expected information attains a maximum value.‡ Let $I_M(\mathbf{s}_0)$ be the expected information at $t_M(\mathbf{s}_0)$. It is reasonable to choose that run (i.e., the value of $\mathbf{s}_0$) whose $I_M(\mathbf{s}_0)$ has the largest value. The actual observations to be made during the run, i.e., the values of the t_μ, are chosen in that portion of the $I(t)$ curve where I is not much below I_M. The problem of determining values of t_μ is further treated by Heineken *et al.* (1967a,b).

Other complications arise when the cost of an experiment depends strongly on the experimental conditions. It may then be cheaper to gain a certain amount of information by performing several cheap though inefficient experiments, rather than a single efficient though expensive one. The simplest solution is to divide the expected information gain in an experiment by the cost of that experiment, and maximize the expected information per unit cost. Design criteria based purely on economic considerations can be derived from decision theory, as shown in Section 10-10.

10-8. Computational Considerations

The problem is to locate the maximum of the design criterion, which is a complicated nonlinear function D of the experimental conditions $\mathbf{x}$. The function is often so complicated that analytic computation of its derivatives is out of the question. Additional factors which contribute to the difficulty of the problem are the following:

1. The maximum is usually located on the boundary of the feasible region.
2. There are usually several local maxima. In the cases that have been studied in detail, the number of local maxima tended to be close to the number of unknown parameters in the model.

‡ It is possible for the maximum to be approached asymptotically as $t \to \infty$. We then select t_M as the time at which $I = I_M - \varepsilon$.

As indicated in Chapter V, maximization of a nonlinear function is easiest when derivatives can be calculated, no constraints apply, and there is a unique local maximum. On all these scores our problem is a difficult one. Furthermore, if we wish to obtain the most information in each experiment, we must repeat the maximization procedure before each experiment is performed. Fortunately, there are some mitigating circumstances:

1. There is no need to locate the maximum with a great deal of precision.
2. The locations of the local maxima do not seem to vary much from one experiment to the next. What do change are the relative heights of the various maxima, so that the conditions chosen for a sequence of experiments cycle among the several local maxima.‡

Indeed, Box (1968) shows that if a sequence of n experiments is designed (nonsequentially) to estimate $l < n$ parameters, then an optimal or near-optimal design is usually obtained if the l best experiments are each replicated n/l (as closely as possible) times.

It seems, therefore, that we need search for the local maxima throughout the entire feasible region only the first time around, i.e., after the initial experiments have been performed. After that, when the results of each new experiment come in, we need only search in the neighborhood of each already established local maximum so as to locate its current position (which may shift slightly after each experiment).

The safest way to conduct the initial thorough search for local maxima is to evaluate the design criterion at all points on a dense grid throughout the feasible region. Those grid points where the design criterion exceeds the values at all neighbors are selected as approximate locations of the local maxima. Further refinement can then be achieved by starting hill climbing procedures (e.g., direct search optimization, see Section 5-19) at these points. A sufficiently fine initial grid makes this step superfluous.

The grid search technique is feasible only when the number of independent experimental conditions is small. With three variables, a ten-level grid in each dimension results in a thousand points, which is not excessive if the model equations are simple. With four or more dimensions, the grid search technique is likely to be impractical. In this case we suggest the following procedure:

1. Select a feasible point at random.
2. Starting from this point, apply a direct search optimization procedure until a local maximum is reached.
3. Repeat 1 and 2 until at least l (= maximum number of parameters in

‡ This statement, like most others in this section, is based solely on a limited amount of experience with computer-simulated experiments.

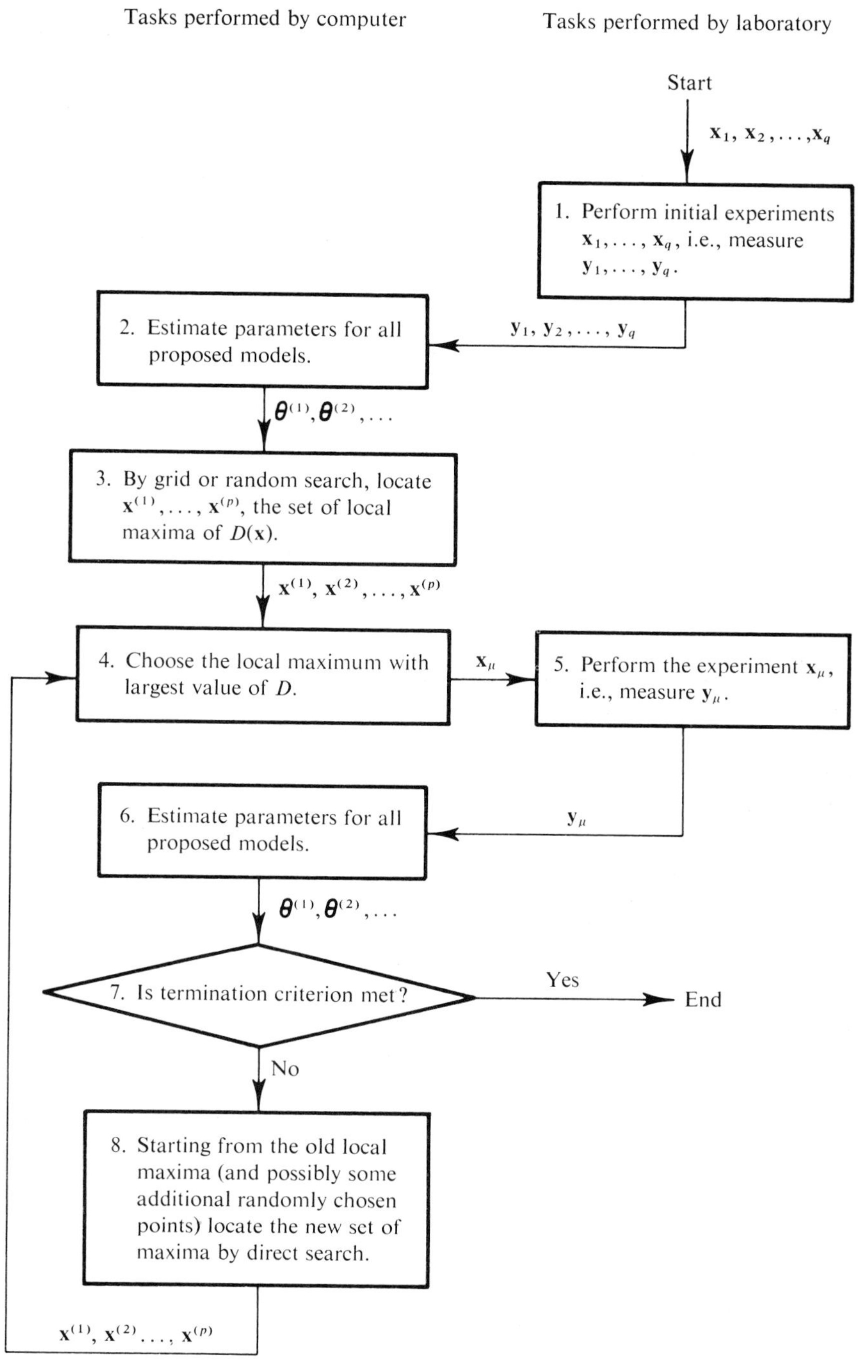

Fig. 10-4. A sequential experimental procedure. Symbols next to arrows indicate transmitted data.

any of the models considered) distinct local maxima have been obtained, or until a certain number of tries has failed to uncover a new local maximum.

Let $\mathbf{x}^{(1)}, \mathbf{x}^{(2)}, \ldots, \mathbf{x}^{(p)}$ be the known local maxima after n experiments have been performed. The $(n+1)$st experiment is, of course, conducted at the highest local maximum, i.e., at the $\mathbf{x}^{(i)}$ whose design criterion is largest. After this experiment has been performed, we establish new values of the $\mathbf{x}^{(i)}$ by applying the direct search technique, starting at each of the old $\mathbf{x}^{(i)}$. It is not uncommon for some of the new $\mathbf{x}^{(i)}$ to coalesce; i.e., searches starting at several of the old $\mathbf{x}^{(i)}$ may lead to the same (within some tolerance ε) new $\mathbf{x}^{(i)}$. To guard against the possibility that some maxima are being overlooked, one may also include after each experiment additional random starting points for searches.

Fig. 10-4 contains a proposed flow diagram for the procedure to be followed. The diagram is divided into two sections, dealing respectively with the functions of the computer (estimation and design), and of the laboratory (execution of specified experiments). This raises the question of how to implement the links between the computer and the laboratory apparatus. The answer depends on the circumstances; if the experiments are of very short duration and suitable instrumentation is available, the computer may be connected directly on-line to the apparatus. Otherwise, manual transfer of data is required. Note that the computer functions described here are quite distinct from actual on-line control of experiments, where the question is not what experiments to perform, but how to insure that the specified experiment is carried out properly. Of course, on-line design cannot perform unless the control function is also implemented, but the latter is outside the scope of this book.

10-9. Computer Simulated Experiments

Before applying our design methods to real experiments it may be wise to test them on computer simulated experiments. In this way we can determine economically whether the method is likely to succeed.

How do we simulate an experiment on the computer? An experiment is, from our point of view, merely a device for generating the value of $\mathbf{y}_\mu$ for a given value of $\mathbf{x}_\mu$. To simulate the experiment, all we need then is a computer routine which accepts a value of $\mathbf{x}_\mu$, and returns a value of $\mathbf{y}_\mu$. Internally, this subroutine should compute $\mathbf{y}_\mu$ using a formula such as

$$\mathbf{y}_\mu = \mathbf{f}^{(0)}(\mathbf{x}_\mu, \boldsymbol{\theta}^{(0)}) + \boldsymbol{\varepsilon}_\mu \tag{10-9-1}$$

where $\mathbf{f}^{(0)}$ is one of the models proposed for the phenomenon under study, and $\boldsymbol{\theta}^{(0)}$ is a specific set of values for the parameters appearing in this model.

The error term $\boldsymbol{\varepsilon}_\mu$ consists of pseudorandom numbers with the proper probability distribution (see Section 3-3). In addition, we may include a systematic error, to test what happens if none of the proposed models is really correct.

The experimental design procedure is tested by applying the procedure of Fig. 10-4, with the functions of boxes (1) and (5) performed by the computer routine just discussed. Note that only this particular routine "knows" which model has been selected, and what parameter values have been assigned; precisely as in nature the laboratory apparatus "knows" the model and the parameters. The only way in which other computer routines (e.g., those performing the functions of boxes (2) and (6)) can guess at the right model and parameter values is by analyzing the data ($\mathbf{y}_\mu$ values) supplied by boxes (1) and (5).

We present, now, a numerical example (Bard and Lapidus, 1968) in which the design method for discrimination among models is applied to computer-simulated experiments. The example clearly illustrates the potential power of the method.

Hougen and Watson (1947, pp. 943–958) have proposed eighteen alternative models for determining the rate of catalytic hydrogenation of mixed isooctenes into isooctane

$$C_8H_{16} + H_2 \rightarrow C_8H_{18} \tag{10-9-2}$$

Blakemore and Hoerl (1963) have attempted to fit all these models, and two additional ones, to data that were available in the literature. They found that all but two of the models could be rejected immediately. There was no conclusive evidence to choose between these two, which have the forms

$$y = \theta_1^{(1)} x_1 x_2 / (1 + \theta_2^{(1)} x_1^{1/2} + \theta_3^{(1)} x_2 + \theta_4^{(1)} x_3)^3 \tag{10-9-3}$$

and

$$y = \theta_1^{(2)} x_1 x_2 / (1 + \theta_2^{(2)} x_1 + \theta_3^{(2)} x_2 + \theta_4^{(2)} x_3)^2 \tag{10-9-4}$$

where y is the rate of reaction and x_1, x_2 and x_3 are the partial pressures of hydrogen, isooctene, and isooctane, respectively. Blakemore and Hoerl conclude, in part

"Carefully designed experiments are necessary . . . there are no fitting techniques which can overcome the deficiencies of poorly-designed experiments . . ."

This system was, therefore, considered a good one for testing the experimental design procedure. To simulate the reaction on the computer, the following relations were used

For experiments $\mu = 1, 2, \ldots, 6$

$$y = 0.0653477 x_1 x_2 [1 + \varepsilon(\sigma)] / (1 + 0.128246 x_1^{1/2} + 0.159038 x_2 + 0.0206618 x_3)^3 \tag{10-9-5}$$

and for experiments $\mu = 7, 8, \ldots$

$$y = 0.0558x_1x_2[1 + \varepsilon(\sigma)]/(1 + 0.104x_1 + 0.264x_2 + 0.0151x_3)^2 \quad (10\text{-}9\text{-}6)$$

where $\varepsilon(\sigma)$ is a pseudorandom number with distribution $N_1(0, \sigma^2)$. Note that σ is the standard deviation of the relative error in y. This choice of model is to be interpreted as follows:

Model Eq. (4) is the correct one, but by chance the first six experiments happen to give wrong results which appear to be closer to model Eq. (3). The aim was to see how soon the experimental design procedure could pick out Eq. (4) as the correct model, in spite of the handicap posed by the first six observations. The parameter values used in Eqs. (5) and (6) were those that gave the best least squares fits to the literature data used by Blakemore and Hoerl. The permitted ranges of the independent variables were the same as in the literature data, i.e.,

$$\begin{aligned} 0.1 &\leqslant x_1 \leqslant 2.5 \\ 0.1 &\leqslant x_2 \leqslant 3 \\ 0.05 &\leqslant x_3 \leqslant 2.7 \end{aligned} \quad (10\text{-}9\text{-}7)$$

The flow chart of Fig. 10-4 was implemented in the following way:

Box 1 The initial experiments, six in number, formed a fractional factorial design. They consisted of the centers of the six surfaces bounding the region defined by Eq. (7). The conditions for these experiments are listed in Table 10-1, along with the results [computed from Eq. (5)] for the case $\sigma = 0.03$, i.e., 3% relative error.

Table 10-1
Initial Experiments

μ	x_1	x_2	x_3	$y\ (\sigma = 0.03)$
1	0.1	1.55	1.375	0.00441
2	2.5	1.55	1.375	0.07932
3	1.3	0.1	1.375	0.00508
4	1.3	3	1.375	0.05633
5	1.3	1.55	0.05	0.04912
6	1.3	1.55	2.7	0.04292

Box 2 The least squares criterion was used to estimate parameters for both models. The fact that the relative rather than absolute error remained constant from experiment to experiment was ignored (i.e., it was assumed that the experimenter did not know that the error standard deviation varied from

experiment to experiment). The parameter estimates with their standard deviations and the residual standard deviations for the data of Table 10-1 are presented in Table 10-2.

Table 10-2
Parameter Estimates for Initial Experiments ($\sigma = 0.03$)

Model	θ_1	θ_2	θ_3	θ_4	Standard deviation of residuals
Eq. (10-9-3)	0.064372 ± 0.000294	0.116329 ± 0.001065	0.160034 ± 0.000544	0.024028 ± 0.000272	0.488055×10^{-4}
Eq. (10-9-4)	0.056738 ± 0.001808	0.071874 ± 0.005881	0.277537 ± 0.009287	0.040166 ± 0.003611	3.75137×10^{-4}

It is not surprising that at this point model Eq. (3) gives much the better fit, and its parameters are the better-determined ones.

Box 3 Since there are only three independent variables, a complete grid search was considered feasible. The design criterion function $J_{1,2}(\mathbf{x})$ of Eq. (10-5-7) was evaluated at all points on an $11 \times 11 \times 11$ grid encompassing the feasible region defined by Eq. (7). Local maxima are taken to be those grid points at which $J_{1,2}$ exceeds values at all direct neighbors. The local maxima after the six preliminary experiments are listed in Table 10-3, with the highest maximum underlined.

Table 10-3
Local Maxima of Design Criterion Function After Initial Experiments ($\boldsymbol{\sigma} = 0.03$)

x_1	x_2	x_3	$J_{1,2}$
2.02	0.68	0.05	1.01445
2.5	2.13	0.05	0.1536827
0.58	2.42	0.05	0.9023107
2.5	3	2.7	<u>1.287048</u>

Box 4 We choose the highest maximum of $J_{1,2}$ for our next experiment. According to Table 10-3, then, we perform the seventh experiment at $\mathbf{x}_7 = (2.5, 3, 2.7)$.

Box 5 Eq. (6) is used to generate y_μ. In our example, y_7 turns out to be 0.09769.

Box 6 Procedure identical to Box 2.

Box 7 The simulation runs were terminated after 30 experiments. However, the likelihood ratio Eq. (10-6-6) was evaluated and printed out after each experiment, so that the number of experiments that would have been required for given confidence levels α, β could be determined easily. Let $R_\mu = L_\mu^{(2)}/L_\mu^{(1)}$ after μ experiments, and assume $\alpha = \beta$. Then quitting after μ experiments would be correct had we set $\beta = 1/(1 + R_\mu)$, and our confidence in preferring the second model after μ experiments is given by

$$C_\mu^{(2)} \equiv 1 - \beta = R_\mu/(1 + R_\mu) = L^{(2)}/(L^{(1)} + L^{(2)}) = (\sigma^{(1)})^\mu/[(\sigma^{(1)})^\mu + (\sigma^{(2)})^\mu] \tag{10-9-8}$$

Box 8 The entire grid search of box 3 was repeated after each experiment. This, of course, would be impractical in larger problems. The procedure described in Fig. 10-4 was also tried, and led to results that were very nearly as good.

Table 10-4 gives the details of experiments 7–30 for the case $\sigma = 0.03$. In addition to $\mathbf{x}_\mu$ and y_μ we list the logarithm of the likelihood ratio and the confidence $C_\mu^{(2)}$ in preferring model Eq. (4) over Eq. (3) after each experiment has been processed.

Similar runs were made with relative error standard deviations of 1%, 3% and 6%. Fig. 10-5 summarizes the results of the three runs. It should be noted that to establish a preference for model Eq. (4) with 95% confidence, we needed 17 experiments with $\sigma = 0.01$, 21 experiments with $\sigma = 0.03$, and by the method of Section 10-6 we predicted that 36 experiments would be required with $\sigma = 0.06$.

For this problem, the use of $\max|y^{(2)} - y^{(1)}|$ as the design criterion worked just as well as using Eq. (10-5-7).

To determine whether the sequential design procedure employed here provides any improvement over classical design procedures, the 27 experiments of a $3 \times 3 \times 3$ factorial design were simulated. These are formed by taking all possible combinations of the independent variables at the following levels:

$$x_1 = 0.1, 1.3, 2.5$$
$$x_2 = 0.1, 1.55, 3$$
$$x_3 = 0.05, 1.375, 2.7$$

These include the six initial experiments of Table 10-1. The results are compared to those obtained in the sequential design procedure in Table 10-5.

Table 10-4
Computer Designed Experiments ($\sigma = 0.03$)

μ	x_1	x_2	x_3	y	$\log(L^{(2)}/L^{(1)})$	Confidence in preferring model Eq. (10-9-4)
1–6		(see Table 10-1)			−12.24	0.00005
7	2.5	3	2.7	0.09769	1.29	0.563
8	1.78	0.68	0.05	0.03709	−0.443	0.391
9	0.58	2.42	0.05	0.02649	0.635	0.654
10	1.78	0.39	0.05	0.02243	1.392	0.801
11	2.5	1.84	1.375	0.08126	1.029	0.737
12	0.34	2.42	1.11	0.01589	1.148	0.759
13	1.54	0.68	0.05	0.03363	1.630	0.836
14	0.34	2.42	3.15	0.01588	1.936	0.874
15	2.02	3	2.17	0.08211	2.209	0.901
16	2.5	1.84	1.64	0.08494	1.825	0.861
17	1.54	0.68	0.05	0.03364	2.325	0.911
18	0.34	2.42	0.05	0.01638	2.503	0.924
19	2.02	3	2.435	0.07961	1.757	0.853
20	2.5	1.84	1.905	0.08108	2.499	0.924
21	1.54	0.68	0.05	0.03411	3.081	0.956
22	0.34	2.42	0.05	0.01525	3.677	0.975
23	1.54	0.68	0.05	0.03146	3.077	0.956
24	0.34	2.42	0.05	0.01627	3.280	0.964
25	2.02	3	2.7	0.08174	3.884	0.980
26	2.5	1.84	2.17	0.07816	4.308	0.987
27	1.54	0.68	0.05	0.03096	3.783	0.978
28	2.5	1.84	1.11	0.07775	3.636	0.974
29	0.34	2.42	1.11	0.01610	3.812	0.978
30	2.5	1.84	2.7	0.08084	4.053	0.983

To interpret the numbers in the table, remember that a 0.5 preference level indicates complete indifference between the two models. Thus, at error levels of 3% or more, the factorial design completely fails to differentiate between the models, whereas the sequential procedure generates 83.3% confidence in the correct model even with a 6% error. At a 1% error level, the factorial design barely prefers the correct model, whereas the sequential design selects the proper model with almost complete certainty.

Admittedly, systematic errors and other complications that may be expected in practice were absent from this study. Still, the benefits of the sequential approach turned out to be very substantial. One has reason to hope that even under less favorable circumstances, at least some of these benefits will be retained. In fact, Hunter and Mezaki (1967) have reported

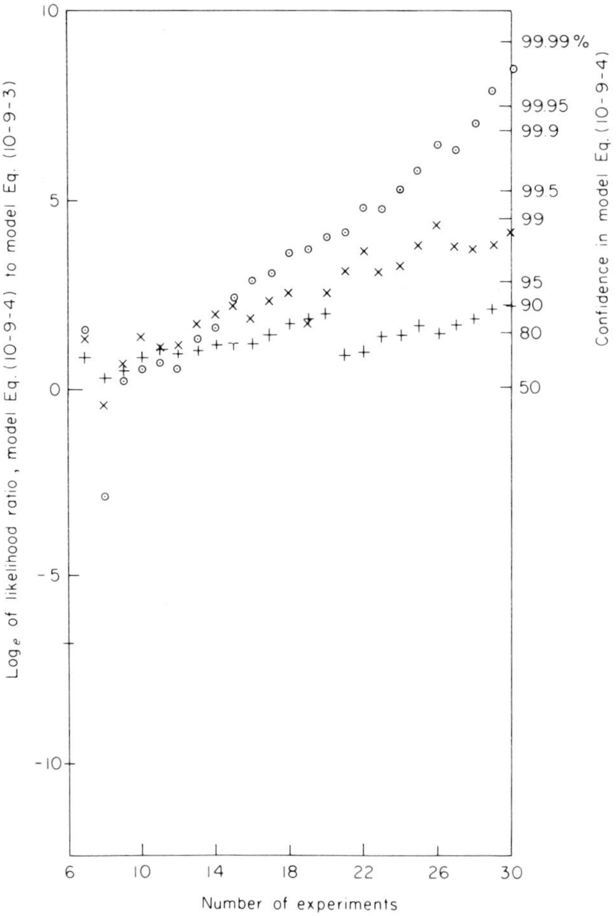

Fig. 10-5. Sequential discrimination between models. Standard deviation of measurement errors: ⊙, 1%; ×, 3%; +, 6%.

Table 10-5
Comparison of Experimental Design Procedures

	Preference for model Eq. (10-9-4) after 27 experiments	
σ	Factorial design	Sequential design
0.01	0.589	0.9983
0.03	0.481	0.978
0.06	0.500	0.833

successful application of the sequential design procedure to the discrimination between two alternative models for the kinetics of the catalytic hydrogenation of propylene. Nine experiments previously performed yielded a likelihood ratio of $L^{(1)}/L^{(2)} = 1.22$. After a mere four additional properly designed experiments a firm preference for model 1 was established with $L^{(1)}/L^{(2)} = 99$.

10-10. Design for Decision Making

So far we have been concerned with the somewhat abstract aim of elucidating the "true" model or parameter values. Consequently, we used an abstract measure of information to select the experiments. When the parameter values are required for some specific purpose it may be more appropriate to minimize the total expected cost of achieving that purpose. We have already (Section 4-16) introduced the loss function $c(\boldsymbol{\theta}^*, \hat{\boldsymbol{\theta}})$ which represents the cost of using the value $\boldsymbol{\theta}^*$ where $\hat{\boldsymbol{\theta}}$ is the true value. Similarly, we introduce a function $d(\mathbf{X})$ which represents the cost of performing the series of experiments $\mathbf{X} \equiv [\mathbf{x}_1, \mathbf{x}_2, \ldots, \mathbf{x}_n]^{\mathrm{T}}$. The outcome of the as yet unperformed experiments is the random variable $\mathbf{Y} \equiv [\mathbf{y}_1, \mathbf{y}_2, \ldots, \mathbf{y}_n]^{\mathrm{T}}$ whose pdf is $p(\mathbf{Y} | \mathbf{X}, \boldsymbol{\theta})$. The latter function is by definition also the likelihood‡ $L(\boldsymbol{\theta} | \mathbf{X}, \mathbf{Y})$ of any hypothetical sample $\mathbf{Y}$. Hence, given a prior density $p_0(\boldsymbol{\theta})$, we can form the posterior density $p^*(\boldsymbol{\theta} | \mathbf{X}, \mathbf{Y}) = kp_0(\boldsymbol{\theta})L(\boldsymbol{\theta} | \mathbf{X}, \mathbf{Y})$ for any possible outcome $\mathbf{Y}$. We can also form the expected (marginal) pdf of $\mathbf{Y}$ by assigning the weight $p_0(\boldsymbol{\theta})$ to each possible value of $p(\mathbf{Y} | \mathbf{X}, \boldsymbol{\theta})$

$$p(\mathbf{Y} | \mathbf{X}) = \int p(\mathbf{Y} | \mathbf{X}, \boldsymbol{\theta}) p_0(\boldsymbol{\theta}) \, d\boldsymbol{\theta} \tag{10-10-1}$$

Using Eq. (4-16-1) we can evaluate the risk associated with using the value $\boldsymbol{\theta}^*$ on the assumption that the outcome of the experiments to be performed will be some specific value $\mathbf{Y}$

$$R(\boldsymbol{\theta}^* | \mathbf{X}, \mathbf{Y}) = \int c(\boldsymbol{\theta}^*, \boldsymbol{\theta}) p^*(\boldsymbol{\theta} | \mathbf{X}, \mathbf{Y}) \, d\boldsymbol{\theta} \tag{10-10-2}$$

Once the outcome $\mathbf{Y}$ becomes known, we shall of course choose $\boldsymbol{\theta}^*$ so as to minimize the risk. We denote this minimum risk $R^*(\mathbf{X}, \mathbf{Y})$

$$R^*(\mathbf{X}, \mathbf{Y}) \equiv \min_{\boldsymbol{\theta}^*} R(\boldsymbol{\theta}^* | \mathbf{X}, \mathbf{Y}) \tag{10-10-3}$$

We cannot yet evaluate R^* because we have not measured $\mathbf{Y}$. However, following Raiffa and Schlaifer (1961), we can find the expected value of R^*

‡ Contrary to previous practice we retain the arguments $\mathbf{X}$ and $\mathbf{Y}$ in the expression for the likelihood because the data have not yet been taken.

by averaging over all possible outcomes of the proposed experiments $\mathbf{X}$, using $p(\mathbf{Y}|\mathbf{X})$ as defined in Eq. (1)

$$\hat{R}(\mathbf{X}) \equiv \int R^*(\mathbf{X}, \mathbf{Y})p(\mathbf{Y}|\mathbf{X})\, d\mathbf{Y} \tag{10-10-4}$$

$\hat{R}(\mathbf{X})$ is the expected risk associated with performing the experiments $\mathbf{X}$. To this we add the cost of experimentation $d(\mathbf{X})$ to obtain the total expected cost of $\mathbf{X}$

$$C(\mathbf{X}) \equiv d(\mathbf{X}) + \hat{R}(\mathbf{X}) \tag{10-10-5}$$

We shall perform the set of experiments $\mathbf{X}$ for which $C(\mathbf{X})$ is minimum. Among the possible sets of experiments is the null set, i.e., no experiments at all. In this case $d(\mathbf{X}) = 0$ and $p^*(\boldsymbol{\theta}|\mathbf{X}, \mathbf{Y}) = p_0(\boldsymbol{\theta})$. Hence R does not depend on $\mathbf{Y}$, and

$$C = \hat{R} = \min_{\boldsymbol{\theta}^*} R(\boldsymbol{\theta}^*) = \min_{\boldsymbol{\theta}^*} \int c(\boldsymbol{\theta}^*, \boldsymbol{\theta})p_0(\boldsymbol{\theta})\, d\boldsymbol{\theta}$$

We now analyze the case in which both $p_0(\boldsymbol{\theta})$ and $p(\mathbf{Y}|\mathbf{X}, \boldsymbol{\theta})$ are normal. Let:

$$p_0(\boldsymbol{\theta}) = N_l(\boldsymbol{\theta}_0, \mathbf{V}_0) \tag{10-10-6}$$

$$p(\mathbf{Y}|\mathbf{X}, \boldsymbol{\theta}) = N_{mn}[\mathbf{F}(\mathbf{X}, \boldsymbol{\theta}), \boldsymbol{\Pi}] \tag{10-10-7}$$

where, as usual, $\mathbf{Y}$ denotes the mn-dimensional vector obtained by adjoining to each other the n rows of $\mathbf{Y}$, and $\boldsymbol{\Pi}$ is the joint covariance matrix of the errors in all projected experiments, usually given by Eq. (10-3-3). We now assume that the model equations $\mathbf{F}(\mathbf{X}, \boldsymbol{\theta})$ can be reasonably approximated in the region of interest by a first-order Taylor series expansion around $\boldsymbol{\theta} = \boldsymbol{\theta}_0$, i.e.,

$$\mathbf{F}(\mathbf{X}, \boldsymbol{\theta}) \approx \mathbf{Y}_0 + \mathbf{B}(\boldsymbol{\theta} - \boldsymbol{\theta}_0) \tag{10-10-8}$$

where $\mathbf{Y}_0 \equiv \mathbf{F}(\mathbf{X}, \boldsymbol{\theta}_0)$ and $\mathbf{B} \equiv \partial\mathbf{F}/\partial\boldsymbol{\theta})_{\boldsymbol{\theta}=\boldsymbol{\theta}_0}$. Note that $\mathbf{B}$ is a function of $\mathbf{X}$. Now Eq. (7) can be rewritten as

$$p(\mathbf{Y}|\mathbf{X}, \boldsymbol{\theta}) = N_{mn}[\mathbf{Y}_0 + \mathbf{B}(\boldsymbol{\theta} - \boldsymbol{\theta}_0), \boldsymbol{\Pi}] \tag{10-10-9}$$

We leave it as an exercise for the reader to show that both the posterior density of $\boldsymbol{\theta}$ and the marginal density of $\mathbf{Y}$ are also normal. Specifically

$$p^*(\boldsymbol{\theta}|\mathbf{X}, \mathbf{Y}) = N_l(\tilde{\boldsymbol{\theta}}, \tilde{\mathbf{V}}) \tag{10-10-10}$$

where

$$\tilde{\boldsymbol{\theta}} = \boldsymbol{\theta}_0 + (\mathbf{V}_0^{-1} + \mathbf{B}^T\boldsymbol{\Pi}^{-1}\mathbf{B})^{-1}\mathbf{B}^T\boldsymbol{\Pi}^{-1}(\mathbf{Y} - \mathbf{Y}_0) \tag{10-10-11}$$

$$\tilde{\mathbf{V}} = (\mathbf{V}_0^{-1} + \mathbf{B}^T\boldsymbol{\Pi}^{-1}\mathbf{B})^{-1} \tag{10-10-12}$$

and

$$p(\mathbf{Y}|\mathbf{X}) = N_{mn}(\mathbf{Y}_0, \tilde{\boldsymbol{\Pi}}) \tag{10-10-13}$$

where

$$\tilde{\boldsymbol{\Pi}} = [\boldsymbol{\Pi}^{-1} - \boldsymbol{\Pi}^{-1}\mathbf{B}\tilde{\mathbf{V}}\mathbf{B}^{\mathrm{T}}\boldsymbol{\Pi}^{-1}]^{-1} \tag{10-10-14}$$

The situation is particularly tractable if the loss function is quadratic, i.e., as in Eq. (4-16-6)

$$c(\boldsymbol{\theta}^*, \hat{\boldsymbol{\theta}}) = (\boldsymbol{\theta}^* - \hat{\boldsymbol{\theta}})^{\mathrm{T}}\mathbf{P}(\boldsymbol{\theta}^* - \hat{\boldsymbol{\theta}}) \tag{10-10-15}$$

where $\mathbf{P}$ is a given positive definite (or at least semidefinite) matrix. As was shown in Section 4-16, this leads to an optimal choice of $\boldsymbol{\theta}^* = \tilde{\boldsymbol{\theta}}$ (the mean of the posterior distribution). Then the minimum risk is the expected value of $(\tilde{\boldsymbol{\theta}} - \boldsymbol{\theta})^{\mathrm{T}}\mathbf{P}(\tilde{\boldsymbol{\theta}} - \boldsymbol{\theta})$ under the posterior distribution i.e.,

$$R^*(\mathbf{X}, \mathbf{Y}) = E[(\tilde{\boldsymbol{\theta}} - \boldsymbol{\theta})^{\mathrm{T}}\mathbf{P}(\tilde{\boldsymbol{\theta}} - \boldsymbol{\theta})] = \mathrm{Tr}\,\mathbf{P}\tilde{\mathbf{V}} \tag{10-10-16}$$

A glance at Eq. (12) indicates that $\tilde{\mathbf{V}}$ and hence R^* are independent of $\mathbf{Y}$, hence $\hat{R}(\mathbf{X}) = R^*(\mathbf{X}, \mathbf{Y})$ and

$$C(\mathbf{X}) = d(\mathbf{X}) + \mathrm{Tr}\,\mathbf{P}(\mathbf{V}_0^{-1} + \mathbf{B}^{\mathrm{T}}\boldsymbol{\Pi}^{-1}\mathbf{B})^{-1} \tag{10-10-17}$$

When Eq. (10-3-3) applies

$$C(\mathbf{X}) = d(\mathbf{X}) + \mathrm{Tr}\,\mathbf{P}\Big(\mathbf{V}_0^{-1} + \sum_{\mu=1}^{n} \mathbf{B}_\mu{}^{\mathrm{T}}\mathbf{V}^{-1}\mathbf{B}_\mu\Big)^{-1} \tag{10-10-18}$$

When minimizing $C(\mathbf{X})$ we can seek to find the optimal number of experiments as well as the conditions under which they are to be performed. If the experiments are to be performed in sequence, it is only necessary at any given time to find the optimal conditions for a single experiment $\mathbf{x}_1$, and compare the associated cost $\min_{\mathbf{x}_1} C(\mathbf{x}_1)$ with the expected cost of performing no experiment at all (which is $\mathrm{Tr}\,\mathbf{P}\mathbf{V}_0^{-1}$ when Eq. (18) holds). If the outcome is favorable to the additional experiment, we perform that experiment, replace $p_0(\boldsymbol{\theta})$ by $p^*(\boldsymbol{\theta})$, and repeat the procedure. The stopping rule is obvious; cease experimentation when the expected cost of no experiment falls below the minimum expected cost of the next experiment.

It must be admitted that while the procedures outlined above are conceptually simple and appealing, their implementation is difficult in most practical situations. While the minimization of Eq. (18) is no more difficult than the minimization of Eq. (10-3-2), almost any other loss function leads to severe computational difficulties which arise from the need to evaluate multiple infinite integrals for all possible values of $\mathbf{X}$, $\mathbf{Y}$, and $\boldsymbol{\theta}$. To this must be added the difficulty of assigning realistic cost functions, a by no means trivial task.

10-11. Problems

1. Verify Eq. (10-10-10)–(10-10-12).

2. Using Eq. (10-2-2), Eq. (10-2-4), and Eq. (10-10-12), show that in the case of normal prior and error distributions and a linear model, the value of I is positive; i.e., one gains information no matter what the outcome of the experiment. For more general results, see Lindley (1956).

3. Derive Eq. (10-5-6).

4. Derive a decision-theoretic design criterion for discriminating between alternative models. Assume that one is given prior probabilities $\pi^{(i)}$ that the ith model is correct, and that the loss function has the form $c_{ij}(\boldsymbol{\theta}, \hat{\boldsymbol{\theta}})$ which represents the cost of assuming that model i holds with parameter values $\boldsymbol{\theta}$, when in fact model j holds with parameter values $\hat{\boldsymbol{\theta}}$.

Appendix

A

Matrix Analysis

A-1. Matrix Algebra

The reader unfamiliar with matrix notation may prefer to write out matrix expressions in full. But he will soon develop facility in manipulating matrices and will no longer need subscripts and summations. This will greatly enhance his insight and enjoyment of the subject.

Throughout the book, boldface normal size capital letters (both latin and greek) denote matrices, e.g.,

$$\mathbf{A} = [A_{ij}] = \begin{bmatrix} A_{11} & A_{12} & \cdots & A_{1n} \\ A_{21} & A_{22} & \cdots & A_{2n} \\ \vdots & & & \\ A_{m1} & A_{m2} & \cdots & A_{mn} \end{bmatrix} \tag{A-1-1}$$

is an $m \times n$ matrix. $\mathbf{A}$ is *square* if $m = n$. A matrix all of whose elements are zero is denoted $\mathbf{0}$ and is called the *null matrix*. Bold face small capital letters denote column vectors obtained by adjoining to each other the rows of the corresponding matrix. Thus, if $\mathbf{A}$ is defined by Eq. (1), then

$$\mathbf{A} \equiv \begin{bmatrix} A_{11} \\ A_{12} \\ \vdots \\ A_{1n} \\ A_{21} \\ A_{22} \\ \vdots \\ A_{mn} \end{bmatrix} \tag{A-1-2}$$

Boldface lower case letters denote column vectors, e.g.,

$$\mathbf{a} = [a_i] = \begin{bmatrix} a_1 \\ a_2 \\ \vdots \\ a_m \end{bmatrix} \tag{A-1-3}$$

is an m-dimensional vector. A vector all of whose elements are zero is denoted $\mathbf{0}$. All nonboldface characters are scalars. Capital or lower case nonboldface letters with subscripts may be elements of the corresponding matrix or vector. A subscripted boldface letter indicates one in a set of vectors or matrices.

The superscript T denotes transposition. if $\mathbf{A}$ is defined by Eq. (1), then

$$\mathbf{A}^{\mathrm{T}} = [A_{ji}] = \begin{bmatrix} A_{11} & A_{21} & \cdots & A_{m1} \\ A_{12} & A_{22} & \cdots & A_{m2} \\ \vdots & & & \\ A_{1n} & A_{2n} & \cdots & A_{mn} \end{bmatrix} \qquad \text{(A-1-4)}$$

is an $n \times m$ matrix. A square matrix $\mathbf{A}$ is *symmetric* if $\mathbf{A}^{\mathrm{T}} = \mathbf{A}$; i.e., $A_{ij} = A_{ji}$ for all i and j.

If $\mathbf{a}$ is defined by Eq. (3), then

$$\mathbf{a}^{\mathrm{T}} = [a_1, a_2, \ldots, a_m] \qquad \text{(A-1-5)}$$

is an m-dimensional row vector.

If $\mathbf{A}$ and $\mathbf{B}$ are both $m \times n$, then $[\mathbf{A} + \mathbf{B}]_{ij} = A_{ij} + B_{ij}$.

We define the following matrix products:

(a) $\mathbf{A}$ is $m \times n$ and $\mathbf{B}$ is $n \times k$. Then $\mathbf{AB}$ is the $m \times k$ matrix whose i, j element is

$$[\mathbf{AB}]_{ij} = \sum_{l=1}^{n} A_{il} B_{lj} \qquad (i = 1, 2, \ldots, m; j = 1, 2, \ldots, k) \qquad \text{(A-1-6)}$$

(b) $\mathbf{A}$ is $m \times n$ and $\mathbf{b}$ is n-dimensional. Then $\mathbf{Ab}$ is the m-dimensional column vector whose ith element is

$$[\mathbf{Ab}]_i = \sum_{l=1}^{n} A_{il} b_l \qquad (i = 1, 2, \ldots, m) \qquad \text{(A-1-7)}$$

(c) $\mathbf{A}$ is $m \times n$ and $\mathbf{b}$ is m-dimensional. Then $\mathbf{b}^{\mathrm{T}}\mathbf{A}$ is the n-dimensional row vector whose ith element is

$$[\mathbf{b}^{\mathrm{T}}\mathbf{A}]_i = \sum_{l=1}^{m} b_l A_{li} \qquad \text{(A-1-8)}$$

(d) $\mathbf{a}$ and $\mathbf{b}$ are m-dimensional. Then the *inner product* $\mathbf{a}^{\mathrm{T}}\mathbf{b} = \mathbf{b}^{\mathrm{T}}\mathbf{a}$ is the scalar

$$\mathbf{a}^{\mathrm{T}}\mathbf{b} = \sum_{i=1}^{m} a_i b_i \qquad \text{(A-1-9)}$$

The inner product of a vector with itself, i.e., $\mathbf{a}^{\mathrm{T}}\mathbf{a}$ is the square of the *length* (also called *norm*) of $\mathbf{a}$. We use the notation $\|\mathbf{a}\|$ to designate the norm of $\mathbf{a}$.

(e) **a** is m-dimensional and **b** is n-dimensional. Then the *outer product* $\mathbf{ab}^{\mathrm{T}}$ is the $m \times n$ matrix whose i, j element is

$$[\mathbf{ab}^{\mathrm{T}}]_{ij} = a_i b_j \qquad (i = 1, 2, \ldots, m; j = 1, 2, \ldots, n) \qquad \text{(A-1-10)}$$

If we regard an m-dimensional column vector as an $m \times 1$ matrix and a similar row vector as a $1 \times m$ matrix, then all the above products become special cases of (a). From these definitions, one can work out the product of any number of terms. For example, the quantity $\mathbf{a}^{\mathrm{T}}\mathbf{Ab}$ is the scalar

$$\mathbf{a}^{\mathrm{T}}\mathbf{Ab} = \sum_{i, j} a_i A_{ij} b_j \qquad \text{(A-1-11)}$$

which may be verified by applying Eq. (7) first, and then Eq. (9). This is permissible because matrix and vector products are associative, i.e.,

$$\mathbf{a}^{\mathrm{T}}\mathbf{Ab} = \mathbf{a}^{\mathrm{T}}(\mathbf{Ab}) = (\mathbf{a}^{\mathrm{T}}\mathbf{A})\mathbf{b}$$

Let **A** be a square $m \times m$ matrix. The *main diagonal* of **A** is the set of elements $A_{11}, A_{22}, \ldots, A_{mm}$. A *diagonal* matrix is one whose only nonzero elements are on the main diagonal. The *identity* matrix **I** is a diagonal matrix, all of whose diagonal elements are unity, i.e.,

$$\mathbf{I} = \begin{bmatrix} 1 & 0 & \cdots & 0 \\ 0 & 1 & \cdots & 0 \\ \vdots & \cdot & & \\ 0 & 0 & \ldots & 1 \end{bmatrix} \qquad \text{(A-1-12)}$$

or

$$I_{ij} = \delta_{ij} \equiv \begin{matrix} 1 & (i = j) \\ 0 & (i \neq j) \end{matrix} \qquad \text{(A-1-13)}$$

The symbol δ_{ij} is called the *Kronecker delta*. Clearly

$$\mathbf{IA} = \mathbf{A}, \qquad \mathbf{BI} = \mathbf{B}, \qquad \mathbf{Ia} = \mathbf{a}, \qquad \mathbf{b}^{\mathrm{T}}\mathbf{I} = \mathbf{b}^{\mathrm{T}}$$

for any suitable matrices **A** and **B**, and vectors **a** and **b**.

If **A** is a square matrix, then $\mathbf{A}^{-1}$ designates a matrix (if one exists) such that

$$\mathbf{A}^{-1}\mathbf{A} = \mathbf{AA}^{-1} = \mathbf{I} \qquad \text{(A-1-14)}$$

$\mathbf{A}^{-1}$ is called the *inverse* of **A**. A matrix **A** can possess at most one inverse. If **A** has no inverse, it is said to be *singular*.

The following relations may be derived easily

$$(\mathbf{Ab})^{\mathrm{T}} = \mathbf{b}^{\mathrm{T}}\mathbf{A}^{\mathrm{T}}, \quad (\mathbf{AB})^{\mathrm{T}} = \mathbf{B}^{\mathrm{T}}\mathbf{A}^{\mathrm{T}}, \quad (\mathbf{AB})^{-1} = \mathbf{B}^{-1}\mathbf{A}^{-1}, \quad (\mathbf{A}^{\mathrm{T}})^{-1} = (\mathbf{A}^{-1})^{\mathrm{T}} \qquad \text{(A-1-15)}$$

A nonzero vector $\mathbf{v}$ is an *eigenvector* of the square matrix $\mathbf{A}$, and λ is the associated *eigenvalue*, if

$$\mathbf{A}\mathbf{v} = \lambda\mathbf{v} \tag{A-1-16}$$

Vectors $\mathbf{a}$ and $\mathbf{b}$ are *orthogonal* to each other if $\mathbf{a}^{\mathrm{T}}\mathbf{b} = \mathbf{0}$. If $\mathbf{A}$ is symmetric $m \times m$, then one can find m mutually orthogonal eigenvectors $\mathbf{v}_1, \mathbf{v}_2, \ldots, \mathbf{v}_m$ of $\mathbf{A}$. Usually, we normalize the vectors so that

$$\mathbf{v}_i{}^{\mathrm{T}}\mathbf{v}_j = \delta_{ij} \qquad (i, j = 1, 2, \ldots, m) \tag{A-1-17}$$

The $\mathbf{v}_i$ then form a set of *orthonormal* eigenvectors of $\mathbf{A}$.

Let $\mathbf{V}$ be the $m \times m$ matrix whose ith column is $\mathbf{v}_i$. In view of Eq. (17), we have $\mathbf{V}^{\mathrm{T}}\mathbf{V} = \mathbf{V}\mathbf{V}^{\mathrm{T}} = \mathbf{I}$, i.e., $\mathbf{V}^{\mathrm{T}} = \mathbf{V}^{-1}$. The matrix $\mathbf{V}$ is said to be *unitary*.

If $\mathbf{A}\mathbf{x} = \mathbf{0}$ ($\mathbf{x} \neq \mathbf{0}$), then $\mathbf{x}$ is called a *null vector* of $\mathbf{A}$. If $\mathbf{A}$ is square, then it can possess null vectors only if it is singular. A singular matrix has at least one zero eigenvalue.

Let $\mathbf{x}$ be a vector and $\mathbf{A}$ a symmetric matrix. The scalar $\mathbf{x}^{\mathrm{T}}\mathbf{A}\mathbf{x}$ may be regarded as a function of $\mathbf{x}$. It is called the *quadratic form* associated with $\mathbf{A}$. The matrix $\mathbf{A}$ is *positive definite* if $\mathbf{x}^{\mathrm{T}}\mathbf{A}\mathbf{x} > 0$ for all $\mathbf{x} \neq \mathbf{0}$, and *positive semidefinite* if $\mathbf{x}^{\mathrm{T}}\mathbf{A}\mathbf{x} \geqslant 0$ for all $\mathbf{x}$. Negative definiteness is defined analogously. All eigenvalues of a positive definite or positive semidefinite matrix are positive or nonnegative, respectively.

The symbol A_{ij}^{-1} is used to denote the i, j element of $\mathbf{A}^{-1}$, and not the reciprocal of A_{ij}.

If $\mathbf{A}$ is a square nonsingular matrix and $\mathbf{y}$ is a known vector, then the solution to the set of simultaneous linear equations

$$\mathbf{A}\mathbf{x} = \mathbf{y} \tag{A-1-18}$$

is given by

$$\mathbf{y} = \mathbf{A}^{-1}\mathbf{x} \tag{A-1-19}$$

Suppose $\mathbf{A}$ is any matrix, not necessarily square. Then there exists [see Penrose (1955)] a unique matrix $\mathbf{A}^{+}$, called the *pseudoinverse* of $\mathbf{A}$, satisfying the relations

$$\mathbf{A}\mathbf{A}^{+}\mathbf{A} = \mathbf{A}, \qquad \mathbf{A}^{+}\mathbf{A}\mathbf{A}^{+} = \mathbf{A}^{+}, \qquad \mathbf{A}^{+}\mathbf{A} = (\mathbf{A}^{+}\mathbf{A})^{\mathrm{T}}, \qquad \mathbf{A}\mathbf{A}^{+} = (\mathbf{A}\mathbf{A}^{+})^{\mathrm{T}} \tag{A-1-20}$$

If $\mathbf{A}$ is square nonsingular, then $\mathbf{A}^{+} = \mathbf{A}^{-1}$. If $\mathbf{A}$ is $m \times n$, then $\mathbf{A}^{+}$ is $n \times m$. If the equations $\mathbf{A}\mathbf{x} = \mathbf{y}$ have a solution, then $\mathbf{x} = \mathbf{A}^{+}\mathbf{y}$ is the solution of minimum length. If $\mathbf{A}\mathbf{x} = \mathbf{y}$ has no solution, then $\mathbf{x} = \mathbf{A}^{+}\mathbf{y}$ minimizes the sum of squares of the deviations $\mathbf{y} - \mathbf{A}\mathbf{x}$; and of all vectors having this property, $\mathbf{x} = \mathbf{A}^{+}\mathbf{y}$ has minimum length.

The *trace* of an $m \times m$ matrix **A** is the scalar

$$\mathrm{Tr}(\mathbf{A}) \equiv \sum_{i=1}^{m} A_{ii} \tag{A-1-21}$$

The trace of a matrix is equal to the sum of its eigenvalues and the *determinant* of a matrix is equal to the product of its eigenvalues.

One verifies easily that

$$\mathrm{Tr}(\mathbf{AB}) = \sum_{i,\, j} A_{ij} B_{ji} = \mathrm{Tr}(\mathbf{BA})$$

Hence

$$\mathrm{Tr}(\mathbf{ab}^{\mathrm{T}}) = \mathbf{b}^{\mathrm{T}}\mathbf{a} = \mathbf{a}^{\mathrm{T}}\mathbf{b}$$

and

$$\mathbf{a}^{\mathrm{T}}\mathbf{A}\mathbf{a} = \mathrm{Tr}(\mathbf{A}\mathbf{a}\mathbf{a}^{\mathrm{T}})$$

If **I** is the $m \times m$ identity matrix, then $\mathrm{Tr}(\mathbf{I}) = m$ and $\det(\mathbf{I}) = 1$.

Let **A** be the $m \times n$ matrix defined by Eq. (1). Suppose k and l are positive integers satisfying $k < m$ and $l < n$. Define the following matrices:

$$\mathbf{B} \equiv \begin{bmatrix} A_{11} & A_{12} & \cdots & A_{1l} \\ A_{21} & A_{22} & \cdots & A_{2l} \\ \vdots & & & \\ A_{k1} & A_{k2} & \cdots & A_{kl} \end{bmatrix}$$

$$\mathbf{C} \equiv \begin{bmatrix} A_{1,\, l+1} & A_{1,\, l+2} & \cdots & A_{1,\, n} \\ A_{2,\, l+1} & A_{2,\, l+2} & \cdots & A_{2,\, n} \\ \vdots & & & \\ A_{k,\, l+1} & A_{k,\, l+2} & \cdots & A_{kn} \end{bmatrix} \tag{A-1-22}$$

$$\mathbf{D} \equiv \begin{bmatrix} A_{k+1,\, 1} & A_{k+1,\, 2} & \cdots & A_{k+1,\, l} \\ A_{k+2,\, 1} & A_{k+2,\, 2} & \cdots & A_{k+2,\, l} \\ \vdots & & & \\ A_{m,\, 1} & A_{m,\, 2} & \cdots & A_{ml} \end{bmatrix}$$

$$\mathbf{E} \equiv \begin{bmatrix} A_{k+1,\, l+1} & A_{k+1,\, l+2} & \cdots & A_{k+1,\, n} \\ A_{k+2,\, l+1} & A_{k+2,\, l+2} & \cdots & A_{k+2,\, n} \\ \vdots & & & \\ A_{m,\, l+1} & A_{m,\, l+2} & \cdots & A_{mn} \end{bmatrix}$$

We write the matrix **A** in *partitioned form* as

$$\mathbf{A} = \begin{bmatrix} \mathbf{B} & \mathbf{C} \\ \mathbf{D} & \mathbf{E} \end{bmatrix} \tag{A-1-23}$$

Matrices in partitioned form may be multiplied as though the submatrices were elements, provided the resulting expressions make sense. For instance, let $\mathbf{x}$ be an n-dimensional vector partitioned as follows

$$\mathbf{x} = \begin{bmatrix} \mathbf{a} \\ \mathbf{b} \end{bmatrix} \tag{A-1-24}$$

where

$$\mathbf{a} \equiv \begin{bmatrix} x_1 \\ x_2 \\ \vdots \\ x_l \end{bmatrix}, \qquad \mathbf{b} \equiv \begin{bmatrix} x_{l+1} \\ x_{l+2} \\ \vdots \\ x_n \end{bmatrix} \tag{A-1-25}$$

Then one may easily verify that

$$\mathbf{Ax} = \begin{bmatrix} \mathbf{B} & \mathbf{C} \\ \mathbf{D} & \mathbf{E} \end{bmatrix} \begin{bmatrix} \mathbf{a} \\ \mathbf{b} \end{bmatrix} = \begin{bmatrix} \mathbf{Ba} + \mathbf{Cb} \\ \mathbf{Da} + \mathbf{Eb} \end{bmatrix} \tag{A-1-26}$$

Note that this makes sense only if $\mathbf{x}$ is partitioned so that the dimension of $\mathbf{a}$ equals the number of columns in $\mathbf{B}$ and $\mathbf{D}$.

The partitioning of a matrix into more than four submatrices proceeds analogously.

The *rank* of a matrix is the maximum number of linearly independent columns or rows in the matrix (it makes no difference whether we take rows or columns). A nonzero vector has rank 1. The rank of a square matrix equals the number of nonzero eigenvalues. We have:

$$\operatorname{rank}(\mathbf{A} + \mathbf{B}) \leqslant \operatorname{rank}\mathbf{A} + \operatorname{rank}\mathbf{B} \tag{A-1-27}$$

$$\operatorname{rank}(\mathbf{AB}) \leqslant \min(\operatorname{rank}\mathbf{A}, \operatorname{rank}\mathbf{B}) \tag{A-1-28}$$

It follows that

$$\operatorname{rank}(\mathbf{ab}^{\mathrm{T}}) = \begin{matrix} 1 & (\mathbf{a} \neq \mathbf{0} \neq \mathbf{b}) \\ 0 & (\mathbf{a} = \mathbf{0} \text{ or } \mathbf{b} = \mathbf{0}) \end{matrix} \tag{A-1-29}$$

and

$$\operatorname{rank} \sum_{i=1}^{n} \mathbf{a}_i \mathbf{b}_i^{\mathrm{T}} \leqslant n \tag{A-1-30}$$

A matrix whose rank equals the number of rows or columns (whichever is less) is said to be of *full rank*. A square matrix of full rank is nonsingular, and vice versa.

A matrix of the form $\mathbf{A} = \mathbf{aa}^{\mathrm{T}}$ is positive semidefinite, because for every vector $\mathbf{x}$

$$\mathbf{x}^{\mathrm{T}}\mathbf{Ax} = (\mathbf{x}^{\mathrm{T}}\mathbf{a})^2 \geqslant 0$$

The sum of positive semidefinite matrices is positive semidefinite. Hence $\sum_{i=1}^{n} \mathbf{a}_i \mathbf{a}_i^{\mathrm{T}}$ is positive semidefinite.

If $\mathbf{A}$ is positive semidefinite, then so is $\mathbf{B}^{\mathrm{T}}\mathbf{A}\mathbf{B}$ where $\mathbf{B}$ is any matrix or vector.

Let $\mathbf{A}$ be a square matrix. Suppose $\lambda_{\min}$ and $\lambda_{\max}$ are the eigenvalues of $\mathbf{A}$ with smallest and largest absolute values, respectively. Then for any vector $\mathbf{b}$:

$$|\lambda_{\min}|\ \|\mathbf{b}\| \leqslant \|\mathbf{A}\mathbf{b}\| \leqslant |\lambda_{\max}|\ \|\mathbf{b}\| \tag{A-1-31}$$

$$|\lambda_{\min}|\ \|\mathbf{b}\|^2 \leqslant |\mathbf{b}^{\mathrm{T}}\mathbf{A}\mathbf{b}| \leqslant |\lambda_{\max}|\ \|\mathbf{b}\|^2 \tag{A-1-32}$$

If $\mathbf{A}$ and $\mathbf{B}$ are $m \times n$ and $n \times m$ matrices, respectively, then (Wilkinson, 1965, p. 54)

$$\det(\mathbf{I}_m + \mathbf{A}\mathbf{B}) = \det(\mathbf{I}_n + \mathbf{B}\mathbf{A}) \tag{A-1-33}$$

A-2. Matrix Differentiation

Let α be a scalar function of a vector $\mathbf{a}$ and a matrix $\mathbf{A}$; let $\mathbf{b}$ be a vector function of a scalar β and a vector $\mathbf{c}$, and let $\mathbf{C}$ be a matrix function of a scalar γ. Table A-1 lists the various derivatives that may be formed. Derivatives of vectors with respect to matrices, and matrices with respect to vectors and matrices, require more than two subscripts. They cannot, therefore, be represented in matrix notation. On the rare occasion when they are needed, subscript notation will be used.

Table A-1
Matrix Derivatives

The symbol	is a	whose elements are
$\partial\alpha/\partial\mathbf{a}$	column vector[a]	$(\partial\alpha/\partial\mathbf{a})_i \equiv \partial\alpha/\partial a_i$
$\partial\alpha/\partial\mathbf{A}$	matrix	$(\partial\alpha/\partial\mathbf{A})_{ij} \equiv \partial\alpha/\partial A_{ij}$
$\partial\mathbf{b}/\partial\beta$	column vector	$(\partial\mathbf{b}/\partial\beta)_i \equiv \partial b_i/\partial\beta$
$\partial\mathbf{b}/\partial\mathbf{c}$	matrix	$(\partial\mathbf{b}/\partial\mathbf{c})_{ij} \equiv \partial b_i/\partial c_j$
$\partial\mathbf{C}/\partial\gamma$	matrix	$(\partial\mathbf{C}/\partial\gamma)_{ij} \equiv \partial C_{ij}/\partial\gamma$

[a] A case may be made for defining $\partial\alpha/\partial\mathbf{a}$ as a row vector, but we prefer to regard all vectors that do not carry the symbol ^{T} as column vectors.

To differentiate a product of vectors and matrices with respect to one term, we proceed as follows (assume we are computing $\partial\alpha/\partial\mathbf{A}$):

1. Write the expression out in terms of subscripts and summations. Do not use the symbols i and j as subscripts.

2. Suppose the term A_{kl} appears in the summation. Remove this term, replace the remaining appearances of subscripts k and l with i and j, respectively, and remove summations with respect to k and l. The result is the derivative with respect to A_{ij}.

3. Reorder the expression so that the term containing i appears first and the term containing j appears last. Reorder the other terms so that any two occurrences of other indices are in consecutive terms. It may happen that some of the terms are left over. These terms can be grouped to form a scalar, which can be placed in front of the remaining matrix expression, as in example (e) below.

4. Drop all summations and indices. Add ^{T} symbols where necessary.

Examples

(a) $\alpha = \mathbf{a}^{\mathrm{T}}\mathbf{A}\mathbf{b}$.

1. $\alpha = \sum_{k,l} a_k A_{kl} b_l$.
2. $\partial\alpha/\partial A_{ij} = a_i b_j$.
3. $\partial\alpha/\partial A_{ij} = a_i b_j$.
4. $\partial\alpha/\partial\mathbf{A} = \mathbf{a}\mathbf{b}^{\mathrm{T}}$.

(b) $\alpha = \mathrm{Tr}(\mathbf{B}\mathbf{A}^{\mathrm{T}}\mathbf{C})$.

1. $\alpha = \sum_{m,k,l} B_{ml} A_{kl} C_{km}$.
2. $\partial\alpha/\partial A_{ij} = \sum_m B_{mj} C_{im}$.
3. $\partial\alpha/\partial A_{ij} = \sum_m C_{im} B_{mj}$.
4. $\partial\alpha/\partial\mathbf{A} = \mathbf{C}\mathbf{B}$.

If the matrix $\mathbf{A}$ appears more than once, each appearance should be treated separately and the results added.

Example

(c) $\alpha = \mathbf{a}^{\mathrm{T}}\mathbf{A}\mathbf{B}\mathbf{A}^{\mathrm{T}}\mathbf{b}$.

1. $\alpha = \sum_{k,l,m,n} a_k A_{kl} B_{lm} A_{nm} b_n$.
2. $\partial\alpha/\partial A_{ij} = \sum_{m,n} a_i B_{jm} A_{nm} b_n + \sum_{k,l} a_k A_{kl} B_{lj} b_i$.
3. $\partial\alpha/\partial A_{ij} = \sum_{m,n} a_i b_n A_{nm} B_{jm} + \sum_{k,l} b_i a_k A_{kl} B_{lj}$.
4. $\partial\alpha/\partial\mathbf{A} = \mathbf{a}\mathbf{b}^{\mathrm{T}}\mathbf{A}\mathbf{B}^{\mathrm{T}} + \mathbf{b}\mathbf{a}^{\mathrm{T}}\mathbf{A}\mathbf{B}$.

The handling of other derivatives is analogous.

Examples

(d) Compute $\partial\alpha/\partial\mathbf{a}$, where $\alpha = \mathbf{a}^{\mathrm{T}}\mathbf{A}\mathbf{a}$.

1. $\alpha = \sum_{k,l} a_k A_{kl} a_l$.
2. $\partial\alpha/\partial a_i = \sum_l A_{il} a_l + \sum_k a_k A_{ki}$.
3. $\partial\alpha/\partial a_i = \sum_l A_{il} a_l + \sum_k A_{ki} a_k$.
4. $\partial\alpha/\partial\mathbf{a} = \mathbf{A}\mathbf{a} + \mathbf{A}^{\mathrm{T}}\mathbf{a}$. If $\mathbf{A}$ is symmetric, $\partial\alpha/\partial\mathbf{a} = 2\mathbf{A}\mathbf{a}$.

(e) Compute $\partial\mathbf{b}/\partial\mathbf{c}$, where $\mathbf{b} = \mathbf{Aca}^{\mathrm{T}}\mathbf{Bc}$.

1. $b_i = \sum_{k,l,m} A_{ik} c_k a_l B_{lm} c_m$.
2. $\partial b_i/\partial c_j = \sum_{l,m} A_{ij} a_l B_{lm} c_m + \sum_{k,l} A_{ik} c_k a_l B_{lj}$.
3. $\partial b_i/\partial c_j = \sum_{l,m}(a_l B_{lm} c_m) A_{ij} + \sum_{k,l} A_{ik} c_k a_l B_{lj}$.
4. $\partial\mathbf{b}/\partial\mathbf{c} = (\mathbf{a}^{\mathrm{T}}\mathbf{Bc})\mathbf{A} + \mathbf{Aca}^{\mathrm{T}}\mathbf{B}$.

(note that the term $\sum_{l,m} a_l B_{lm} c_m = \mathbf{a}^{\mathrm{T}}\mathbf{Bc}$ is a scalar and can be placed anywhere in a product).

We shall also need the following derivatives:

(f) We wish to compute $\partial A_{kl}^{-1}/\partial A_{ij}$, where A_{kl}^{-1} is the k, l element of $\mathbf{A}^{-1}$. By definition

$$\partial A_{kl}^{-1}/\partial A_{ij} = \lim_{\varepsilon\to 0}(1/\varepsilon)[(\mathbf{A} + \varepsilon\mathbf{B})^{-1} - \mathbf{A}^{-1}]_{kl} \tag{A-2-1}$$

where $\mathbf{B}$ is a matrix whose m, n element is $\delta_{mi}\delta_{nj}$; i.e., the i, j element is unity, and all other elements are zero. Now

$$(\mathbf{A} + \varepsilon\mathbf{B})^{-1} = [\mathbf{A}(\mathbf{I} + \varepsilon\mathbf{A}^{-1}\mathbf{B})]^{-1} = (\mathbf{I} + \varepsilon\mathbf{A}^{-1}\mathbf{B})^{-1}\mathbf{A}^{-1} \tag{A-2-2}$$

For sufficiently small ε the following series expansion is valid

$$\begin{aligned}(\mathbf{I} + \varepsilon\mathbf{A}^{-1}\mathbf{B})^{-1}\mathbf{A}^{-1} &= (\mathbf{I} - \varepsilon\mathbf{A}^{-1}\mathbf{B} + \varepsilon^2\mathbf{A}^{-1}\mathbf{BA}^{-1}\mathbf{B} - \cdots)\mathbf{A}^{-1}\\ &= \mathbf{A}^{-1} - \varepsilon\mathbf{A}^{-1}\mathbf{BA}^{-1} + \varepsilon^2\mathbf{A}^{-1}\mathbf{BA}^{-1}\mathbf{BA}^{-1} - \cdots\end{aligned} \tag{A-2-3}$$

and we can prove easily that

$$\lim_{\varepsilon\to 0}(1/\varepsilon)[(\mathbf{A} + \varepsilon\mathbf{B})^{-1} - \mathbf{A}^{-1}] = -\mathbf{A}^{-1}\mathbf{BA}^{-1} \tag{A-2-4}$$

Therefore

$$\begin{aligned}\partial A_{kl}^{-1}/\partial A_{ij} &= -[\mathbf{A}^{-1}\mathbf{BA}^{-1}]_{kl} = -\sum_{m,n} A_{km}^{-1} B_{mn} A_{nl}^{-1}\\ &= -\sum_{m,n} A_{km}^{-1}\,\delta_{mi}\,\delta_{nj} A_{nl}^{-1} = -A_{ki}^{-1} A_{jl}^{-1}\end{aligned} \tag{A-2-5}$$

which is the desired result.

(g) Now we can evaluate, for example, $\partial\alpha/\partial\mathbf{A}$ where $\alpha \equiv \mathbf{x}^{\mathrm{T}}\mathbf{A}^{-1}\mathbf{x}$. Indeed:

$$\alpha = \sum_{k,l} x_k A_{kl}^{-1} x_l$$

$$\partial\alpha/\partial A_{ij} = \sum_{k,l} x_k(\partial A_{kl}^{-1}/\partial A_{ij})x_l = -\sum_{k,l} x_k A_{ki}^{-1} A_{jl}^{-1} x_l = -\sum_{k,l} A_{ki}^{-1} x_k x_l A_{jl}^{-1} \tag{A-2-6}$$

so that

$$\partial(\mathbf{x}^{\mathrm{T}}\mathbf{A}^{-1}\mathbf{x})/\partial\mathbf{A} = -(\mathbf{A}^{-1})^{\mathrm{T}}\mathbf{xx}^{\mathrm{T}}(\mathbf{A}^{-1})^{\mathrm{T}} \tag{A-2-7}$$

(h) Let $\alpha = \det \mathbf{A}$. We wish to evaluate $\partial\alpha/\partial\mathbf{A}$. Let us expand the determinant in cofactors of the ith row, i.e.,

$$\det \mathbf{A} = \sum_k A_{ik} A_{ik}^* \tag{A-2-8}$$

where A_{ik}^* is the cofactor of A_{ik}. A_{ik}^* does not depend on any of the elements in the ith row. Therefore

$$\partial \det \mathbf{A}/\partial A_{ik} = A_{ik}^* \tag{A-2-9}$$

As is well known

$$A_{mn}^{-1} = A_{nm}^*/\det \mathbf{A} \tag{A-2-10}$$

Hence, $A_{ik}^* = A_{ki}^{-1} \det \mathbf{A}$ and

$$\partial \det \mathbf{A}/\partial\mathbf{A} = (\mathbf{A}^{-1})^{\mathrm{T}} \det \mathbf{A} \tag{A-2-11}$$

Furthermore

$$\partial \log \det \mathbf{A}/\partial\mathbf{A} = (1/\det \mathbf{A})\, \partial \det \mathbf{A}/\partial\mathbf{A} = (\mathbf{A}^{-1})^{\mathrm{T}} \tag{A-2-12}$$

A-3. Pivoting and Sweeping

Many computations involving matrices may be viewed as a sequence of operations called *pivoting*. It is useful to examine the pivoting operation in detail, and list some of its applications. In the sequel we always assume that we start with a given matrix $\mathbf{B}$ which is progressively modified by successive pivotings. Unless otherwise stated, whenever we refer to $\mathbf{B}$ or to any of its elements we mean the current, rather than the original values.

Definition Suppose $B_{ij} \neq 0$ for some pair of indices i, j. Then performing a *Gauss-Jordan pivot*, or simply *pivoting* on (i, j) means changing the elements of $\mathbf{B}$ according to the following scheme:

1. Replace B_{pq} by $B_{pq} - B_{iq}B_{pj}/B_{ij}$ for all $p \neq i$, $q \neq j$.
2. Replace B_{iq} by B_{iq}/B_{ij} for all $q \neq j$.
3. Replace B_{pj} by $-B_{pj}/B_{ij}$ for all $p \neq i$.
4. Replace B_{ij} by $1/B_{ij}$.

The element B_{ij} (before pivoting) is referred to as the *pivot*. Pivoting on (i, i), i.e., with a pivot on the main diagonal, is referred to as *sweeping* (Beaton, 1964) row i. Two pivots are *unrelated* if they differ in both row and column, i.e., B_{ij} and B_{kl} are unrelated if $i \neq k$ and $j \neq l$. The following properties are easily verified:

1. Pivoting is reversible, i.e., pivoting on (i, j) twice restores the original matrix.

2. Pivoting on unrelated pivots is commutative, i.e., pivoting first on (i, j) and then on (k, l) produces the same matrix as pivoting first on (k, l) and then on (i, j), provided $i \neq k$ and $j \neq l$. Since different elements on the main diagonal are unrelated, it follows that sweeps are always commutative.

3. From 1 and 2 we deduce that pivoting in sequence on (i, j), (k, l), and (i, j) is equivalent to pivoting on (k, l) alone if (i, j) and (k, l) are unrelated.

The following applications will motivate the definition of pivoting:

(a) Exchange of Variables. Suppose $\mathbf{B}$ is $m \times n$, $\mathbf{x}$ is an n-vector, and $\mathbf{y}$ is an m-vector satisfying

$$\mathbf{y} + \mathbf{Bx} = \mathbf{0} \tag{A-3-1}$$

The elements of $\mathbf{x}$ and $\mathbf{y}$ may be regarded as independent and dependent variables, respectively. Suppose we wish to interchange the roles of, say, x_1 and y_1. That is, we wish to express the variables $x_1, y_2, y_3, \ldots, y_m$ as functions of $y_1, x_2, x_3, \ldots, x_n$. The first row of Eq. (1) reads

$$y_1 + B_{11}x_1 + B_{12}x_2 + \cdots = 0 \tag{A-3-2}$$

If $B_{11} \neq 0$, then this is equivalent to

$$x_1 + B_{11}^{-1}y_1 + B_{12}B_{11}^{-1}x_2 + \cdots = 0 \tag{A-3-3}$$

Solving for x_1 and substituting in the ith row of Eq. (1) we find, after collecting terms

$$y_i - B_{i1}B_{11}^{-1}y_1 + (B_{i2} - B_{i1}B_{12}/B_{11})x_2 + \cdots = 0 \tag{A-3-4}$$

Consider the following tableau as a schematic representation of Eq. (1)

	x_1	x_2	$\cdots$	x_n
y_1	B_{11}	B_{12}		B_{1n}
y_2	B_{21}	B_{22}		B_{2n}
$\vdots$				
y_m	B_{m1}	B_{m2}		B_{mn}

(A-3-5)

Then, after exchanging x_1 with y_1 we can represent Eq. (3) and Eq. (4) in a new tableau

	y_1	x_2	$\cdots$	x_n
x_1	$1/B_{11}$	B_{12}/B_{11}		B_{1n}/B_{11}
y_2	$-B_{21}/B_{11}$	$B_{22} - B_{21}B_{12}/B_{11}$		$B_{2n} - B_{21}B_{1n}/B_{11}$
$\vdots$				
y_m	$-B_{m1}/B_{11}$	$B_{m2} - B_{m1}B_{12}/B_{11}$		$B_{mn} - B_{m1}B_{1n}/B_{11}$

(A-3-6)

It is evident that the elements of **B** have been transformed as by pivoting on (1, 1). Generally, exchanging y_i for x_j is accomplished by pivoting on (i, j).

(b) Partial Elimination. Instead of interchanging just one pair of variables, we may wish to interchange several. Let the equations of Eq. (1) be partitioned as follows

$$\mathbf{y}_1 + \mathbf{B}_{11}\mathbf{x}_1 + \mathbf{B}_{12}\mathbf{x}_2 = \mathbf{0}, \qquad \mathbf{y}_2 + \mathbf{B}_{21}\mathbf{x}_1 + \mathbf{B}_{22}\mathbf{x}_2 = \mathbf{0} \qquad \text{(A-3-7)}$$

The corresponding tableau is

	$\mathbf{x}_1^{\mathrm{T}}$	$\mathbf{x}_2^{\mathrm{T}}$
$\mathbf{y}_1$	$\mathbf{B}_{11}$	$\mathbf{B}_{12}$
$\mathbf{y}_2$	$\mathbf{B}_{21}$	$\mathbf{B}_{22}$

(A-3-8)

Let $\mathbf{B}_{11}$ be a $k \times k$ nonsingular submatrix of **B**. Then we can solve the first k equations in Eq. (7) for $\mathbf{x}_1$, and substitute in the remaining equations to obtain

$$\mathbf{x}_1 + \mathbf{B}_{11}^{-1}\mathbf{y}_1 + \mathbf{B}_{11}^{-1}\mathbf{B}_{12}\mathbf{x}_2 = \mathbf{0}, \qquad \mathbf{y}_2 - \mathbf{B}_{21}\mathbf{B}_{11}^{-1}\mathbf{y}_1 + (\mathbf{B}_{22} - \mathbf{B}_{21}\mathbf{B}_{11}^{-1}\mathbf{B}_{12})\mathbf{x}_2 = \mathbf{0} \qquad \text{(A-3-9)}$$

Suppose it is possible to exchange, in sequence, y_1 for x_1, y_2 for $x_2, \ldots, y_k$ for x_k. The result is the same as exchanging the entire vector $\mathbf{y}_1$ for $\mathbf{x}_1$. According to Eq. (9), then, sweeping (if possible) rows $1, 2, \ldots, k$ of tableau Eq. (8) produces

	$\mathbf{y}_1^{\mathrm{T}}$	$\mathbf{x}_2^{\mathrm{T}}$
$\mathbf{x}_1$	$\mathbf{B}_{11}^{-1}$	$\mathbf{B}_{11}^{-1}\mathbf{B}_{12}$
$\mathbf{y}_2$	$-\mathbf{B}_{21}\mathbf{B}_{11}^{-1}$	$\mathbf{B}_{22} - \mathbf{B}_{21}\mathbf{B}_{11}^{-1}\mathbf{B}_{12}$

(A-3-10)

This property is used in the projection method (Sections 6-2 and 6-3). It can be shown that if $\mathbf{B}_{11}$ is positive definite, then the required sweeps can always be executed, i.e., no B_{ii} ever turns zero.

(c) Matrix Inversion. When $\mathbf{B}_{11}$ is the entire matrix **B**, then sweeping all rows transforms **B** into $\mathbf{B}^{-1}$, since $\mathbf{y} + \mathbf{B}\mathbf{x} = \mathbf{0}$ is changed into $\mathbf{x} + \mathbf{B}^{-1}\mathbf{y} = \mathbf{0}$. This procedure cannot be carried out if zero diagonal elements are encountered. For instance

$$\begin{bmatrix} 0 & 1 \\ 1 & 0 \end{bmatrix}$$

cannot be swept though it is nonsingular. However, we can always proceed as follows:

1. Write out the tableau Eq. (5). The y_i and x_j are symbolic headings, whereas the B_{ij} are numerical values.
2. Among all the elements whose rows are headed by a y_i and whose columns are headed by an x_j, find the one, say B_{pq}, with largest absolute value. If no such elements exist, proceed to step 5. Otherwise:
3. If $B_{pq} = 0$ the matrix $\mathbf{B}$ is singular, and the process is terminated. Otherwise:
4. Pivot on (p, q) and interchange the headings y_p and x_q. Return to step 2.
5. Rearrange the rows so that their headings appear in the natural order $x_1, x_2, \ldots, x_m$.
6. Rearrange the columns so that their headings appear in the natural order $y_1, y_2, \ldots, y_m$. Our tableau now contains $\mathbf{B}^{-1}$.

(d) Simultaneous Linear Equations. We wish to solve for $\mathbf{x}$ the set of simultaneous equations

$$\mathbf{A}\mathbf{x} = \mathbf{b} \tag{A-3-11}$$

where $\mathbf{A}$ is $m \times m$. Let us define $\mathbf{B}$ as the $m \times (m + 1)$ matrix $[\mathbf{A}, \mathbf{b}]$ and let us apply to $\mathbf{B}$ the algorithm of the preceding section, except that no pivots are allowed in the last column. If $\mathbf{A}$ is nonsingular, one ends up with the matrix $[\mathbf{A}^{-1}, \mathbf{A}^{-1}\mathbf{b}]$, i.e., the solution $\mathbf{x}$ is found in the last column. If one is only interested in $\mathbf{x}$, then step 6 may be omitted. Also, in step 5 only the elements of the last column need be rearranged. This method of solving equations is known as *Gauss–Jordan elimination.* Ordinary Gaussian elimination is faster, but Gauss–Jordan elimination is very convenient and economical in storage space when the inverse too is desired.

If $\mathbf{A}$ is singular, the process terminates in step 3 with all eligible pivots equal to zero. Let us partition $\mathbf{x}$ and $\mathbf{y}$ into vectors $\mathbf{x}_1$, $\mathbf{x}_2$ and $\mathbf{y}_1$, $\mathbf{y}_2$, respectively, where subscript 1 refers to those elements which have been exchanged, and subscript 2 to those elements which have not been exchanged. For instance, $\mathbf{x}_2$ consists of elements of $\mathbf{x}$ which appear as column headings in the final tableau. The final tableau takes the form (we have rearranged rows and columns suitably)

	$\mathbf{y}_1^{\mathrm{T}}$	$\mathbf{x}_2^{\mathrm{T}}$	
$\mathbf{x}_1$	$\mathbf{C}_{11}$	$\mathbf{C}_{12}$	$\mathbf{c}_1$
$\mathbf{y}_2$	$\mathbf{C}_{21}$	$\mathbf{C}_{22} = \mathbf{0}$	$\mathbf{c}_2$

(A-3-12)

The matrix $\mathbf{C}_{22}$ must vanish, for otherwise we could have continued pivoting. Let the partitioning of Eq. (11) that corresponds to the partitioning of $\mathbf{x}$ and $\mathbf{y}$ be

$$\mathbf{A}_{11}\mathbf{x}_1 + \mathbf{A}_{12}\mathbf{x}_2 = \mathbf{b}_1, \qquad \mathbf{A}_{21}\mathbf{x}_1 + \mathbf{A}_{22}\mathbf{x}_2 = \mathbf{b}_2 \tag{A-3-13}$$

Then, from Eq. (10) and Eq. (12) we must have:

$$\begin{aligned} &\mathbf{C}_{11} = \mathbf{A}_{11}^{-1}, && \mathbf{C}_{12} = \mathbf{A}_{11}^{-1}\mathbf{A}_{12}, && \mathbf{c}_1 = \mathbf{A}_{11}^{-1}\mathbf{b}_1 \\ &\mathbf{C}_{21} = -\mathbf{A}_{21}\mathbf{A}_{11}^{-1}, && \mathbf{C}_{22} = \mathbf{A}_{22} - \mathbf{A}_{21}\mathbf{A}_{11}^{-1}\mathbf{A}_{12} = \mathbf{0}, && \mathbf{c}_2 = \mathbf{b}_2 - \mathbf{A}_{21}\mathbf{A}_{11}^{-1}\mathbf{b}_1 \end{aligned} \tag{A-3-14}$$

Now, if we eliminate $\mathbf{x}_1$ directly from Eq. (13) we find

$$\mathbf{x}_1 = \mathbf{A}_{11}^{-1}\mathbf{b}_1 - \mathbf{A}_{11}^{-1}\mathbf{A}_{12}\mathbf{x}_2, \qquad (\mathbf{A}_{22} - \mathbf{A}_{21}\mathbf{A}_{11}^{-1}\mathbf{A}_{12})\mathbf{x}_2 = \mathbf{b}_2 - \mathbf{A}_{21}\mathbf{A}_{11}^{-1}\mathbf{b}_1 \tag{A-3-15}$$

which, in view of Eq. (14), can be written as

$$\mathbf{x}_1 = \mathbf{c}_1 - \mathbf{C}_{12}\mathbf{x}_2, \qquad \mathbf{0}\mathbf{x}_2 = \mathbf{c}_2 \tag{A-3-16}$$

From this we deduce the following:

1. If $\mathbf{c}_2 \neq \mathbf{0}$ then the equations $\mathbf{Ax} = \mathbf{b}$ have no solution. Note that $\mathbf{c}_2$ is the set of elements in the last column which belong to rows with y headings.
2. If $\mathbf{c}_2 = \mathbf{0}$, then the equations $\mathbf{Ax} = \mathbf{0}$ have infinitely many solutions. These can be obtained by assigning arbitrary values to $\mathbf{x}_2$, and letting $\mathbf{x}_1 = \mathbf{c}_1 - \mathbf{C}_{12}\mathbf{x}_2$.

(e) Rank of Matrix and Linear Independence of Vectors. Let $\mathbf{a}_1, \mathbf{a}_2, \ldots, \mathbf{a}_m$ be a set of n-vectors, and let $\mathbf{A}$ be the $m \times n$ matrix whose ith row is $\mathbf{a}_i^{\mathrm{T}}$. We wish to determine the rank of $\mathbf{A}$, or what is the same, the number of linearly independent $\mathbf{a}_i$. We write down $\mathbf{A}$ in tableau form, and proceed to apply the algorithm of (c) above. The number of pivots executed before the process had to be halted equals the rank of the matrix. Referring to Eq. (14), the condition $\mathbf{C}_{22} = \mathbf{0}$ may be rewritten as $\mathbf{A}_{22} = -\mathbf{C}_{21}\mathbf{A}_{12} = \mathbf{A}_{21}\mathbf{C}_{12}$. But also $\mathbf{A}_{21} = -\mathbf{C}_{21}\mathbf{A}_{11}$ and $\mathbf{A}_{12} = \mathbf{A}_{11}\mathbf{C}_{12}$. Combining these, we find:

$$\begin{bmatrix} \mathbf{A}_{12} \\ \mathbf{A}_{22} \end{bmatrix} = \begin{bmatrix} \mathbf{A}_{11} \\ \mathbf{A}_{21} \end{bmatrix} \mathbf{C}_{12} \tag{A-3-17}$$

$$[\mathbf{A}_{21}, \ \mathbf{A}_{22}] = -\mathbf{C}_{21}[\mathbf{A}_{11}, \ \mathbf{A}_{12}] \tag{A-3-18}$$

Thus, $\mathbf{C}_{12}$ and $\mathbf{C}_{21}$ contain the coefficients for the linear dependence among the columns and rows of $\mathbf{A}$, respectively. The rows of $[\mathbf{A}_{11}, \mathbf{A}_{12}]$ form a maximal linearly independent subset of the $\mathbf{a}_i$. The columns of

$$\begin{bmatrix} \mathbf{A}_{11} \\ \mathbf{A}_{21} \end{bmatrix}$$

form a maximal linearly independent subset of the columns of $\mathbf{A}$. When the rank of $\mathbf{A}$ equals the number of rows, then $\mathbf{A}_{21}$, $\mathbf{A}_{22}$, $\mathbf{C}_{21}$, and $\mathbf{C}_{22}$ are vacuous; when the rank of $\mathbf{A}$ equals the number of columns, then $\mathbf{A}_{12}$, $\mathbf{A}_{22}$, $\mathbf{C}_{12}$, and $\mathbf{C}_{22}$ are vacuous. When $\mathbf{A}$ is square and nonsingular, only $\mathbf{A}_{11}$ and $\mathbf{C}_{11} = \mathbf{A}_{11}^{-1}$ exist.

(f) Determinant. To compute the determinant of $\mathbf{B}$, we follow the procedure of (c); step 6 may be omitted. If the process cannot be completed, then $\det \mathbf{B} = 0$. Otherwise, the determinant equals the product of all pivots times $(-1)^r$, where r is the number of row interchanges required in step 5.

(g) Stepwise Linear Regression. We wish to find the l-vector $\boldsymbol{\theta}$ which minimizes

$$\Phi(\boldsymbol{\theta}) = (\mathbf{Y} - \mathbf{B}\boldsymbol{\theta})^{\mathrm{T}}\mathbf{V}^{-1}(\mathbf{Y} - \mathbf{B}\boldsymbol{\theta})$$

Let us form the $(l+1) \times (l+1)$ matrix

$$\mathbf{A} \equiv \begin{bmatrix} \mathbf{B}^{\mathrm{T}}\mathbf{V}^{-1}\mathbf{B} & \mathbf{B}^{\mathrm{T}}\mathbf{V}^{-1}\mathbf{Y} \\ (\mathbf{B}^{\mathrm{T}}\mathbf{V}^{-1}\mathbf{Y})^{\mathrm{T}} & \mathbf{Y}^{\mathrm{T}}\mathbf{V}^{-1}\mathbf{Y} \end{bmatrix}$$

Suppose we sweep some of the first l rows of $\mathbf{A}$, producing a modified matrix $\mathbf{A}$ (whenever we speak of $\mathbf{A}$ we are referring to its current form). Let I denote the set of indices of the swept rows, and J the set of indices of the unswept rows (excluding row $l+1$). Let $\mathbf{a}$ be the last column of $\mathbf{A}$. Then $a_\alpha (\alpha \in I)$ is the optimal value of θ_α provided all θ_β $(\beta \in J)$ are restricted to vanish. Furthermore, a_{l+1} is the minimum of $\Phi(\boldsymbol{\theta})$ under the above restriction, and

$$t_\beta^{\,2} \equiv a_\beta^{\,2}/A_{\beta\beta} \quad (\beta \in J)$$

is the reduction in $\Phi(\boldsymbol{\theta})$ that would ensue if θ_β were to be included in the regression, i.e., if row β were to be swept. Therefore, the following algorithm is suggested for forward stepwise regression:

1. Choose a small positive number ε such that a change ε in $\Phi(\boldsymbol{\theta})$ is considered insignificant.
2. Construct the matrix $\mathbf{A}$. Let I be the empty set, and let $J = \{1, 2, \ldots, l\}$.
3. Of the elements $\beta \in J$ for which $t_\beta^{\,2} > \varepsilon$, find the one, say β^*, for which $t_\beta^{\,2}$ is largest. Sweep row β^*, and transfer β^* from J to I.
4. Repeat step 3 until no $t_\beta^{\,2}$ exceeds ε. At this point, the model is represented by the equation

$$f_\mu = \sum_{\alpha \in I} B_{\mu\alpha}\theta_\alpha, \qquad \theta_\alpha = \begin{matrix} a_\alpha = A_{\alpha,\,l+1} & (\alpha \in I) \\ 0 & (\alpha \in J) \end{matrix}$$

In backward stepwise regression, we start with

$$\mathbf{A} \equiv \begin{bmatrix} (\mathbf{B}^T\mathbf{V}^{-1}\mathbf{B})^{-1} & \boldsymbol{\theta}^* \\ -\boldsymbol{\theta}^{*T} & \mathbf{Y}^T\mathbf{V}^{-1}\mathbf{Y} - \boldsymbol{\theta}^{*T}\mathbf{B}^T\mathbf{V}^{-1}\mathbf{B}\boldsymbol{\theta}^* \end{bmatrix}$$

where $\boldsymbol{\theta}^* = (\mathbf{B}^T\mathbf{V}^{-1}\mathbf{B})^{-1}\mathbf{B}^T\mathbf{V}^{-1}\mathbf{Y}$. We let $I = \{1, 2, \ldots, l\}$ and J is the empty set. Step 3 above becomes:

3′. Of the elements $\alpha \in I$ for which $t_\alpha^2 \leqslant \varepsilon$, find the one, say α^*, for which t_α^2 is smallest. Sweep row α^*, and transfer α^* from I to J.

A-4. Eigenvalues and Vectors of a Real Symmetric Matrix

Presented below are the computational details for an algorithm which combines Givens-Householder reduction to tridiagonal form, the QR algorithm with origin shifts for diagonalizing the tridiagonal matrix, and successive orthogonal transformations of the unit matrix to obtain the eigenvectors. For explanations, see Wilkinson (1965) and Ortega and Kaiser (1963). Steps which are starred (*) can be omitted if only eigenvalues are to be computed.

A is the $n \times n$ symmetric matrix whose eigenvalues are to be computed. Step 2 is omitted if $n = 2$. Two constants ε_1 and ε_2 are used in termination tests [steps 8 and 9 below]. The following method is suggested for selecting the values of these constants: Let ε be the desired relative accuracy of the largest eigenvalue (this cannot exceed the precision of the computer. For instance, if a k-bit word length is used, we must have $\varepsilon > 2^{-k}$). Let $S = \sum_{i,j=1}^{n} A_{ij}^2$. Then let $\varepsilon_1 = \varepsilon(2S)^{1/2}$ and $\varepsilon_2 = \varepsilon_1/n^2$.

1*. Set $\mathbf{V} = \mathbf{I}_n$.
2. For $i = 1, 2, \ldots, n-2$ in turn, perform the following steps:
   a. Let $a = 1$ if $A_{i+1,i} \geqslant 0$, $a = -1$ otherwise.
   b. Let $c = \sum_{j=i+1}^{n} A_{j,i}^2$, $s = ac^{1/2}$.
   c. Let $b_i = -s$. If $s = 0$, proceed to step i.
   d. Let $\alpha = 1/(c + |A_{i+1,i}s|)$.
   e. Let $w_{i+1} = A_{i+1,i} + s$, $w_j = A_{j,i}$ $(j = i+2, i+3, \ldots, n)$.
   f. Let $u_j = \alpha w_j$ $(j = i+1, i+2, \ldots, n)$.
   g*. Let $p_k = \sum_{j=i+1}^{n} V_{k,j} w_j$ $(k = 1, 2, \ldots, n)$.
   h*. Replace V_{kj} with $V_{kj} - p_k u_j (k = 1, 2, \ldots, n;\ j = i+1,\ i+2, \ldots, n)$.
   i. Let $q_k = \sum_{j=i+1}^{n} A_{kj} u_j$ $(k = i+1, i+2, \ldots, n)$.
   j. Let $\beta = \frac{1}{2} \sum_{k=i+1}^{n} q_k u_k$.
   k. Let $q_k = q_k - \beta w_k$ $(k = i+1, i+2, \ldots, n)$.
   l. Replace A_{jk} with $A_{jk} - q_j w_k - w_j q_k$ $(j, k = i+1, i+2, \ldots, n)$.

3. Let $b_{n-1} = A_{n-1,n}$, $a_i = A_{i,i}$ $(i = 1, 2, \ldots, n)$.
4. Let $m = n$, $g = 0$.
5. Let $c_1 = 1$, $\alpha = a_m$, $p = a_1$.
6. For $i = 1, 2, \ldots, m-1$ in turn, perform the following steps:
    a. If $|b_i| \leqslant \varepsilon_2$, replace b_i by zero, set $s = 0$ and $c = \text{sign}(p)$, and proceed to step e.
    b. Let $x = (p^2 + b_i^2)^{1/2}$.
    c. Let $s = b_i/x$, $c = p/x$.
    d*. For $j = 1, 2, \ldots, n$ in turn, let $\beta = cV_{j,i+1} - sV_{j,i}$. Replace $V_{j,i}$ with $sV_{j,i+1} + cV_{j,i}$ and $V_{j,i+1}$ with β.
    e. Let $r = cp + sb_i$, $d = cc_1$.
    f. Let $q = db_i + sa_{i+1}$.
    g. Replace a_i with $dr + sq$.
    h. If $i > 1$, replace b_{i-1} with $s_1 r$.
    i. Let $s_1 = s$, $p = ca_{i+1} - sc_1 b_i$, $c_1 = c$.
7. Replace b_{m-1} with $s_1 p$ and a_m with $c_1 p$.
8. If $\sum_{i=1}^{m-1} |b_i| \leqslant \varepsilon_1$ proceed to step 16.
9. If $|b_{m-1}| \leqslant \varepsilon_2$ proceed to step 13.
10. If $||a_m/\alpha| - 1| > \frac{1}{2}$, return to step 5.
11. Replace g with $g + a_m$.
12. Replace a_i with $a_i - a_m$ $(i = 1, 2, \ldots, m)$ and return to step 5.
13. Replace a_m with $a_m + g$.
14. Replace m with $m - 1$.
15. If $m \geqslant 2$, return to step 9.
16. Replace a_i with $a_i + g$ $(i = 1, 2, \ldots, m)$.
17. At this point, a_i is the ith eigenvalue of **A**, and (*) V_{ji} $(j = 1, 2, \ldots, n)$ is the ith eigenvector of **A** $(i = 1, 2, \ldots, n)$. These eigenvectors form an orthonormal set, i.e., $\sum_i V_{ij} V_{ik} = \delta_{jk}$ $(j, k = 1, 2, \ldots, n)$.

A-5. Spectral Decompositions

Let **A** be a symmetric $l \times l$ matrix. Suppose **D** and **E** are, respectively, diagonal and nonsingular $l \times l$ matrices, satisfying

$$\mathbf{A} = \mathbf{E}\mathbf{D}\mathbf{E}^{\mathrm{T}} \tag{A-5-1}$$

Then $\mathbf{EDE}^{\mathrm{T}}$ is referred to as a *spectral decomposition* of **A**. In component form, a spectral decomposition is given by

$$A_{ij} = \sum_{k=1}^{l} d_k E_{ik} E_{jk} \tag{A-5-2}$$

where

$$d_k \equiv D_{kk} \tag{A-5-3}$$

Let $\mathbf{x}$ be any l-dimensional column vector. The quantity

$$A(\mathbf{x}) \equiv \mathbf{x}^{\mathrm{T}}\mathbf{A}\mathbf{x} = \sum_{i,j=1}^{l} A_{ij} x_i x_j$$

is called the *quadratic form* defined by $\mathbf{A}$. From Eq. (2) we have

$$A(\mathbf{x}) = \mathbf{x}^{\mathrm{T}}\mathbf{A}\mathbf{x} = \mathbf{x}^{\mathrm{T}}\mathbf{E}\mathbf{D}\mathbf{E}^{\mathrm{T}}\mathbf{x} = \mathbf{y}^{\mathrm{T}}\mathbf{D}\mathbf{y} = \sum_{i=1}^{l} d_i y_i^2 \tag{A-5-4}$$

where

$$\mathbf{y} \equiv \mathbf{E}^{\mathrm{T}}\mathbf{x} \tag{A-5-5}$$

The matrix $\mathbf{A}$ is positive (negative) definite if $A(\mathbf{x}) > 0$ (<0) for all $\mathbf{x} \neq \mathbf{0}$. It follows from Eq. (4) that $\mathbf{A}$ is positive (negative) definite if and only if all d_i are positive (negative).

If none of the d_i are zero, the matrix $\mathbf{A}$ is nonsingular, and we can form

$$\mathbf{A}^{-1} = (\mathbf{E}^{\mathrm{T}})^{-1}\mathbf{D}^{-1}\mathbf{E}^{-1} \tag{A-5-6}$$

since $\mathbf{E}$ was assumed nonsingular, and $\mathbf{D}^{-1}$ is a diagonal matrix with $(\mathbf{D}^{-1})_{ii} = d_i^{-1}$.

Any symmetric matrix possesses infinitely many spectral decompositions. Of these, the following play important roles:

(a) The Eigenvalue Decomposition. Suppose $\mathbf{E}$ is a unitary matrix $\mathbf{V}$ satisfying.

$$\mathbf{V}^{\mathrm{T}} = \mathbf{V}^{-1} \tag{A-5-7}$$

In this case we denote $\mathbf{D}$ by $\mathbf{\Lambda}$ and d_i by λ_i. Then we have from Eq. (1) and Eq. (7)

$$\mathbf{A}\mathbf{V} = \mathbf{V}\mathbf{\Lambda}\mathbf{V}^{\mathrm{T}}\mathbf{V} = \mathbf{V}\mathbf{\Lambda} \tag{A-5-8}$$

Let $\mathbf{v}_i$ denote the ith column of $\mathbf{V}$. Then Eq. (8) is equivalent to

$$\mathbf{A}\mathbf{v}_i = \lambda_i \mathbf{v}_i \qquad (i = 1, 2, \ldots, l) \tag{A-5-9}$$

which states that the λ_i and $\mathbf{v}_i$ are, respectively, the eigenvalues and eigenvectors of $\mathbf{A}$. The equation

$$\mathbf{A} = \mathbf{V}\mathbf{\Lambda}\mathbf{V}^{\mathrm{T}} = \sum_{i=1}^{l} \lambda_i \mathbf{v}_i \mathbf{v}_i^{\mathrm{T}} \tag{A-5-10}$$

represents the *eigenvalue decomposition* of $\mathbf{A}$. Inverting Eq. (10) we find

$$\mathbf{A}^{-1} = (\mathbf{V}^{\mathrm{T}})^{-1}\mathbf{\Lambda}^{-1}\mathbf{V}^{-1} = \mathbf{V}\mathbf{\Lambda}^{-1}\mathbf{V}^{\mathrm{T}} = \sum_{i=1}^{l} \lambda_i^{-1} \mathbf{v}_i \mathbf{v}_i^{\mathrm{T}} \tag{A-5-11}$$

provided all $\lambda_i \neq 0$. If we omit from the summation in Eq. (11) all the terms for which $\lambda_i = 0$, we obtain a matrix $\mathbf{A}^{+}$, called the *pseudoinverse* of $\mathbf{A}$. This definition of the pseudoinverse applies only to symmetric matrices; for the general case see Eq. (A-1-20).

Equations (10) and (11) show how both a matrix and its inverse can be reconstituted when the eigenvalues and vectors are known. We now consider the quadratic form $A(\mathbf{x})$. We have

$$\mathbf{A}(\mathbf{x}) = \mathbf{x}^{\mathrm{T}}\mathbf{V}\boldsymbol{\Lambda}\mathbf{V}^{\mathrm{T}}\mathbf{x} = \mathbf{y}^{\mathrm{T}}\boldsymbol{\Lambda}\mathbf{y} = \sum_{i=1}^{l} \lambda_i {y_i}^2 \tag{A-5-12}$$

where

$$\mathbf{y} = \mathbf{V}^{\mathrm{T}}\mathbf{x} \tag{A-5-13}$$

Since $\mathbf{V}$ is unitary, the transformation of coordinates given by Eq. (13) does not affect the shape of the contours of the function $A(\mathbf{x})$, i.e., the shape of the surfaces on which $A(\mathbf{x}) =$ constant. From Eq. (12) it is evident that these surfaces are quadratics whose ith principal axis is inversely proportional to $|\lambda_i|^{1/2}$ and lies in the direction of $\mathbf{v}_i$. For instance, if $l = 2$ and λ_1 and λ_2 are positive, the contours are ellipses whose principal axes are proportional in length to $\lambda_1^{-1/2}$ and $\lambda_2^{-1/2}$. If the eigenvalues are nearly equal, the contours are nearly circles; if they differ widely, the contours are very elongated.

The most extensive analysis of methods for computation of eigenvalues and vectors can be found in Wilkinson (1965). A summary of a fast and convenient method based on the QR algorithm appears in the previous section. If computations are carried to n digits of precision, then the error in any computed eigenvalue is about $\pm 10^{-n}\lambda_{\max}$, where $\lambda_{\max}$ is the eigenvalue of largest absolute value. It follows that eigenvalues much smaller than $\lambda_{\max}$ cannot be computed with great precision. We define the *condition number* of a matrix as the ratio of largest to smallest (in absolute value) eigenvalues. The computation of the small eigenvalues (corresponding to long principal axes) of a matrix with a large condition number poses a serious problem. Fortunately, this problem can usually be eliminated if we use a different spectral decomposition, as described below.

(b) The Scaled and Inverse Scaled Decompositions. An apparent *ill-conditioning* (large condition number) of matrices encountered in practice is often due to the scaling of the variables. For instance, consider the function $\Phi(\boldsymbol{\theta}) = \frac{1}{2}({\theta_1}^2 + {\theta_2}^2)$. This has the Hessian matrix

$$\mathbf{H} = \begin{bmatrix} 1 & 0 \\ 0 & 1 \end{bmatrix}$$

which is very well-conditioned indeed, having both eigenvalues equal to one. Let us rescale the first variable by substituting $\eta_1 = 10^5\theta_1$. We leave the second variable unchanged, setting $\eta_2 = \theta_2$. In terms of the new coordinates, $\Phi = \frac{1}{2}[(\eta_1/10^5)^2 + {\eta_2}^2]$

$$\mathbf{H} = \begin{bmatrix} 10^{-10} & 0 \\ 0 & 1 \end{bmatrix}$$

The condition number has been increased from 1 to 10^{10}. This suggests that before computing eigenvalues and vectors we should scale the matrix properly.

The simplest scaling is one which reduces all diagonal elements to unit magnitude. If our matrix is a Hessian, this means that we are rescaling all variables so that the curvature of the objective function at the minimum is unity along all coordinate axes. If our matrix is the inverse of a Hessian, we are scaling all variables to possess unit standard deviation (see Section 7-5). If our matrix is positive or negative definite, the proposed scaling sets the magnitude of all off-diagonal elements to less than unity. If the matrix is not positive definite, this scaling method may fail by leaving very large off-diagonal elements. On the whole, however, the method has given very good results.

Given a matrix $\mathbf{A}$, we define a diagonal matrix $\mathbf{B}$ with

$$B_{ii} \equiv \begin{cases} |A_{ii}|^{1/2} & (A_{ii} \neq 0) \\ 1 & (A_{ii} = 0) \end{cases} \tag{A-5-14}$$

then the matrix

$$\mathbf{C} = \mathbf{B}^{-1}\mathbf{A}\mathbf{B}^{-1} \tag{A-5-15}$$

has elements $C_{ij} = A_{ij}/|A_{ii}A_{jj}|^{1/2}$ (except when A_{ii} or $A_{jj} = 0$), and, in particular, $C_{ii} = 1$. We refer to $\mathbf{C}$ as the *scaled version* of $\mathbf{A}$. If $\mathbf{A}$ is a covariance matrix then $\mathbf{C}$ is the correlation matrix. Let the eigenvalue decomposition of $\mathbf{C}$ be given by

$$\mathbf{C} = \mathbf{U}\mathbf{\Pi}\mathbf{U}^{\mathrm{T}} \tag{A-5-16}$$

where $\mathbf{\Pi}$ is diagonal; $\Pi_{ii} = \pi_i$, the ith eigenvalue of $\mathbf{C}$; $\mathbf{U}$ is the matrix whose ith column is $\mathbf{u}_i$, the ith eigenvector of $\mathbf{C}$; and $\mathbf{U}^{\mathrm{T}} = \mathbf{U}^{-1}$.

Eqs. (15) and (16) may be combined to yield

$$\mathbf{A} = \mathbf{B}\mathbf{U}\mathbf{\Pi}\mathbf{U}^{\mathrm{T}}\mathbf{B} = \mathbf{F}\mathbf{\Pi}\mathbf{F}^{\mathrm{T}} \tag{A-5-17}$$

where

$$\mathbf{F} \equiv \mathbf{B}\mathbf{U}, \text{ i.e., } F_{ij} = B_{ii}U_{ij} \tag{A-5-18}$$

We call the relation $\mathbf{A} = \mathbf{F}\mathbf{\Pi}\mathbf{F}^{\mathrm{T}}$ the *scaled decomposition* of $\mathbf{A}$. Inverting Eq. (17) yields

$$\mathbf{A}^{-1} = \mathbf{B}^{-1}\mathbf{U}\mathbf{\Pi}^{-1}\mathbf{U}^{\mathrm{T}}\mathbf{B}^{-1} = \mathbf{G}\mathbf{\Pi}^{-1}\mathbf{G}^{\mathrm{T}} \tag{A-5-19}$$

where

$$\mathbf{G} = \mathbf{B}^{-1}\mathbf{U}, \text{ i.e., } G_{ij} = B_{ii}^{-1}U_{ij} \tag{A-5-20}$$

We call the relation $\mathbf{A}^{-1} = \mathbf{G}\mathbf{\Pi}^{-1}\mathbf{G}^{\mathrm{T}}$ the *inverse scaled decomposition* of $\mathbf{A}$.

The following is a summary of the steps required to compute the scaled decompositions:

1. Divide each element A_{ij} by $|A_{ii}A_{jj}|^{1/2}$, forming the matrix $\mathbf{C}$.
2. Obtain the eigenvalues π_i and eigenvectors $\mathbf{u}_i$ of $\mathbf{C}$.

3. Multiply the jth element of $\mathbf{u}_i$ by $|A_{jj}|^{1/2}$ to form a vector $\mathbf{f}_i$, which is the ith column of $\mathbf{F}$.

4. The scaled decomposition of $\mathbf{A}$ is given by

$$\mathbf{A} = \sum_{i=1}^{l} \pi_i \mathbf{f}_i \mathbf{f}_i^{\mathrm{T}} \tag{A-5-21}$$

This is equivalent to Eq. (17).

5. Divide the jth element of $\mathbf{u}_i$ by $|A_{jj}|^{1/2}$ to form a vector $\mathbf{g}_i$, which is the ith column of $\mathbf{G}$.

6. The inverse scaled decomposition of $\mathbf{A}^{-1}$ is given by

$$\mathbf{A}^{-1} = \sum_{i=1}^{l} \pi_i^{-1} \mathbf{g}_i \mathbf{g}_i^{\mathrm{T}} \tag{A-5-22}$$

provided all $\pi_i \neq 0$. This is equivalent to Eq. (19).

Note: Replace any zero A_{ii} by one in the above computations.

A numerical example appears in Section 5-21.

The above procedure [omitting steps 3 and 4] can be regarded as a method for computing the inverse of a symmetric matrix. As such it is unlikely to win any prizes for speed, but it is quite accurate and stable. It provides insight into the nature of the matrix, and lets us generate "almost inverses" of $\mathbf{A}$. By this we mean matrices which (like the pseudoinverse) differ from Eq. (22) only in the values of the π_i, the latter being chosen so as to confer certain desirable properties (e.g., positive definiteness, or well-conditioning) on the matrix. For examples, see Sections 5-7–5-8.

(c) The Square Root Decomposition. If $\mathbf{A}$ is positive definite, it is possible to obtain spectral decompositions in which $\mathbf{D} = \mathbf{I}$, the identity matrix; i.e., $\mathbf{A} = \mathbf{E}\mathbf{E}^{\mathrm{T}}$. Of particular interest is the decomposition in which $\mathbf{E}$ is a symmetric matrix $\mathbf{S}$, whence $\mathbf{A} = \mathbf{S}^2$. The matrix $\mathbf{S}$ is called the *square root* of $\mathbf{A}$. If $\mathbf{A} = \mathbf{V}\mathbf{\Lambda}\mathbf{V}^{\mathrm{T}}$ is the eigenvalue decomposition of $\mathbf{A}$, then we have, because $\mathbf{V}^{\mathrm{T}}\mathbf{V} = \mathbf{I}$

$$\mathbf{A} = (\mathbf{V}\mathbf{\Lambda}^{1/2}\mathbf{V}^{\mathrm{T}})^2 \tag{A-5-23}$$

so that $\mathbf{A}^{1/2} = \mathbf{S} = \mathbf{V}\mathbf{\Lambda}^{1/2}\mathbf{V}^{\mathrm{T}}$. Here $\mathbf{\Lambda}^{1/2}$ is a diagonal matrix with elements $\lambda_i^{1/2}$.

(d) The Cholesky Decomposition. [See, e.g., Fox (1964)]. Again we assume that $\mathbf{A}$ is positive definite, and choose $\mathbf{D} = \mathbf{I}$. Now, however, we specify that $\mathbf{E}$ should be a lower diagonal matrix $\mathbf{L}$, that is, a matrix whose elements above the main diagonal are all zero

$$L_{ij} = 0 \qquad (j > i) \tag{A-5-24}$$

Since $\mathbf{A} = \mathbf{L}\mathbf{L}^{\mathrm{T}}$ we have $\mathbf{A}_{ij} = \sum_{k=1}^{l} L_{ik} L_{jk}$, which in view of Eq. (24) becomes:

$$A_{ij} = \sum_{k=1}^{j} L_{ik} L_{jk} \qquad (j < i) \tag{A-5-25}$$

$$A_{ii} = \sum_{k=1}^{j} L_{ik}^2 \tag{A-5-26}$$

These equations may be solved recursively for the L_{ij}. From Eq. (26)

$$L_{11} = A_{11}^{1/2} \tag{A-5-27}$$

From Eq. (25)

$$L_{i1} = A_{i1}/L_{11} \quad (i = 2, 3, \ldots, l) \tag{A-5-28}$$

Then, using Eq. (25) and Eq. (26) alternately for $i = 2, 3, \ldots, l$:

$$L_{ij} = \left(A_{ij} - \sum_{k=1}^{j-1} L_{ik} L_{jk}\right) \Big/ L_{jj} \quad (j = 2, 3, \ldots, i-1;\ \text{skip for } i = 2) \tag{A-5-29}$$

$$L_{ii} = \left(A_{ii} - \sum_{k=1}^{i-1} L_{ik}^2\right)^{1/2} \tag{A-5-30}$$

This procedure can be carried through provided all of the square root arguments are positive. This occurs if and only if **A** is positive definite. Of all the decompositions discussed, the last is the only one that can be accomplished in a finite procedure, which requires approximately $l^3/6$ multiplications. All the other decompositions depend on the evaluation of eigenvalues, which requires an iterative procedure.

The Cholesky decomposition is particularly useful in solving for **x** the set of linear equations

$$\mathbf{A}\mathbf{x} = \mathbf{b} \tag{A-5-31}$$

These can be rewritten as

$$\mathbf{L}\mathbf{y} = \mathbf{b} \tag{A-5-32}$$

where

$$\mathbf{y} = \mathbf{L}^{\mathrm{T}}\mathbf{x} \tag{A-5-33}$$

Now Eq. (32) on account of the triangular nature of **L**, has the form:

$$\begin{aligned} L_{11} y_1 &= b_1 \\ L_{21} y_1 + L_{22} y_2 &= b_2 \\ L_{31} y_1 + L_{32} y_2 + L_{33} y_3 &= b_3 \\ &\vdots \end{aligned} \tag{A-5-34}$$

These equations can easily be solved in turn for $y_1, y_2, \ldots, y_l$. Then Eq. (33), which has the form:

$$\begin{aligned} L_{11}x_1 + L_{21}x_2 + \cdots + L_{l1}x_l &= y_1 \\ L_{22}x_2 + \cdots + L_{l2}x_l &= y_2 \\ &\vdots \\ L_{ll}x_l &= y_l \end{aligned} \tag{A-5-35}$$

can be solved in turn for $x_l, x_{l-1}, \ldots, x_1$. This is the fastest method for solving Eq. (31) when $\mathbf{A}$ is positive definite.

We conclude this section with a computational note. In most applications the matrix $\mathbf{A}$ is given only in decomposed form [Eq. (1)]. We are interested in computing a vector $\mathbf{y} = \mathbf{Ax}$, where $\mathbf{x}$ is a known vector, but have no use for $\mathbf{A}$ itself. The following procedure is much more economical than generating $\mathbf{A}$ first and then computing $\mathbf{Ax}$:

1. Suppose $\mathbf{A} = \mathbf{EDE}^{\mathrm{T}}$. Compute the vector $\mathbf{z} = \mathbf{E}^{\mathrm{T}}\mathbf{x}$.
2. Compute the vector $\mathbf{u} = \mathbf{Dz}$.

This is done simply by applying the formula $u_j = d_j z_j$ to each component of $\mathbf{z}$.

3. Compute $\mathbf{y} = \mathbf{Eu}$.

In summary, the proper order of the calculations is given by

$$z_j = \sum_k E_{kj} x_k, \qquad u_j = d_j z_j, \qquad y_i = \sum_j E_{ij} u_j \tag{A-5-36}$$

A numerical example appears in Section 5-21.

If the distribution of $\boldsymbol{\xi}$ depends on some other variables $\mathbf{y}$, we write the pdf as $p(\mathbf{x}|\mathbf{y})$. If $\mathbf{y}$ is a possible value of some other vector random variable $\boldsymbol{\eta}$, then we call $p(\mathbf{x}|\mathbf{y})$ the *conditional pdf of* $\boldsymbol{\xi}$ *given* $\boldsymbol{\eta} = \mathbf{y}$,

The following equation relates the joint, marginal, and conditional distributions

$$p(\mathbf{x}, \mathbf{y}) = p(\mathbf{x}|\mathbf{y})p(\mathbf{y}) \tag{B-15}$$

hence

$$p(\mathbf{x}|\mathbf{y}) = p(\mathbf{x}, \mathbf{y}) \Big/ \int p(\mathbf{u}, \mathbf{y})\, d\mathbf{u} \tag{B-16}$$

provided the denominator does not vanish. If $\boldsymbol{\xi}$ and $\boldsymbol{\eta}$ are independent, we find, in view of Eq. (13) and Eq. (15), that $p(\mathbf{x}|\mathbf{y}) = p(\mathbf{x})$, i.e., knowledge of $\boldsymbol{\eta}$ does not affect the distribution of $\boldsymbol{\xi}$.

If $\mathbf{x}$ is an m-dimensional vector random variable with pdf $\hat{p}(\mathbf{x})$, and $\mathbf{f}(\mathbf{x})$ is an m-dimensional vector of continuous, differentiable single-valued functions such that $\mathbf{f}(\mathbf{x}_1) = \mathbf{f}(\mathbf{x}_2)$ only if $\mathbf{x}_1 = \mathbf{x}_2$, then the vector $\mathbf{y} = f(\mathbf{x})$ is a random variable with pdf

$$p(\mathbf{y}) = \hat{p}(\mathbf{x})|\det^{-1} \partial\mathbf{f}/\partial\mathbf{x}| \tag{B-17}$$

The *Jacobian matrix* of the transformation from $\mathbf{x}$ to $\mathbf{y}$, is defined as $\partial\mathbf{f}/\partial\mathbf{x}$, and its determinant (which under the above conditions must be nonzero) is the *Jacobian*.

Appendix

C

The Rao–Cramer Theorem

Let $p(\mathbf{Y}|\boldsymbol{\phi})$ be the pdf of the sample $\mathbf{Y}$. Then from the definition of a pdf

$$\int p(\mathbf{Y}|\boldsymbol{\phi})\, d\mathbf{Y} = 1 \tag{C-1}$$

Let $\mathbf{t}^*$ be a vector-valued statistic of the sample, i.e.,

$$\mathbf{t}^* = \mathbf{t}^*(\mathbf{Y}) \tag{C-2}$$

and let $\mathbf{t}$ be the expected value of $\mathbf{t}^*$, i.e.,

$$\mathbf{t}(\boldsymbol{\phi}) = \int \mathbf{t}^*(\mathbf{Y}) p(\mathbf{Y}|\boldsymbol{\phi})\, d\mathbf{Y} \tag{C-3}$$

From Eq. (1) we have

$$\frac{\partial}{\partial \boldsymbol{\phi}} \int p\, d\mathbf{Y} = \int \left(\frac{\partial p}{\partial \boldsymbol{\phi}}\right)^{\mathrm{T}} d\mathbf{Y} = \mathbf{0} \tag{C-4}$$

Also, from Eq. (3)

$$\frac{\partial}{\partial \boldsymbol{\phi}} \int \mathbf{t}^* p\, d\mathbf{Y} = \int \mathbf{t}^* \left(\frac{\partial p}{\partial \boldsymbol{\phi}}\right)^{\mathrm{T}} d\mathbf{Y} = \mathbf{P} \tag{C-5}$$

where

$$\mathbf{P} \equiv \partial \mathbf{t}/\partial \boldsymbol{\phi} \tag{C-6}$$

Thus, using Eq. (4) and Eq. (5)

$$\int (\mathbf{t}^* - \mathbf{t})(\partial p/\partial \boldsymbol{\phi})^{\mathrm{T}}\, d\mathbf{Y} = \int \mathbf{t}^*(\partial p/\partial \boldsymbol{\phi})^{\mathrm{T}}\, d\mathbf{Y} - \mathbf{t} \int (\partial p/\partial \boldsymbol{\phi})^{\mathrm{T}}\, d\mathbf{Y} = \partial \mathbf{t}/\partial \boldsymbol{\phi} = \mathbf{P} \tag{C-7}$$

Now

$$\partial p/\partial \boldsymbol{\phi} = p\, \partial \log p/\partial \boldsymbol{\phi}$$

hence Eq. (7) may be rewritten as

$$\int \mathbf{u}\mathbf{v}^{\mathrm{T}}\, d\mathbf{Y} = \mathbf{P} \tag{C-8}$$

where we define

$$\mathbf{u} \equiv p^{1/2}(\mathbf{t}^* - \mathbf{t}), \qquad \mathbf{v} \equiv p^{1/2}\, \partial \log p/\partial \boldsymbol{\phi} \tag{C-9}$$

The covariance matrix of the statistic $\mathbf{t}^*$ is

$$\mathbf{V}_{\mathbf{t}^*} \equiv \int \mathbf{u}\mathbf{u}^{\mathrm{T}}\, d\mathbf{Y} = \int p(\mathbf{t}^* - \mathbf{t})(\mathbf{t}^* - \mathbf{t})^{\mathrm{T}}\, d\mathbf{Y} = \mathbf{E}(\mathbf{t}^* - \mathbf{t})(\mathbf{t}^* - \mathbf{t})^{\mathrm{T}} \tag{C-10}$$

Let

$$\begin{aligned}\mathbf{R} &\equiv \int \mathbf{v}\mathbf{v}^{\mathrm{T}}\, d\mathbf{Y} = \int p(\partial \log p/\partial \boldsymbol{\phi})(\partial \log p/\partial \boldsymbol{\phi})^{\mathrm{T}}\, d\mathbf{Y} \\ &= E(\partial \log p/\partial \boldsymbol{\phi})(\partial \log p/\partial \boldsymbol{\phi})^{\mathrm{T}}\end{aligned} \tag{C-11}$$

Let $\mathbf{A}(\boldsymbol{\phi})$ be an arbitrary matrix function of $\boldsymbol{\phi}$ such that $\mathbf{Av}$ is a column vector of the same dimension as $\mathbf{u}$. The matrix $(\mathbf{u} + \mathbf{Av})(\mathbf{u} + \mathbf{Av})^{\mathrm{T}}$ is clearly positive semidefinite, and so is the sum of any number of matrices of this form. Hence, if

$$\mathbf{B} \equiv \int (\mathbf{u} + \mathbf{Av})(\mathbf{u} + \mathbf{Av})^{T}\, d\mathbf{Y} \tag{C-12}$$

then $\mathbf{B}$ is positive semidefinite. But

$$\begin{aligned}\mathbf{B} &= \int (\mathbf{u}\mathbf{u}^{\mathrm{T}} + \mathbf{A}\mathbf{v}\mathbf{v}^{\mathrm{T}}\mathbf{A}^{\mathrm{T}} + \mathbf{A}\mathbf{v}\mathbf{u}^{\mathrm{T}} + \mathbf{u}\mathbf{v}^{\mathrm{T}}\mathbf{A})\, d\mathbf{Y} = \mathbf{V}_{\mathbf{t}^*} + \mathbf{ARA}^{\mathrm{T}} + \mathbf{AP}^{\mathrm{T}} + \mathbf{PA}^{\mathrm{T}} \\ &= \mathbf{V}_{\mathbf{t}^*} - \mathbf{PR}^{-1}\mathbf{P}^{\mathrm{T}} + (\mathbf{A} + \mathbf{PR}^{-1})\mathbf{R}(\mathbf{A}^{\mathrm{T}} + \mathbf{R}^{-1}\mathbf{P}^{\mathrm{T}})\end{aligned} \tag{C-13}$$

Now, $\mathbf{B}$ must be positive semidefinite for any $\mathbf{A}$; in particular it must be so for $\mathbf{A} = -\mathbf{PR}^{-1}$, in which case $\mathbf{B} = \mathbf{V}_{\mathbf{t}^*} - \mathbf{PR}^{-1}\mathbf{P}^{\mathrm{T}}$. Hence, $\mathbf{V}_{\mathbf{t}^*} - \mathbf{PR}^{-1}\mathbf{P}^{\mathrm{T}}$ must be positive semidefinite.

From Eq. (12), $\mathbf{B} = \mathbf{0}$ if and only if $\mathbf{u} = -\mathbf{Av}$. In this case, from Eq. (8)

$$\mathbf{P} = \int -\mathbf{A}\mathbf{v}\mathbf{v}^{\mathrm{T}}\, d\mathbf{Y} = -\mathbf{AR} \tag{C-14}$$

and therefore Eq. (13) reduces to

$$\mathbf{0} = \mathbf{V}_{\mathbf{t}^*} - \mathbf{PR}^{-1}\mathbf{P}^{\mathrm{T}} \tag{C-15}$$

Conversely, suppose $\mathbf{V}_{\mathbf{t}^*} - \mathbf{PR}^{-1}\mathbf{P} = \mathbf{0}$. Then, from Eq. (13)

$$\mathbf{B} = (\mathbf{A} + \mathbf{PR}^{-1})\mathbf{R}(\mathbf{A}^{\mathrm{T}} + \mathbf{R}^{-1}\mathbf{P}^{\mathrm{T}})$$

In particular, if we choose $\mathbf{A} = -\mathbf{PR}^{-1}$, we have $\mathbf{B} = \mathbf{0}$. Hence, from Eq. (12)

$$\int (\mathbf{u} - \mathbf{PR}^{-1}\mathbf{v})(\mathbf{u} - \mathbf{PR}^{-1}\mathbf{v})^{\mathrm{T}}\, d\mathbf{Y} = \mathbf{0}$$

and it follows that $\mathbf{u} = \mathbf{P}\mathbf{R}^{-1}\mathbf{v}$. Associating the estimate $\boldsymbol{\phi}^*$ with $\mathbf{t}^*$ and $\bar{\boldsymbol{\Phi}}$ with $\mathbf{t}$, we obtain the results stated in Section 3-2.

Note that the proof is valid only if p satisfies regularity conditions which permit the differentiations under the integral sign in Eq. (4) and Eq. (5). We have also assumed that $\mathbf{R}$ was nonsingular.

Appendix

D

Generating a Sample from a Given Multivariate Normal Distribution

We wish to generate on a computer a sample from the distribution $N_k(\mathbf{a}, \mathbf{V})$. That is, we need a vector $\mathbf{z}$ of numbers $z_1, z_2, \ldots, z_k$ derived from a normal distribution with mean $\mathbf{a}$ and covariance matrix $\mathbf{V}$. We proceed as follows:

(a) Let $m = k/2$ if k is even, or $m = (k+1)/2$ if k is odd.

(b) Generate $2m$ pseudorandom numbers $x_1, x_2, \ldots, x_{2m}$ uniformly and independently distributed on the interval zero to one. For a discussion of methods to accomplish this see Moshman (1967) and Lewis *et al.* (1969).

(c) From the x_i we generate y_j $(j = 1, 2, \ldots, k)$ which are normally and independently distributed with zero means and unit variances. For this transformation many methods have been proposed, but only the following two are easy to program and reasonably fast, yet produce the required distribution exactly:

1. Method of Box and Muller (1958). Compute:

$$\begin{aligned} y_{2i-1} &= (-2 \log x_{2i-1})^{1/2} \cos 2\pi x_{2i} \\ y_{2i} &= (-2 \log x_{2i-1})^{1/2} \sin 2\pi x_{2i} \end{aligned} \qquad (i = 1, 2, \ldots, m) \qquad \text{(D-1)}$$

If k is odd, y_{2m} need not be computed.

2. Method of Marsaglia and Bray (1964). Compute $u_i = 2(x_i - 1)$ for $i = 1, 2, \ldots, 2m$. If for any $j = 1, 2, \ldots, m$ it happens that $u_{2j-1}^2 + u_{2j}^2 > 1$, replace x_{2j-1} and x_{2j} by a new pair of uniform random numbers, recompute u_{2j-1} and u_{2j}, and repeat until $u_{2j-1}^2 + u_{2j}^2 \leq 1$. Compute:

$$\begin{aligned} y_{2i-1} &= u_{2i-1}[-2 \log (u_{2i-1}^2 + u_{2i}^2)/(u_{2i-1}^2 + u_{2i}^2)]^{1/2} \\ y_{2i} &= u_{2i}[-2 \log (u_{2i-1}^2 + u_{2i}^2)/(u_{2i-1}^2 + u_{2i}^2)]^{1/2} \end{aligned} \qquad (i = 1, 2, \ldots, m)$$

$$\text{(D-2)}$$

The second method is probably faster, since it requires no evaluation of trigonometric functions.

(d) Compute the eigenvalues λ_i and eigenvectors $\mathbf{v}_i$ ($i = 1, 2, \ldots, k$) of $\mathbf{V}$. Generate the matrix $\mathbf{U}$ whose ith column is $\lambda_i^{1/2}\,\mathbf{v}_i$.

Note that the λ_i must all be nonnegative. A faster method, useful if $\mathbf{V}$ is known to be nonsingular, is to find a lower triangular matrix $\mathbf{U}$ such that $\mathbf{U}\mathbf{U}^{\mathrm{T}} = \mathbf{V}$ by means of the Cholesky decomposition (see Section 5-5).

(e) Compute

$$\mathbf{z} = \mathbf{a} + \mathbf{U}\mathbf{y} \tag{D-3}$$

to obtain the desired sample $\mathbf{z}$.

If many samples are required from the same distribution, step (d) should be performed once for all samples at the beginning.

Appendix

E

The Gauss–Markov Theorem

Suppose we have a model

$$\mathbf{y} - \mathbf{B}\boldsymbol{\theta} = \boldsymbol{\varepsilon} \tag{E-1}$$

where $\boldsymbol{\varepsilon}$ is a random vector with mean $\mathbf{0}$ and nonsingular covariance matrix $\mathbf{V}$, $\mathbf{y}$ is a vector of observations, and $\mathbf{B}$ a known matrix. We wish to find the least-variance linear unbiased estimator $\boldsymbol{\theta}^*$ for $\boldsymbol{\theta}$. Linearity implies the existence of a matrix $\mathbf{A}$ independent of $\mathbf{y}$ such that

$$\boldsymbol{\theta}^* = \mathbf{A}\mathbf{y} \tag{E-2}$$

Therefore

$$E(\boldsymbol{\theta}^*) = E(\mathbf{A}\mathbf{y}) = \mathbf{A}E(\mathbf{y}) = \mathbf{A}\mathbf{B}\boldsymbol{\theta} \tag{E-3}$$

If the estimate is unbiased, we must have $E(\boldsymbol{\theta}^*) = \boldsymbol{\theta}$ for any $\boldsymbol{\theta}$, hence

$$\mathbf{A}\mathbf{B} = \mathbf{I} \tag{E-4}$$

The covariance matrix of $\boldsymbol{\theta}^*$ is given by

$$\begin{aligned}\mathbf{V}_{\boldsymbol{\theta}} &= E(\boldsymbol{\theta}^* - \boldsymbol{\theta})(\boldsymbol{\theta}^* - \boldsymbol{\theta})^{\mathrm{T}} = E(\mathbf{A}\mathbf{y} - \boldsymbol{\theta})(\mathbf{A}\mathbf{y} - \boldsymbol{\theta})^{\mathrm{T}} \\ &= E[\mathbf{A}(\mathbf{y} - \mathbf{B}\boldsymbol{\theta}) + (\mathbf{A}\mathbf{B}\boldsymbol{\theta} - \boldsymbol{\theta})][\mathbf{A}(\mathbf{y} - \mathbf{B}\boldsymbol{\theta}) + (\mathbf{A}\mathbf{B}\boldsymbol{\theta} - \boldsymbol{\theta})]^{T} \\ &= E[\mathbf{A}(\mathbf{y} - \mathbf{B}\boldsymbol{\theta})(\mathbf{y} - \mathbf{B}\boldsymbol{\theta})^{\mathrm{T}}\mathbf{A}^{\mathrm{T}}] = E(\mathbf{A}\boldsymbol{\varepsilon}\boldsymbol{\varepsilon}^{\mathrm{T}}\mathbf{A}^{\mathrm{T}}) = \mathbf{A}\mathbf{V}\mathbf{A}^{\mathrm{T}}\end{aligned} \tag{E-5}$$

We wish to find the matrix $\mathbf{A}$ which minimizes some measure of $\mathbf{V}_{\boldsymbol{\theta}}$. The following are possible measures:

1. The so-called "generalized variance," i.e., det $\mathbf{V}_{\boldsymbol{\theta}}$.
2. Some weighted average of the elements of $\mathbf{V}_{\boldsymbol{\theta}}$, e.g., $\mathrm{Tr}(\mathbf{G}\mathbf{V}_{\boldsymbol{\theta}})$, where $\mathbf{G}$ is an arbitrary positive definite matrix.
3. The spectral norm (largest eigenvalue) of $\mathbf{V}$.

All measures lead to the same answer. We shall use the measure (1) here, and the reader may derive the result for the other measures as an exercise.

We wish, then, to determine the matrix $\mathbf{A}$ which minimizes det $\mathbf{AVA}^{\mathrm{T}}$ while satisfying $\mathbf{AB} = \mathbf{I}$. We introduce a matrix of Lagrange multipliers $\mathbf{\Lambda}$, and construct the Lagrangian

$$\mathscr{L}(\mathbf{A}, \mathbf{\Lambda}) = \det \mathbf{AVA}^{\mathrm{T}} + \mathrm{Tr}[\mathbf{\Lambda}(\mathbf{AB} - \mathbf{I})] \qquad \text{(E-6)}$$

We must find the stationary point of Eq. (6). By the methods of Section A-2 we find

$$\partial \mathscr{L}/\partial \mathbf{A} = 2(\mathbf{AVA}^{\mathrm{T}})^{-1}\mathbf{AV} + \mathbf{\Lambda}^{\mathrm{T}}\mathbf{B}^{\mathrm{T}} = \mathbf{0} \qquad \text{(E-7)}$$

Postmultiplying Eq. (7) by $\mathbf{A}^{\mathrm{T}}$, we obtain in view of Eq. (4)

$$2\mathbf{I} + \mathbf{\Lambda}^{\mathrm{T}} = \mathbf{0} \qquad \text{(E-8)}$$

Substituting $\mathbf{\Lambda}^{\mathrm{T}} = -2\mathbf{I}$ in Eq. (7), we obtain

$$(\mathbf{AVA}^{\mathrm{T}})^{-1}\mathbf{AV} = \mathbf{B}^{\mathrm{T}} \qquad \text{(E-9)}$$

Postmultiplying by $\mathbf{V}^{-1}$ and then by $\mathbf{B}$ we find, successively:

$$(\mathbf{AVA}^{\mathrm{T}})^{-1}\mathbf{A} = \mathbf{B}^{\mathrm{T}}\mathbf{V}^{-1} \qquad \text{(E-10)}$$

$$(\mathbf{AVA}^{\mathrm{T}})^{-1}\mathbf{AB} = (\mathbf{AVA}^{\mathrm{T}})^{-1} = \mathbf{B}^{\mathrm{T}}\mathbf{V}^{-1}\mathbf{B} \qquad \text{(E-11)}$$

Substituting Eq. (11) in Eq. (10)

$$\mathbf{B}^{\mathrm{T}}\mathbf{V}^{-1}\mathbf{BA} = \mathbf{B}^{\mathrm{T}}\mathbf{V}^{-1} \qquad \text{(E-12)}$$

So that finally

$$\mathbf{A} = (\mathbf{B}^{\mathrm{T}}\mathbf{V}^{-1}\mathbf{B})^{-1}\mathbf{B}^{\mathrm{T}}\mathbf{V}^{-1} \qquad \text{(E-13)}$$

and

$$\boldsymbol{\theta}^* = (\mathbf{B}^{\mathrm{T}}\mathbf{V}^{-1}\mathbf{B})^{-1}\mathbf{B}^{\mathrm{T}}\mathbf{V}^{-1}\mathbf{y} \qquad \text{(E-14)}$$

in agreement with Eq. (4-4-7). It is not difficult to verify that this solution is indeed a minimum of det $\mathbf{AVA}^{\mathrm{T}}$.

A treatment of the general case where singular matrices may occur is given by Price (1964). The results call for substitution of pesudoinverses for inverses whenever needed.

Appendix

F

A Convergence Theorem for Gradient Methods

Theorem Given a continuous function $\Phi(\boldsymbol{\theta})$ with continuous differentiable first derivatives. Let $\Phi_1 = \Phi(\boldsymbol{\theta}_1)$, and let $\mathscr{D}$ be the set of all points $\boldsymbol{\theta}$ such that $\Phi(\boldsymbol{\theta}) \leqslant \Phi_1$. Define a sequence of points $\boldsymbol{\theta}_2, \boldsymbol{\theta}_3, \ldots$, by

$$\boldsymbol{\theta}_{i+1} = \boldsymbol{\theta}_i - \rho_i \mathbf{R}_i \mathbf{q}_i \qquad \text{(F-1)}$$

where

$$\mathbf{q}_i \equiv \partial\Phi/\partial\boldsymbol{\theta})_{\boldsymbol{\theta}=\boldsymbol{\theta}_i}$$

We make the following further assumptions.

1. There exists a number M such that no eigenvalue of the Hessian $\mathbf{H}(\boldsymbol{\theta})$ exceeds M in absolute value for all $\boldsymbol{\theta} \in \mathscr{D}$.
2. All $\mathbf{R}_i$ are positive definite matrices whose eigenvalues fall between two positive numbers $0 < \beta < \gamma$.
3. All ρ_i are chosen so that

$$\min(\rho_0, \alpha\mu_i) \leqslant \rho_i \leqslant \mu_i \qquad \text{(F-2)}$$

where ρ_0 is a positive constant, α a constant satisfying $0 < \alpha < 1$, and μ_i is the smallest nonnegative ρ at which $\Phi(\boldsymbol{\theta}_i - \rho\mathbf{R}_i\mathbf{q}_i)$ is a stationary function of ρ.

Let $\boldsymbol{\theta}^*$ be a limit point of the sequence $\{\boldsymbol{\theta}_i\}$. Then $\boldsymbol{\theta}^*$ is a stationary point of Φ, i.e., $\mathbf{q}^* \equiv \mathbf{q}(\boldsymbol{\theta}^*) = \mathbf{0}$.

Comment: Such a limit point (not necessarily unique) must exist if $\mathscr{D}$ is bounded.

Proof Clearly $\{\Phi_i\} \equiv \{\Phi(\boldsymbol{\theta}_i)\}$ is a monotone nonincreasing sequence. Because of continuity, we must have

$$\Phi^* \equiv \Phi(\boldsymbol{\theta}^*) \leqslant \Phi_i \qquad (i = 1, 2, \ldots) \qquad \text{(F-3)}$$

Suppose $\boldsymbol{\theta}^*$ is not stationary. Then $\|\mathbf{q}^*\| = a > 0$. Due to continuity of Φ and $\mathbf{q}$, and the definition of a limit point, we can find an integer j such that

$$2a \geqslant \|\mathbf{q}_j\| \geqslant \tfrac{1}{2}a \tag{F-4}$$

and

$$\Phi_j \leqslant \Phi^* + \min\left[\frac{(2-\alpha)\alpha a^2\beta^2}{256\gamma^2 M}, \frac{a^2\beta^2}{256\gamma^2 M}, \frac{a^2\beta\rho_0}{16}\right] \tag{F-5}$$

Consider the function

$$\Psi(\rho) \equiv \Phi(\boldsymbol{\theta}_j - \rho\mathbf{R}_j\mathbf{q}_j) \tag{F-6}$$

we have

$$d\Psi/d\rho = (\partial\Phi/\partial\boldsymbol{\theta})(\partial\boldsymbol{\theta}/\partial\rho) = -\mathbf{q}^{\mathrm{T}}\mathbf{R}_j\mathbf{q}_j \tag{F-7}$$

and

$$d\Psi/d\rho)_{\rho=0} = -\mathbf{q}_j{}^{\mathrm{T}}\mathbf{R}_j\mathbf{q}_j \leqslant -\beta\|\mathbf{q}_j\|^2 \leqslant -\beta a^2/4 \tag{F-8}$$

which follows from Eq. (A-1-32) and Eq. (4).

Also

$$\begin{aligned}|d^2\Psi/d\rho^2| &= |(\partial\boldsymbol{\theta}/\partial\rho)^T(\partial^2\Phi/\partial\boldsymbol{\theta}\,\partial\boldsymbol{\theta})(\partial\boldsymbol{\theta}/\partial\rho)| = |\mathbf{q}_j{}^{\mathrm{T}}\mathbf{R}_j\mathbf{H}_j\mathbf{R}_j\mathbf{q}_j| \\ &\leqslant M\|\mathbf{R}_j\mathbf{q}_j\|^2 \leqslant 4a^2\gamma^2 M\end{aligned} \tag{F-9}$$

which follows from Eq. (A-1-31) and Eq. (A-1-32).

In view of Eq. (8) and Eq. (9) we have, for $\rho > 0$

$$d\Psi/d\rho \leqslant -\beta a^2/4 + 4a^2\gamma^2 M\rho \tag{F-10}$$

At $\rho = \mu_j$ we have $d\Psi/d\rho = 0$. Hence, from Eq. (10)

$$0 \leqslant -\beta a^2/4 + 4a^2\gamma^2 M\mu_j \tag{F-11}$$

or, $\mu_j \geqslant (\beta/16\gamma^2 M)$ and $\alpha\mu_j \geqslant (\alpha\beta/16\gamma^2 M)$. In view of Eq. (10) we have, for $\rho > 0$

$$\Psi(\rho) \leqslant \Phi_j - (\beta a^2/4)\rho + 2a^2\gamma^2 M\rho^2 \tag{F-12}$$

Suppose ρ_j has been chosen so that $\alpha\mu_j \leqslant \rho_j \leqslant \mu_j$ [see (Eq. 2)]. Since $\Psi(\rho)$ is monotonically nonincreasing for $0 \leqslant \rho \leqslant \mu_j$, we have because of Eq. (12)

$$\begin{aligned}\Phi_{j+1} &= \Psi(\rho_j) \leqslant \Psi(\alpha\mu_j) \leqslant \Psi(\alpha\beta/16\gamma^2 M) \\ &\leqslant \Phi_j - (\beta a^2/4)(\alpha\beta/16\gamma^2 M) + 2a^2\gamma^2 M(\alpha\beta/16\gamma^2 M)^2 \\ &= \Phi_j - [(2-\alpha)\alpha a^2\beta^2/128\gamma^2 M]\end{aligned} \tag{F-13}$$

Employing Eq. (5) we find, then

$$\Phi_{j+1} \leqslant \Phi^* - [(2-\alpha)\alpha a^2\beta^2/256\gamma^2 M] < \Phi^* \tag{F-14}$$

This contradicts Eq. (3).

The other alternative is that $\rho_0 \leqslant \rho_j \leqslant \mu_j$. But then

$$\Phi_{j+1} \leqslant \Phi_j - (\beta a^2/4)\rho_0 + 2a^2\gamma^2 M\rho_0{}^2 \tag{F-15}$$

Now there are two possibilities

(a) $\rho_0 \leqslant \beta/16\gamma^2 M$. Therefore

$$\begin{aligned}\Phi_{j+1} &\leqslant \Phi_j - a^2\rho_0(\beta/4 - 2\gamma^2 M\rho_0) \leqslant \Phi_j - a^2\rho_0[\beta/4 - 2\gamma^2 M(\beta/16\gamma^2 M)] \\ &= \Phi_j - a^2\beta\rho_0/8 \leqslant \Phi^* - a^2\beta\rho_0/16 < \Phi^* \end{aligned}\tag{F-16}$$

In contradiction with Eq. (3).

(b) $\beta/16\gamma^2 M \leq \rho_0 \leqslant \mu_j$. Therefore, because $\Psi(\rho)$ is monotonically non-increasing at $\rho = \rho_0$

$$\begin{aligned}\Phi_{j+1} &= \Phi(\rho_0) \leqslant \Phi(\beta/16\gamma^2 M) \leqslant \Phi_j - (\beta a^2/4)(\beta/16\gamma^2 M) + 2a^2\gamma^2 M(\beta/16\gamma^2 M)^2 \\ &= \Phi_j - (a^2\beta^2/128\gamma^2 M) \leqslant \Phi^* - (a^2\beta^2/256\gamma^2 M) < \Phi^* \end{aligned}\tag{F-17}$$

again contradicting Eq. (3). Hence $\boldsymbol{\theta}^*$ must be stationary.

Appendix

G

Some Estimation Programs

It is impossible to list all existing estimation programs. The ones listed below are either of historical interest, possess special features, or are in widespread use. The list is in chronological order.

1. G. W. Booth and T. I. Peterson (1958), Nonlinear estimation (IBM-Princeton), 704 G2 3226 NLI (Previously designated SHARE 687 WLNLI). First generally available computer program for nonlinear parameter estimation. Written in IBM 704 Assembly Language. Uses Gauss method with finite difference approximation to solve single equation least squares problems.
2. M. A. Efroymson (1961), Nonlinear regression with differential equations, 7090 G2 3146 NLR. Written in FORTRAN II for the IBM 7090 specifically to handle models which are in differential equation form. Uses Gauss method with finite difference approximations.
3. L. Lapidus and T. I. Peterson (1964), Chemical reaction analysis by nonlinear estimation, 7090 T2 IBM0014. A package combining program 1 above with a FORTRAN II interface for estimating parameters in chemical-kinetics models.
4. D. W. Marquardt (1965), Least squares estimation of nonlinear parameters, 7040 G2 3094 NLIN. Written in FORTRAN IV for the IBM 7040. Uses Marquardt's method with analytic derivatives or finite difference approximations to solve weighted least squares problems.
5. H. Eisenpress, A. Bomberault, and J. Greenstadt (1966), Nonlinear regression equations and systems, estimation and prediction, 7090 G2 IBM0035. Written in FORTRAN IV-FORMAC for the IBM 7090. Performs maximum likelihood estimation of multiple equation econometric models. The FORMAC system evaluates automatically analytic derivatives of all orders required for the full Newton method with rotational discrimination.
6. D. F. Shanno, 1967, CREEP—Constrained nonlinear estimation package, 7094 G2 3492. Written in FORTRAN IV for the IBM 7094. Estimates parameters in least squares models with constraints, using a

modified Marquardt method combined with gradient projection. Requires analytic derivatives.

7. Y. Bard, 1968, Nonlinear estimation and programming, 360D 13.6.003. Written in FORTRAN IV for the IBM System/360. Solves least squares and multiple equation maximum likelihood problems with known or unknown covariance matrix. Uses the Gauss method with analytic derivatives. Includes provisions for constraints (penalty function method), prior distributions, models in standard dynamic form (sensitivity equations), and chemical kinetics models.

8. H. Eisenpress, 1968, Nonlinear regression equations and systems, estimation and prediction, 360D 13.6.005. Program 5 above, rewritten in PL/I-FORMAC for the IBM System/360.

9. F. S. Wood, 1971, Nonlinear least squares curve-fitting program, 360D 13.6.007. Written in FORTRAN IV for the IBM System/360. Solves least squares problems, using a modified Marquardt method.

Note: At the time of writing, several of the above programs were available from the SHARE Program Library Agency, Triangle Universities Computation Center, P.O. Box 12076, Research Triangle Park, North Carolina 27709.

References

Abadie, J., ed. (1967a). "Nonlinear Programming." Wiley, New York.

Abadie, J. (1967b). "Generalization of the Wolfe Reduced Gradient Method to the Case of Nonlinear Constraints." Electricité de France, Direction des Etudes et Recherches, Clamart, France.

Abadie, J., and Carpentier, J. (1966). Generalisation de la methode du gradient reduit de Wolfe au cas de contraintes non lineaires. *Internat. Congr. Operations Research, 4th, Boston.*

Acton, F. S. (1959). "Analysis of Straight Line Data." Wiley, New York.

Afifi, A. A., and Elashoff, R. M. (1966). Missing observations in multivariate statistics: I. Review of the literature. *J. Amer. Statist. Assoc.* **61,** 595–604.

Albert, A. E., and Gardner, L. A. (1967). "Stochastic Approximation and Nonlinear Regression." MIT Press, Cambridge, Massachusetts.

Anderson, T. W. (1958). "An Introduction to Multivariate Statistical Analysis." Wiley, New York.

Anscombe, F. J. (1960). Rejection of outliers. *Technometrics* **2,** 123–147.

Arndt, R. A., and MacGregor, M. H. (1966). Nucleon–nucleon phase shift analyses by chi-squared minimization. *In* "Methods in Computational Physics," (B. Alder, F. Fernbach, and M. Rotenberg, eds.), Vol. 6. Academic Press, New York.

Atkinson, A. C., and Hunter, W. G. (1968). The design of experiments for parameter estimation. *Technometrics.* **10,** 271–289.

Bard, Y. (1967). "A Function Maximization Method with Application to Parameter Estimation." New York Scientific Center Report 322.0902, IBM, New York.

Bard, Y. (1968). On a numerical instability of Davidon-like methods. *Math. Comp.* **22,** 665–666.

Bard, Y. (1970). Comparison of gradient methods for the solution of nonlinear parameter estimation problems. *SIAM J. Numer. Anal.* **7,** 157–186.

Bard, Y. (1971). An eclectic approach to nonlinear programming. *Proc. ANU Sem. Optimization, Canberra,* Austral. Nat. Univ.

Bard, Y., and Lapidus, L. (1968). Kinetics analysis by digital parameter estimation. *Catal. Rev.* **2,** 67–112.

Barnett, V. D. (1967). A note on linear structural relationships when both residual variances are known. *Biometrika.* **54,** 670–672.

Bartels, R. H., and Golub, G. H. (1968). Chebyshev solution to an overdetermined linear system. *Comm. ACM.* **11,** 428.

Bayes, T. (1763). Essay towards solving a problem in the doctrine of chances. *Philos. Trans. Roy. Soc.* **53,** 370–418. [Reprinted in *Biometrika.* **45,** 293–315 (1958)].

Beale, E. M. L. (1960). Confidence regions in nonlinear estimation. *J. Roy. Statist. Soc. Ser. B.* **22,** 41–76.

Beaton, A. E. (1964). "The Use of Special Matrix Operators in Statistical Calculus." Research Bulletin RB-64-51, Educational Testing Service, Princeton, New Jersey.

Beauchamp, J. J. and Cornell, R. G. (1966). Simultaneous nonlinear estimation. *Technometrics.* **8,** 319–326.

Behnken, D. W. (1964). Estimation of copolymer reactivity ratios: an example of nonlinear estimation. *J. Polym. Sci. Part A.* **2,** 645–668.

Bellman, R., Collier, C., Kagiwada, H., Kalaba, R., and Selvester, R. (1964). Estimation of heart parameters using skin potential measurements. *Comm. ACM.* **7,** 666–668.

Bellman, R., Jacquez, J., Kalaba, R., and Schwimmer, S. (1967). Quasilinearization and the estimation of chemical rate constants from raw kinetic data. *Math. Biosci.* **1,** 71–76.

Bellman, R. E., Kagiwada, H. H., Kalaba, R. E., and Sridhar, R. (1964). "Invariant Imbedding and Nonlinear Filtering Theory." Memorandum RM-4374-PR, The Rand Corporation, Santa Monica, California.

Berman, M., Weiss, M. F., and Shahn, E. (1962). Some formal approaches to the analysis of kinetic data in terms of linear compartmental systems. *Biophys. J.* **2,** 289–316.

Blakemore, J. W., and Hoerl, A. E. (1963). Fitting nonlinear reaction rate equations to data. *Chem. Eng. Progr. Symp. Ser.* **59** (42), 14–27.

Bodkin, R. G., and Klein, L. R. (1967). Nonlinear estimation of aggregate production functions. *Rev. Econom. Statist.* **49,** 28–44.

Bond, E., Auslander, M., Grisoff, S., Kenney, R., Myszewski, M., Sammet, J., Tobey, R., and Zilles, S. (1964). FORMAC, An experimental FORmula MAnipulation Compiler. *Proc. Nat. Conf. Ass. Comput. Mach. 19th.*

Booth, G. W., and Peterson, T. I. (1958). "Nonlinear Estimation." IBM SHARE Program Pa. No. 687 WLNLI.

Box, G. E. P. (1957). Use of statistical methods in the elucidation of basic mechanisms. *Bull. Inst. Internat. Statist.* **36,** 215–225.

Box, G. E. P., and Hill, W. J. (1967). Discrimination among mechanistic models. *Technometrics.* **9,** 57–71.

Box, G. E. P., and Hunter, W. G. (1962). A useful method for model-building. *Technometrics.* **4,** 301–318.

Box, G. E. P., and Hunter, W. G. (1963). Sequential design of experiments for nonlinear models. *Proc. IBM Sci. Comput. Symp. Statist., IBM, White Plains, New York.*

Box., G. E. P., and Hunter, W. G. (1965). The experimental study of physical mechanisms. *Technometrics.* **7,** 23–42.

Box, G. E. P., and Lucas, H. L. (1959). Design of experiments in nonlinear situations. *Biometrika.* **46,** 77–90.

Box, G. E. P., and Muller, M. E. (1958). A note on the generation of random normal deviates. *Ann. Math. Statist.* **29,** 610–611.

Box, G. E. P., and Youle, P. V. (1955). The exploration and exploitation of response surfaces: an example of the link between the fitted surface and the basic mechanism of the system. *Biometrics.* **11,** 287–323.

Box, M. J. (1966). A comparison of several current optimization methods, and the use of transformations in constrained problems. *Comput. J.* **9,** 67–77.

Box, M. J. (1968). The occurrence of replications in optimal designs of experiments to estimate parameters in nonlinear models. *J. Roy. Statist. Soc. Ser. B.* **30,** 290–302.

Brent, R. P. (1971). "Algorithms for Finding Zeros and Extrema of Functions without Calculating Derivatives." Computer Science Report STANS-CS-71-198, Stanford University, Palo Alto, California.

Broyden, C. G. (1967). Quasi-Newton methods and their application to function minimization. *Math. Comp.* **21,** 368–381.

Buzzi Ferraris, G. (1968). Metodo automatico per trovare l'ottimo di una funzione. *Ing. Chim. Ital.* **4,** 171–192.

Carney, T. M., and Goldwyn, R. M. (1967). Numerical experiments with various optimal estimators. *J. Optimization Theory Appl.* **1,** 113–130.

Carroll, C. W. (1961). The created response surface technique for optimizing nonlinear, restrained, systems. *Operations Res.* **9,** 169–184.

Chow, G. C. (1964). A comparison of alternative estimators for simultaneous equations. *Econometrica.* **32,** 532–553.

Colville, A. R. (1968). "A Comparative Study of Nonlinear Programming Codes." IBM N.Y. Scientific Center Report No. 320-2949, New York.

Cornfield, J. (1967). Bayes Theorem. *Rev. Inst. Internat. Statist.* **35,** 34–49.

Cottle, R. W., and Dantzig, G. B. (1968). Complementary pivot theory of mathematical programming. *Linear Algebra and Appl.* **1,** 103–125.

Cragg, J. G. (1967). On the relative small sample properties of several structural-equation estimators. *Econometrica.* **35,** 89–110.

Cramér, H. (1946). "Mathematical Methods of Statistics." Princeton Univ. Press, Princeton, New Jersey.

Daniel, J. W. (1971). "The Approximate Minimization of Functionals." Prentice-Hall, Englewood Cliffs, New Jersey.

Dantzig, G. B., and Cottle, R. W. (1967). Positive (semi-) definite programming. *In* "Nonlinear Programming," (J. Abadie, ed.), pp. 55–73. Wiley, New York.

Davidon, W. C. (1959). "Variable Metric Method for Minimization." A.E.C. Research and Development Report ANL-5990 (Rev.).

Davidon, W. C. (1968). Variance algorithm for minimization. *Comput. J.* **10,** 406–410.

Davies, D. (1970). Some practical methods of optimization. *In* "Integer and Nonlinear Programming" (J. Abadie, ed.), North-Holland Publ., Amsterdam.

Davies, O. L. D. (1954). "The Design and Analysis of Industrial Experiments." Oliver & Boyd, Edinburgh.

Deming, W. E. (1943). "Statistical Adjustment of Data." Wiley, New York.

Deutsch, R. (1965). "Estimation Theory." Prentice-Hall, Englewood Cliffs, New Jersey.

Draper, N. R., and Hunter, W. G. (1966). Design of experiments for parameter estimation in multiresponse situations. *Biometrika.* **53,** 525–533.

Draper, N. R., and Hunter, W. G. (1967a). The use of prior distributions in the design of experiments for parameter estimation in nonlinear situations. *Biometrika.* **54,** 147–153.

Draper, N. R., and Hunter, W. G. (1967b). The use of prior distributions in the design of experiments for parameter estimation in nonlinear situations: multiresponse case. *Biometrika.* **54,** 662–665.

Draper, N. R., and Smith, H. (1966). "Applied Regression Analysis." Wiley, New York.

Eisenpress, H., Bomberault, A., and Greenstadt, J. (1966). "Nonlinear Regression Equations and Systems, Estimation and Prediction (IBM) 7090." Computer program 7090-G2 IBM0035 G2, IBM, Hawthorne, New York.

Eisenpress, H., and Greenstadt, J. (1966). The estimation of nonlinear econometric systems. *Econometrica.* **34,** 851–861.

Eisenpress, H., and Surkan, A. (1966). "Fitting Ore Deposit Models to Geophysical Survey Data." Personal Communication.

Fariss, R. H., and Law, V. J. (1967). Practical tactics for overcoming difficulties in nonlinear regression and equation solving. *AIChE Meet., Houston, Feb. 1967.*

Faure, P., and Huard, P. (1965). Resolution des programmes mathematiques à fonction nonlinéaire par la méthode du gradient reduit. *Rev. Française Recherche Opérationelle.* **9,** 167–205.

Feller, W. (1966). "An Introduction to Probability Theory and Its Applications," Vol. II. Wiley, New York.

Ferguson, T. S. (1967). "Mathematical Statistics, A Decision Theoretic Approach." Academic Press, New York.

Fiacco, A. V. (1968). Second-order sufficient conditions for weak and strict constrained minima. *SIAM J. Appl. Math.* **16,** 105–108.

Fiacco, A. V., and McCormick, G. P. (1964). The sequential unconstrained minimization technique for nonlinear programming: A primal-dual method. *Management Sci.* **10,** 360–366.

Fiacco, A. V., and McCormick, G. P. (1965). "The Sequential Unconstrained Minimization Technique for Convex Programming with Equality Constraints." RAC-TP-155, Research Analysis Corporation, McLean, Virginia.

Fiacco, A. V., and McCormick, G. P. (1967). The slacked unconstrained minimization technique for convex programming. *SIAM J. Appl. Math.* **15,** 505–515.

Fiacco, A. V., and McCormick, G. P. (1968). "Nonlinear Programming: Sequential Unconstrained Minimization Techniques." Wiley, New York.

Fisher, R. A. (1935). "The Design of Experiments." Oliver & Boyd, Edinburgh.

Fisher, R. A. (1950). "Contributions to Mathematical Statistics" (Collection of papers published 1920–1943.) Wiley, New York.

Flanagan, P. D., Vitale, P. A., and Mendelsohn, J. (1969). A numerical investigation of several one-dimensional search procedures in nonlinear regression problems. *Technometrics* **11,** 265–284.

Fletcher, R. (1965). Function minimization without evaluating derivatives—a review. *Comput. J.* **8,** 33–41.

Fletcher, R. (1970). A new approach to variable metric algorithms. *Comput. J.* **13,** 317–322.

Fletcher, R., and Powell, M. J. D. (1963). A rapidly convergent descent method for minimization. *Comput. J.* **6,** 163–168.

Fox, L. (1964). "An Introduction to Numerical Linear Algebra." Oxford Univ. Press (Clarendon), London and New York.

Freudenstein, F., and Woo, L. S. (1968). "Kinematics of the Human Knee Joint." IBM New York Scientific Center Report No. 320-2928, New York.

Galambos, J. T., and Cornell, R. G. (1962). Mathematical models for the study of the metabolic pattern of sulfate. *J. Lab. Clin. Med.* **60,** 53–63.

Gauss, K. F. (1809). "Theoria Motus Corporum Coelestium." *In* "Werke," Vol. 7, 240–254.

Goldfarb, D., and Lapidus, L. (1968). A conjugate gradient method for nonlinear programming problems with linear constraints. *Ind. Eng. Chem. Fundam.* **7,** 142–151.

Goldfeld, S. M., Quandt, R. E., and Trotter, H. F. (1966). Maximization by quadratic hill climbing. *Econometrica.* **34,** 541–551.

Goldstein, A. A., and Price, J. F. (1967). An effective algorithm for minimization. *Numer. Math.* **10,** 184–189.

Golub, G. (1965). Numerical methods for solving linear least squares problems. *Numer. Math.* **7,** 206–216.

Golub, G. H. (1969). Matrix decompositions and statistical calculations. *In* "Statistical Computation" (C. Milton and J. A. Nelder, eds.). Academic Press, New York.

Golub, G. H., and Pereyra, V. (1972). The Differentiation of Pseudoinverses and Nonlinear Least Squares Problems Whose Variables Separate." Rep. No. STAN-CS-72-261, Computer Science Dept., Stanford University, Palo Alto, California.

Grant, F. S., and West, G. F. (1965). "Interpretation Theory in Applied Geophysics." McGraw-Hill, New York.

Greenstadt, J. (1967). On the relative efficiencies of gradient methods. *Math. Comp.* **21,** 360–367.

Greenstadt, J. (1970). Variations on variable metric methods. *Math. Comp.* **24,** 1–22.

Guttman, I., and Meeter, D. A. (1965). On Beale's measures of nonlinearity. *Technometrics* **7,** 623–637.

Hadley, G., (1964). "Nonlinear and Dynamic Programming." Addison-Wesley, Reading, Massachusetts.

Hammersley, J. M., and Hanscomb, D. C. (1964). "Monte Carlo Methods." Methuen, London.

Hartley, H. O. (1961). The modified Gauss-Newton method for the fitting of nonlinear regression functions by least squares. *Technometrics* **3,** 269–280.

Hartley, H. O. (1964). Exact confidence regions for the parameters in nonlinear regression laws. *Biometrika.* **51,** 347–353.

Healy, M. J. R. (1968). Multiple regression with a singular matrix. *J. Roy. Statist. Soc. Ser. C Appl. Statist.* **17,** 110–117.

Heineken, F. G., Tsuchiya, H. M., and Aris, R. (1967a). On the mathematical status of the pseudosteady state hypothesis of biochemical kinetics. *Math. Biosci.* **1,** 95–113.

Heineken, F. G., Tsuchiya, H. M., and Aris, R. (1967b). On the accuracy of determining rate constants in enzymatic reactions. *Math. Biosci.* **1,** 115–141.

Hicks, J. S., and Wei, J. (1967). Numerical solution of parabolic partial differential equations with two-point boundary conditions by use of the method of lines. *J. Assoc. Comput. Mach.* **14,** 549–562.

Hill, W. J., Hunter, W. G. and Wichern, D. W. (1968). A joint design criterion for the dual problem of model discrimination and parameter estimation. *Technometrics.* **10,** 145–160.

Himmelblau, D. M., Jones, C. R., and Bischoff, K. B. (1967). Determination of rate constants for complex kinetics models. *Ind. Eng. Chem. Fundam.* **6,** 539–543.

Hoerl, A. E. (1962). Application of ridge analysis to regression problems. *Chem. Eng. Progr.* **58,** 54–59.

Hoerl, A. E., and Kennard, R. W. (1970). Ridge regression: biased estimation for nonorthogonal problems. *Technometrics.* **12,** 55–67.

Hood, W. C., and Koopmans, T. C., eds. (1953). "Studies in Econometric Method." Wiley, New York.

Hooke, R., and Jeeves, T. A. (1961). "Direct search" solution of numerical and statistical problems. *J. Assoc. Comput. Mach.* **8,** 212–229.

Hougen, O. A., and Watson, K. M. (1947). "Chemical Process Principles, Part Three: Kinetics and Catalysis." Wiley, New York.

Howland, J. L., and Vaillancourt, R. (1961). A generalized curve fitting method. *SIAM J. Appl. Math.* **9,** 165–168.

Hunter, W. G., and Mezaki, R. (1964). A model building technique for chemical engineering kinetics. *AIChE. J.* **10,** 315–322.

Hunter, W. G., and Mezaki, R. (1967). An experimental design strategy for distinguishing among rival mechanistic models-an application to the catalytic hydrogenation of propylene. *Can. J. Chem. Eng.* **45,** 247–249.

Jeffreys, H. (1961). "Theory of Probability," 3rd ed. Oxford Univ. Press, London and New York.

Jennrich, R. I., and Sampson, P. F. (1968). Application of stepwise regression to nonlinear estimation. *Technometrics.* **10,** 63–72.

John, F. (1948). Extremum problems with inequalities as subsidiary conditions. *In* "Studies and Essays," 187–204. Wiley (Interscience), New York.

Johnston, J. (1963). "Econometric Methods," McGraw-Hill, New York.

Kalman, R. E. (1960). A new approach to linear filtering and prediction problems. *J. Basic Eng.* **82,** 33–45.
Kelley, H. J., and Denham, W. F. (1966). Orbit determination with the Davidon Method. *Joint Automat. Control Conf. Seattle, Washington.*
Kelley, H. J., and Denham, W. F. (1969). Modeling and adjoints for continuous systems. *J. Optimization Theory Appl.* **3,** 174–183.
Kelley, Jr., J. E., (1958). An application of linear programming to curve fitting. *SIAM J. Appl. Math.* **6,** 15–22.
Kittrell, J. R., Hunter, W. G., and Mezaki, R. (1966). The use of diagnostic parameters for kinetic model building. *AIChE J.* **12,** 1014–1017.
Kittrell, J. R., Hunter, W. G., and Watson, C. C. (1966). Obtaining precise parameter estimates for nonlinear catalytic rate models. *AIChE J.* **12,** 5–10.
Kittrell, J. R., Mezaki, R., and Watson, C. C. (1965). Estimation of parameters for nonlinear least squares analysis. *Ind. Eng. Chem.* **57,** 18–27.
Koopmans, T. C., and Hood, W. C. (1953). The estimation of simultaneous linear economic relationships. *In* "Studies in Econometric Method" (W. C. Hood and T. C. Koopmans, eds.). Wiley, New York.
Korin, B. P. (1968). On the distribution of a statistic used for testing a covariance matrix. *Biometrika* **55,** 171–178.
Kowalik, J., and Osborne, M. R. (1968). "Methods for Unconstrained Optimization Problems." American Elsevier, New York.
Kuhn, H. W., and Tucker, A. W. (1951). Nonlinear programming. *In Proc. Berkeley Symp. Math. Statist. and Probability, 2nd* (J. Neyman, ed). Univ. of California Press, Berkeley, California.
Kullback, S. (1959). "Information Theory and Statistics." Wiley, New York.
Kullback, S., and Leibler, R. A. (1951). On information and sufficiency. *Ann. Math. Statist.* **22,** 79–86.
Künzi, H. P., and Krelle, W. (1966). "Nonlinear Programming." Ginn (Blaisdell), Boston, Massachusetts.
Lawton, W. H., and Sylvestre, E. A. (1971). Elimination of linear parameters in nonlinear regression. *Technometrics.* **13,** 461–467.
Legendre, A. M. (1805). "Nouvelles Méthodes pour la Determination des Orbites de Comètes." Paris.
Lehman, E. L. (1959). "Testing Statistical Hypotheses." Wiley, New York.
Levenberg, K. (1944). A method for the solution of certain nonlinear problems in least squares. *Quart. Appl. Math.* **2,** 164–168.
Lewis, P. A. W., Goodman, A. S., and Miller, J. M. (1969). A pseudorandom number generator for the System/360. *IBM Systems J.* **8,** 136–145.
Lindley, D. V. (1956). On a measure of the information provided by an experiment. *Ann. Math. Statist.* **27,** 986–1005.
Longley, J. W. (1967). An appraisal of least squares programs for the electronic computer from the point of view of the user. *J. Amer. Statist. Assoc.* **62,** 819–841.
Mangasarian, O. L. (1969). "Nonlinear Programming." McGraw-Hill, New York.
Marquardt, D. W. (1963). An algorithm for least squares estimation of nonlinear parameters. *SIAM J.* **11,** 431–441.
Marsaglia, G., and Bray, T. A. (1964). A convenient method for generating normal variables. *SIAM Rev.* **6,** 260–264.
McCormick, G. P. (1967). Second-order conditions for constrained minima. *SIAM J. Appl. Math.* **15,** 641–652.
McGhee, R. B. (1963). "Identification of nonlinear dynamic systems by regression analysis

methods." Doctoral dissertation, Univ. Southern California, Los Angeles, California (University Microfilms 64–2588, Ann Arbor, Mich.).

Melkanoff, M. A., Sawada, T., and Raynal, J. (1966). Nuclear optical model calculations. *In* "Methods in Computational Physics" (B. Alder, S. Fernbach, and M. Rotenberg, eds.), Vol. 6. Academic Press, New York.

von Mises, R. (1919). Fundamentalsätze der Wahrscheinlichkeitsrechnung. *Math. Zeitschrift* **4,** 1–97.

Moshman, J. (1967). Random number generation. *In* "Mathematical Methods for Digital Computers" (A. Ralston and H. S. Wilf, eds.), Volume II. Wiley, New York.

Murtagh, B. A., and Sargent, R. W. H. (1969). A constrained minimization method with quadratic convergence. *In* "Optimization" (R. Fletcher ed.). Academic Press, New York.

Nelder, J. A., and Mead, R. (1965). A simplex method for function minimization. *Comput. J.* **7,** 308–313.

Neyman, J. (1937). Outline of a theory of statistical estimation based on the classical theory of probability. *Phil. Trans. Roy. Soc. London Ser. A*, **231,** 333–380.

Neyman, J. (1962). Two breakthroughs in the theory of statistical decision making. *Rev. Inst. Internat. Statist.* **30,** 11–27.

Ortega, J. M., and Kaiser, H. F. (1963). The LL^T and QR methods for symmetric tri-diagonal matrices. *Comput. J.* **6,** 99–101.

Osborne, M. R., and Watson, G. A. (1969). An algorithm for minimax approximation in the nonlinear case. *Comput. J.* **12,** 63–68.

Pearson, J. D. (1969). Variable metric methods of minimization *Comput. J.* **12,** 171–178.

Penrose, R. (1955). A generalized inverse for matrices. *Proc. Cambridge Philos. Soc.* **51,** 406–413.

Perdreauville, F. J., and Goodson, R. E. (1966). Identification of systems described by partial differential equations. *Trans. ASME, Ser. D*, **88,** 463–468.

Peterson, T. I. (1962). Kinetics and mechanism of naphthalene oxidation by nonlinear estimation. *Chem. Eng. Sci.* **17,** 203–219.

Pontryagin, L. S., Bolyanskii, V. G., Gamkrelidze, R. V., and Mishchenko, E. F. (1962). "The Mathematical Theory of Optimal Processes," K. N. Trigoroff (transl.). Wiley (Interscience), New York.

Powell, M. J. D. (1964). An efficient method for finding the minimum of a function of several variables without calculating derivatives. *Comput. J.* **7,** 155–162.

Powell, M. J. D. (1965). A method for minimizing a sum of squares of nonlinear functions without calculating derivatives. *Comput. J.* **7,** 303–307.

Powell, M. J. D. (1969). A theorem on rank one modification to a matrix and its inverse. *Comput. J.* **12,** 288–290.

Price, C. M. (1964). The matrix pseudoinverse and minimal variance estimates. *SIAM Rev.* **6,** 115–120.

Quenouille, M. H. (1956). Notes on bias in estimation. *Biometrika.* **43,** 353–360.

Raiffa, H., and Schlaifer, R. (1961). "Applied Statistical Decision Theory." Graduate School of Business Administration, Harvard Univ., Boston.

Rao, C. R. (1957). Theory of the method of estimation by minimum chi-square. *Bull. Internat. Statist. Inst.* **35,** 25–32.

Robbins, H. (1955). An empirical Bayes' approach to statistics. *In Proc. Berkeley Symp. Statist. and Probability, 3rd* **1,** 157–164. Univ. of California Press, Berkeley, California.

Robbins, H. (1964). The empirical Bayes' approach to statistical decision problems. *Ann. Math. Statist.* **35,** 1–20.

Rosen, J. B. (1960). The gradient projection method for nonlinear programming: I. Linear constraints. *SIAM J.* **8,** 181–217.

Rosen, J. B. (1961). The gradient projection method for nonlinear programming: II. Nonlinear constraints. *SIAM J.* **9,** 514–532.
Rosenbrock, H. H. (1960). An automatic method for finding the greatest or least value of a function. *Comput. J.* **3,** 175–184.
Rosenbrock, H. H., and Storey, C. (1966). "Computational Methods for Chemical Engineers." Pergamon, Oxford.
Rutemiller, H. C., and Bowers, D. A. (1968). Estimation in a heteroscedastic regression model. *J. Amer. Statist. Assoc.* **63,** 552–557.
Sammet, J. E. (1966). Survey of formula manipulation. *Comm. ACM.* **9,** 555–569.
Savage, L. J. (1954). "The Foundations of Statistics." Wiley, New York.
Scheffé, H. (1959). "The Analysis of Variance." Wiley, New York.
Seal, H. L. (1967). The historical development of the Gauss linear model. *Biometrika.* **54,** 1–24.
Shannon, C. E. (1948). A mathematical theory of communication. *Bell System Tech. J.* **27,** 379–423 and 623–656.
Shinbrot, M. (1954). "On the Analysis of Linear and Nonlinear Dynamical Systems from Transient-Response Data." NACA Technical Notes, TN 3288.
Smith, Jr. F. B., and Shanno, D. F. (1971). An improved Marquardt procedure for nonlinear regressions. *Technometrics.* **13,** 63–74.
Solow, R. M. (1957). Technical change and the aggregate production function, *Rev. Econom. Statist.* **39,** 312–320.
Sorenson, H. W. (1966). Kalman filtering techniques. *In* "Advances in Control Systems" (C. T. Leondes, ed.), Vol. 3 Academic Press, New York.
Spendley, W. (1969). Nonlinear least squares fitting using a modified simplex minimization method. *In* "Optimization" (R. Fletcher, ed.). Academic Press, New York.
Stewart III, G. W. (1967). A modification of Davidon's minimization method to accept difference approximations of derivatives. *J. Assoc. Comput. Mach.* **14,** 72–83.
Swed, F. S., and Eisenhart, C. (1943). Tables for testing randomness of grouping in a sequence of alternatives. *Ann. Math. Statist.* **14,** 66–87.
Tomović, R. (1963). "Sensitivity Analysis of Dynamic Systems." McGraw-Hill, New York.
Turner, M. E., Monroe, R. J., and Homer, L. D. (1963). Generalized kinetic regression analysis: hypergeometric kinetics. *Biometrics.* **19,** 406–428.
Wagner, H. M. (1959). Linear programming techniques for regression analysis. *J. Amer. Statist. Assoc.* **54,** 206–212.
Wald, A. (1947). "Sequential Analysis." Wiley, New York.
Wiener, N. (1949). "Extrapolation, Interpolation and Smoothing of Stationary Time Series." MIT Press, Cambridge, Massachusetts and Wiley, New York.
Wilde, D. J., and Beightler, C. S. (1967). "Foundations of Optimization." Prentice-Hall, Englewood Cliffs, New Jersey.
Wilkinson, J. H. (1965). "The Algebraic Eigenvalue Problem." Oxford Univ. Press (Clarendon), London and New York.
Winkler, R. L. (1967). The assessment of prior distributions in Bayesian analysis, *J. Amer. Statist. Assoc.* **62,** 776–800.
Wolfe, P. (1963). Methods of nonlinear programming. *In* "Recent Advances in Mathematical Programming" (R. L. Graves and P. Wolfe, eds.). McGraw-Hill, New York.
Zangwill, W. I. (1967a). Nonlinear programming via penalty functions. *Management Sci.* **5,** 344–358.
Zangwill, W. I. (1967b). Minimizing a function without calculating derivatives. *Comput. J.* **10,** 293–296.
Zoutendijk, G. (1960). "Methods of Feasible Directions." Elsevier, Amsterdam.

Author Index

Numbers in italics refer to the pages on which the complete references are listed.

A

Abadie, J., 83, 146, *325*
Acton, F. S., 201, *325*
Afifi, A. A., 245, *325*
Albert, A. E., 251, *325*
Anderson, T. W., 170, 200, *325*
Anscombe, F. J., 202, *325*
Aris, R., 273, *329*
Arndt, R. A., 15, *325*
Atkinson, A. C., 265, 272, *325*
Auslander, M., 116, *326*

B

Bard, Y., 91, 96, 107, 109, 110, 148, 151, 277, *324*, *325*
Barnett, V. D., 81, *325*
Bartels, R. H., 77, *325*
Bayes, T., 36, *325*
Beale, E. M. L., 170, 191, *325*
Beaton, A. E., 296, *362*
Beauchamp, J. J., 16, 253, *326*
Behnken, D. W., 265, *326*
Beightler, C. S., 83, *332*
Bellman, R., 16, 226, 230, 242, *326*
Berman, M., 16, *326*
Bischoff, K. B., 220, *329*
Blakemore, J. W., 277, *326*
Bodkin, R. G., 25, 133, 138, *326*
Bolyanskii, V. G., 225, *331*
Bomberault, A., 116, 133, 323, *327*
Bond, E., 116, *326*
Booth, G. W., 7, 323, *326*
Bowers, D. A., 247, *332*
Box, G. E. P., 100, 123, 204, 261, 265, 267, 269, 316, *326*
Box, M. J., 119, 120, 153, 274, *326*
Bray, T. A., 316, *330*
Brent, R. P., 120, *326*
Broyden, C. G., 107, 108, *326*
Buzzi Ferraris, G., 120, *327*

C

Carney, T. M., 61, *327*
Carpentier, J., 146, *325*
Carroll, C. W., 141, *327*
Chow, G. C., 61, *327*
Collier, C., 16, *326*
Colville, A. R., 117, *327*
Cornell, R. G., 16, 253, *326*, *328*
Cornfield, J., 33, *327*
Cottle, R. W., 148, *327*
Cragg, J. G., 61, *327*
Cramér, H., 19, 80, 178, 185, 186, 188, 201, *327*

D

Daniel, J. W., 83, *327*
Dantzig, G. B., 148, *327*
Davidon, W. C., 106, 107, 108, 110, *327*
Davies, D., 146, *327*
Davies, O. L. D., 260, *327*
Deming, W. E., 154, *327*
Denham, W. F., 16, 242, *330*
Deutsch, R., 77, 225, *327*
Draper, N. R., 201, 265, *327*

E

Efroymson, M. A., *323*
Eisenhart, C., 201, *332*
Eisenpress, H., 15, 116, 133, 323, *324*, *327*
Elashoff, R. M., 245, *325*

F

Fariss, R. H., 91, *327*
Faure, P., 146, *327*
Feller, W., 19, *328*
Ferguson, T. S., 33, *328*
Fiacco, A. V., 52, 107, 108, 142, 145, 159, *328*
Fisher, R. A., 7, 260, *328*
Flanagan, P. D., 91, *328*
Fletcher, R., 107, 110, 120, *328*
Fox, L., 307, *328*
Freudenstein, F., 16, *328*

G

Galambos, J. T., 253, *328*
Gamkrelidze, R. V., 225, *331*
Gardner, L. A., 251, *325*
Gauss, K. F., 6, 97, *328*
Goldfarb, D., 110, 146, *328*
Goldfeld, S. M., 94, *328*
Goldstein, A. A., 90, *328*
Goldwyn, R. M., 61, *327*
Golub, G. H., 77, 102, 103, 122, *325*, *328*
Goodman, A. S., 316, *330*
Goodson, R. E., 221, *331*
Grant, F. S., 15, *328*
Greenstadt, J., 89, 92, 107, 116, 133, 323, *327*, *329*
Grisoff, S., 116, *326*
Guttman, I., 170, 191, *329*

H

Hadley, G., 84, *329*
Hammersley, J. M., 46, *329*
Hanscomb, D. C., 46, *329*
Hartley, H. O., 100, 170, 191, *329*
Healy, M. J. R., 102, *329*
Heineken, F. G., 273, *329*
Hicks, J. S., 224, *329*
Hill, W. J., 267, 269, *326*, *329*
Himmelblau, D. M., 220, *329*
Hoerl, A. E., 60, 277, *326*, *329*
Homer, L. D., 16, *332*
Hood, W. C., 7, 64, *329*, *330*
Hooke, R., 119, 120, *329*
Hougen, O. A., 277, *329*
Howland, J. L., 226, *329*
Huard, P., 146, *327*
Hunter, W. G., 123, 204, 265, 269, 272, 281, *325*, *326*, *327*, *329*, *330*

J

Jacquez, J., 226, 230, *326*
Jeeves, T. A., 119, 120, *329*
Jeffreys, H., 35, *329*
Jennrich, R. I., 91, 94, 102, *329*
John, F., 52, *329*
Jonnston, J., 16, *329*
Jones, C. R., 220, *329*

K

Kagiwada, H., 16, 242, *326*
Kaiser, H. F., 302, *331*
Kalaba, R., 16, 226, 230, 242, *326*
Kalman, R. E., 225, *330*
Kelley, H. J., 16, 242, *330*
Kelley, Jr., J. E., 71, *330*
Kennard, R. W., 60, *329*
Kenney, R., 116, *326*
Kittrell, J. R., 120, 123, 265, *330*
Klein, L. R., 25, 133, 138, *326*
Koopmans, T. C., 7, 64, *329*, *330*
Korin, B. P., 200, *330*
Kowalik, J., 84, *330*
Krelle, W., 83, *330*
Kuhn, H. W., 52, *330*
Kullback, S., 267, *330*
Künzi, H. P., 83, *330*

L

Lapidus, L., 110, 146, 277, *323*, *325*, *328*
Law, V. J., 91, *327*
Lawton, W. H., 122, *330*
Legendre, A. M., 6, *330*
Lehman, E. L., 170, *330*
Leibler, R. A., 267, *330*
Levenberg, K., 94, *330*

Lewis, P. A. W., 316, *330*
Lindley, D. V., 261, 286, *330*
Longley, J. W., 103, *330*
Lucas, H. L., 261, 265, *326*

Mc

McCormick, G. P., 52, 107, 108, 142, 145, 159, *328*, *330*
McGhee, R. B., 100, 226, *330*
MacGregor, M. H., 15, *325*

M

Mangasarian, O. L., 53, *330*
Marquardt, D. W., 94, 114, *323*, *330*
Marsaglia, G., 316, *330*
Mead, R., 120, *331*
Meeter, D. A., 170, 191, *329*
Melkanoff, M. A., 15, *331*
Mendelsohn, J., 91, *328*
Mezaki, R., 120, 123, 204, 281, *329*, *330*
Miller, J. M., 316, *330*
von Mises, R., 73, *331*
Mishchenko, E. F., 225, *331*
Monroe, R. J., 16, *332*
Moshman, J., 316, *331*
Muller, M. E., 316, *326*
Murtagh, B. A., 146, *331*
Myszewski, M., 116, *326*

N

Nelder, J. A., 120, *331*
Neyman, J., 35, 185, *331*

O

Ortega, J. M., 302, *331*
Osborne, M. R., 84, 154, *330*, *331*

P

Pearson, J. D., 107, *331*
Penrose, R., 290, *331*
Perdreauville, F. J., 221, *331*
Pereyra, V., 122, *328*
Peterson, T. I., 7, 123, *323*, *326*, *331*
Pontryagin, L. S., 225, *331*
Powell, M. J. D., 107, 110, 120, 251, *328*, *331*
Price, C. M., 319, *331*
Price, J. F., 90, *328*

Q

Quandt, R. E., 94, *328*
Quenouille, M. H., 187, *331*

R

Raiffa, H., 33, 35, 77, 283, *331*
Rao, C. R., 80, *331*
Raynal, J., 15, *331*
Robbins, H., 35, *331*
Rosen, J. B., 146, *331*, *332*
Rosenbrock, H. H., 120, 224, 226, *332*
Rutemiller, H. C., 247, *332*

S

Sammet, J. E., 116, *326*, *332*
Sampson, P. F., 91, 94, 102, *329*
Sargent, R. W. H., 146, *331*
Savage, L. J., 33, *332*
Sawada, T., 15, *331*
Scheffé, H., 190, *332*
Schlaifer, R., 33, 35, 77, 283, *331*
Schwimmer, S., 226, 230, *326*
Seal, H. L., 7, *332*
Selvester, R., 16, *326*
Shahn, E., 16, *326*
Shanno, D. F., 95, *323*, *332*
Shannon, C. E., 19, 261, *332*
Shinbrot, M., 220, *332*
Smith, H., 201, *327*
Smith, F. B., Jr., 95, *332*
Solow, R. M., 134, *332*
Sorenson, H. W., 225, *332*
Spendley, W., 120, *332*
Sridhar, R., 242, *326*
Stewart, III, G. W., 119, *332*

Storey, C., 120, 224, 226, *332*
Surkan, A., 15, *327*
Swed, F. S., 201, *332*
Sylvestre, E. A., 122, *330*

T

Tobey, R., 116, *326*
Tomović, R., 226, *332*
Trotter, H. F., 94, *328*
Tsuchiya, H. M., 273, *329*
Tucker, A. W., 52, *330*
Turner, M. E., 16, *332*

V

Vaillancourt, R., 226, *329*
Vitale, P. A., 91, *328*

W

Wagner, H. M., 71, *332*
Wald, A., 260, 269, 270, 271, *332*
Watson, C. C., 120, 265, *330*
Watson, G. A., 154, *331*
Watson, K. M., 277, *329*
Wei, J., 224, *329*
Weiss, M. F., 16, *326*
West, G. F., 15, *328*
Wichern, D. W., 269, *329*
Wiener, N., 16, *332*
Wilde, D. J., 83, *332*
Wilkinson, J. H., 293, 302, 305, *332*
Winkler, R. L., 35, *332*
Wolfe, P., 146, *332*
Woo, L. S., 16, *328*
Wood, F. S., *324*

Y

Youle, P. V., 123, *326*

Z

Zangwill, W. I., 120, 145, *332*
Zilles, S., 116, *326*
Zoutendijk, G., 148, *332*

Subject Index

B

Bayes' theorem, 36–37
Bayesian estimation, 72–77
Bias, 40–41
  estimation of, 47
  reduction of, 187
Bienaymé–Chebyshev inequality
  for multiple parameters, 188–189
  for single parameter, 186
Bounds on parameters, 151–153
  effect on sampling distribution, 182–183
  need for, 141
Bounds on state variables, 232

C

Canonical form, 174–175
Chebyshev estimate, *see* Minimax deviation estimate
Chemical kinetics models, 15, 222, 229–230, 233–241
Chi square distribution, 21
Cholesky decomposition, 307–309
  modified for Marquardt's method, 95
Complementary pivot problem, 148
Complementary slackness, 52
Computer role in experiments, 275
Computer programs for parameter estimation, 323–324
Conditional distribution, 312
Confidence interval, 6, 184–187
Confidence region, 187–191
  illustration, 208
  for linearized model, 190–191
  minimizing volume of, 263
Constraints, 49–53, *see also* Bounds on parameters, Bounds on state variables, Equality constraints, Inequality constraints
  arising from prior information, 32–33
  effect on sampling distribution, 180–183
Control theory, problems of, 225
Correlation, 311
  test for, 200–201, 216
Curve fitting, 1-2
COVARIANCE MATRIX 178-179, 212-216

D

Data
  errors in, *see* Errors
  randomness of, 18
  requirements for estimation, 69–70
Data matrix, 17
Davidon–Fletcher–Powell method, 110
Decision theory
  applied to design of experiments, 283–285
  applied to parameter estimation, 74–76
Deming's method, 154–159
Dependent variables, 13
Derivative-free methods, 117–120
Design criteria, locating maximum of, 273–276
Design of experiments, 258
  for decision making, 283–285
  for discriminating among models, 266–269
  for parameter estimation, 262–265
  for prediction, 265–266
  termination criteria, 269–271

Determinant, 291
  computation of, 301
Deterministic model, 11–12
Differential equations, *see also* Dynamic models
  models formulated as, 218
  numerical integration of, 230–231
  stability of, 231–232
Differentiation
  analytic, by computer, 116, 323
  of dynamic model objective function, 226–228
  importance of accuracy in, 116
  of matrix functions, 293–296
  numerical, 117–119
Direct search methods, 119–120
Directional discrimination methods, 91–94
Discrimination among models
  design of experiments for, 266–269
  illustration, 277–283
Dynamic models
  computation of objective function, 225–230
  difficulties associated with, 231–233
  gradient of objective function, 226–230
  illustration, 233–238
  methods of solution, 218–221
  reduction to standard form, 223–224
  standard, 221–223

E

Econometric models, 25–26, 133–138, 167–168, 213–216
Eigenvalue decomposition, 304–305
Eigenvalues and vectors, 290
  computation of, for real symmetric matrix, 302–303
Equality constraints, 49–51
  linear, application of projection method to, 160
  model equations viewed as, *see* Exact structural model
  penalty functions for, 159–160
Errors, 54
  distribution of, 22–23
  estimating parameters of distribution, 195
Estimate
  asymptotically efficient, 43
  consistent, 42
  efficient, 41
  ill-determined, 172, 203
  linear, 44
  robust, 44
  statistical properties of, *see* Sampling distribution
  sufficient, 44
  unbiased, 40
  well determined, 172
Estimation procedures
  desirable properties for, 44–45
  reasons for failure, 202–204
Exact structural model, 24
  computation of estimates for, 154–159
  covariance matrix of parameter estimates, 179, 212–214
  estimating parameters of error distribution, 196–197
  illustration, 163–168
  maximum likelihood method for, 68–69
Experimental conditions, 17, 258
Experiments, 17
  cost of, 273, 284–285
  design of, *see* Design of experiments
  simulated by computer, 46, 276–277

F

F distribution, 21, 190
Farris–Law method, 93
Feasible region, 48
Finite differences, 117–119
  central, 119
  determination of optimum length, 118
  for dynamic models, 226
  one-sided, 117–118

G

Gauss–Jordan pivot, 296
Gauss–Markov theorem, 59, 318–319
Gauss method, 97–106
  implementation of, 101–106
  with penalty functions, 106, 144
  with prior distribution, 101, 106, 131–133

as sequence of linear regressions, 99–100
single-equation least squares, 97
illustration, 124–130
Gaussian distribution, *see* Normal distribution
Givens–Householder transformation, 302
Goodness of fit criteria, 198–202
Gradient methods, 86
convergence of, 87–88, 115–117, 320–322
efficiency of, 89
step length determination in, 110–113

H

Hessian matrix, 88
Gauss approximation for, 97–99
variable metric approximations for, 106–110

I

Independent variables, 14, 221
subject to error, *see* Exact structural model
Indifference region, 171–173
illustration, 207
Inequality constraints, 51–53
linear, application of projection method to, 146–153
penalty functions for, 141–145
Inexact structural model, 24–25
Information
for discrimination, 267
in a distribution, 19, 261
gained from an experiment, 261
in normal distribution, 262
prior, 32–35
Inhomogeneous covariance, 246–248
Initial conditions, 221
Initial guess, 120–123
Interpolation-extrapolation methods, 111–113
illustrations, 127–129
Interval estimate, 6
Iterative methods, 84–88, *see also* Derivative-free methods, Gradient methods
acceptable, 85–86
initial guess for, 120–123
termination criteria for, 114–115

K

Kuhn–Tucker condition, 52
for quadratic program, 147

L

Lagrange multipliers, 50–53
Least squares method, 55–61, *see also* Regression
unweighted, 55
weighted 56–57
Levenberg method, *see* Marquardt method
Likelihood, 26–29
concentrated, 65–66
standard reduced model, 27
structural models, 27–29
Likelihood equations, 62
Likelihood ratio, 269
Linear equations, solution of, 299
Linearity, 5
Linearizing transformations, 78–80, 122
illustration, 131
Linearly dependent equations, 238–241

M

Marginal distribution, 311
Marquardt method, 94–96
illustration, 130–131
Matrix
condition of, 89, 305
improved by scaling, 306
rank of, 292, 300–301
spectral decompositions of, 303–309
square root of, 307
trace of, 291
Matrix algebra, 287–293
Matrix functions, differentiation of, 293–296
Matrix inverse, 289, 298–299
updating of, 250
Matrix pseudoinverse, 290, 304

Maximization, *see* Optimization
Maximum likelihood method, 61–71
  exact structural model, 68–69
  illustrations, 133–138
  independent variables subject to error, 67–68
  normal distribution, 63–70
  two-sided exponential distribution, 70–71
  uniform distribution, 70
  unknown covariance matrix, 64–66
Measurement errors, *see* Errors
Minimax deviation estimate, 77
  computational method for, 154
Minimization, *see* Optimization
Minimum chi-square method, 80
Minimum risk estimate, 74–76
Minimum variance bound, 41
Missing observations, 244–245
  illustrations, 251–255
Mode-of-posterior-distribution estimate, 73–74
  illustration, 131–133
  sampling distribution of, 192
Model
  deterministic form, 11–12
  estimation of parameters of, 2–4
  formulation of, illustration, 29–31
  stochastic form, 24–26
Moment matrix, 64
  likelihood expressed as function of, 97–99
Monte Carlo method, 46
  illustration, 210–212

N

Newton–Greenstadt method, 92
Newton's method, 89–91
Nonlinear programming, 83
Normal distribution, 18–21
  generating pseudorandom sample from, 316–317
  information in, 262
  maximum likelihood method for, 63–70
  multivariate, 20–21
  univariate, 20
Normal equations, 49

O

Objective function, 47
  computation for standard dynamic model, 225–230
  as function of moment matrix, 97–99
Observed variables, 221
Optimality conditions
  constrained problems, 49–53
  unconstrained problems, 48–49
Optimization, 47
Orthogonalization, 103–106

P

Parameter estimation, 4
  applications of, 14–16
  computer programs for, 323–324
  design of experiments for, 262–265
  history of, 6–7
  in a probability distribution, 3, 80
  problem formulation, 37
Parameters, 11–12
Penalty functions
  equality constraints, 159–160
  illustration, 160–162
  inequality constraints, 141–145
  as a prior distribution, 145
Pivoting, 296
Point estimate, 6, 39
Positive definite matrix, 290
  role in gradient methods, 86, 116
Posterior distribution, 36–37
  mode of, 73–74
Prediction, 13–14
  design of experiments for, 265–266
  errors in, 204–205
Principal components, 183–184, 208
Prior distribution, 33–35
  informative, 35
  noninformative, 34–35
Probability distribution, 310
  estimating parameters of, 3, 80
Projection method, 146–153
  for bounded parameters, 151–153
  illustration, 162–163
  for linear equality constraints, 160
  step length determination, 149
Pseudomaximum likelihood method, 78
Pseudorandom numbers, 316–317

Q

QR method, 302–303
Quadratic programming, 147
Quasilinearization, 230

R

Rank one correction method, 107–109
Rao–Cramér theorem, 41, 313–315
Reduced model, 13–14
  standard, 26
Regression
  backward selection, 302
  forward selection, 301
  multiple linear, 58–61
    methods of solution, 102–106
  ridge, 60
  stepwise, 59, 101, 301–302
Reparametrization, *see* Transformation of variables
Residuals, 54
  analysis of, illustration, 209–210, 214–216
  outliers, 202
  run tests, 201–202
  statistical properties of, 193–196
  statistical tests on, 199–202
Risk, 74, 283–284
Rotational discrimination methods, 91

S

Sampling distribution, 39–45
  covariance matrix of, 176–179, 207
  effect of constraints on, 180–183
  estimation of statistical properties of, 175–183
  evaluation by Monte Carlo method, 46, 210–212
  evaluation of statistical properties of, 45–47
  statistics of, 40
Scaled decomposition, 305–307
Scientific investigation, goals of, 258–260
Sensitivity equations, 226–230
Sequential reestimation, 248–251
  illustration, 255–257
Serial correlation, 247–248
Simulation of experiments, 46, 276–277
Spectral decompositions, 303–309
State variables, 221
  bounds on, 232
Steepest descent method, 88
Stochastic approximation, 251
Stochastic model, 24–26
Structural model, 12–13, *see also* Exact structural model, Inexact structural model
Sufficient statistic, 43, 62
SUMT method, 412
Sweeping, 296

T

Termination criteria
  for iterative methods, 114–115
  for sequential experiments, 269–271
Transformation of variables, *see also* Linearizing transformations
  effect on sampling distribution, 205–206
  to eliminate constraints, 153
  invariance of estimates under, 44
  to simplify model, 133

U

Uncertainty, 261
Uniform distribution, 21–22
  maximum likelihood method for, 70

V

Variable metric methods, 106–110
Vectors, linear independence of, 300–301

6
B 7
C 8
D 9
E 0
F 1
G 2
H 3
I 4
J 5

Lloyd R Smith
1978